Lecture Notes in Electrical Engineering

Volume 268

W0007855

For further volumes:
http://www.springer.com/series/7818

Corrado Di Natale • Vittorio Ferrari
Andrea Ponzoni • Giorgio Sberveglieri
Marco Ferrari

Editors

Sensors and Microsystems

Proceedings of the 17th National Conference,
Brescia, Italy, 5-7 February 2013

 Springer

Editors
Corrado Di Natale
Department of Electronic Engineering
University of Rome Tor Vergata
Rome, Italy

Vittorio Ferrari
Department of Information Engineering
University of Brescia
Brescia, Italy

Andrea Ponzoni
SENSOR Laboratory, IDASC
National Research Council (CNR)
Brescia, Italy

Giorgio Sberveglieri
Department of Information Engineering
University of Brescia
Brescia, Italy

Marco Ferrari
Department of Information Engineering
University of Brescia
Brescia, Italy

ISSN 1876-1100   ISSN 1876-1119 (electronic)
ISBN 978-3-319-00683-3   ISBN 978-3-319-00684-0 (eBook)
DOI 10.1007/978-3-319-00684-0
Springer Cham Heidelberg New York Dordrecht London

Library of Congress Control Number: 2013953711

Printed on acid-free paper

Springer is part of Springer Science+Business Media (www.springer.com)

Contents

Part I
Physical Sensors

Investigation of Seebeck Effect in Metal Oxide Nanowires for Powering Autonomous Microsystems

Simone Dalola, Vittorio Ferrari, Guido Faglia, Elisabetta Comini,
Matteo Ferroni, Caterina Soldano, Dario Zappa, and Giorgio Sberveglieri

1 Introduction

Quasi-monodimensional metal oxide nanostructures have been recently considered as promising candidates to develop high-efficiency and high-temperature thermo-electric devices due to their reduced dimensionality and excellent durability at high temperature [1]. The fabrication of a thermoelectric generator (TEG) requires both n- and p-type elements. To this purpose, bundles of ZnO (n-type) and CuO (p-type) nanowires have been grown by thermal evaporation process [2] and thermal oxidation techniques [3], respectively, and a planar thermoelectric generator has been fabricated.

2 Material Preparation

The deposition technique for zinc oxide nanowires consists in the thermally driven evaporation of a powder of bulk metal oxides followed by condensation on a substrate in an alumina tubular furnace capable to reach high temperatures ($\approx 1,500\ ^\circ C$), needed to initiate the decomposition and the evaporation of metal oxide powder [2]. The nanowires have been fabricated on alumina 20×20 mm^2 substrates with Au nanoparticles deposited by RF sputtering as catalyst.

S. Dalola (✉) • V. Ferrari
Dipartimento di Ingegneria dell'Informazione, Università degli
Studi di Brescia, Via Branze 38, 25123 Brescia, Italy
e-mail: simone.dalola@ing.unibs.it

G. Faglia • E. Comini • M. Ferroni • C. Soldano • D. Zappa • G. Sberveglieri
SENSOR Lab, Università degli Studi di Brescia and CNR-IDASC,
Via Valotti 9, 25133 Brescia, Italy

C. Di Natale et al. (eds.), *Sensors and Microsystems: Proceedings of the 17th National Conference, Brescia, Italy, 5-7 February 2013*, Lecture Notes in Electrical Engineering 268, DOI 10.1007/978-3-319-00684-0_1, © Springer International Publishing Switzerland 2014

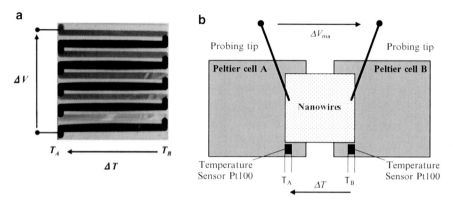

Fig. 1 (a) Picture of the fabricated planar TEG. (b) Schematic diagram of the experimental setup for the measurement of the thermoelectric response of nanowires samples

Copper oxide nanowires have been grown by thermal oxidation of metallic Cu thin film layer, previously deposited by RF magnetron sputtering on 20×20 mm^2 alumina substrates [3].

By combining n- and p-type nanostructured elements, a planar TEG consisting of five ZnO and CuO thermocouples, electrically connected in series and thermally in parallel, has been designed and fabricated. Thermoelectric material depositions have been patterned via shadow mask technique on alumina 20×20 mm^2 substrates to form an array of elements. Each element consists of an S-shaped strip 20 mm in length and 1 mm in width. The electrical contact is provided by the overlap of adjacent strips. ZnO nanowires have been grown by thermal evaporation technique before copper deposition, to avoid contaminations; then the entire device has been oxidized in a furnace. A top-view picture of the planar thermoelectric generator is shown in Fig. 1a.

3   Measurement of the Thermoelectric Response

The thermoelectric response of the fabricated samples has been experimentally investigated by measuring the thermoelectric voltage as a function of the temperature difference $\Delta T = T_A - T_B$ applied by means of an experimental setup, based on two Peltier cells, which provides the temperatures T_A and T_B at the edges of the sample, as shown in Fig. 1b.

The thermoelectric voltages on the ZnO and CuO nanowire bundles have been measured by means of a pair of Chromel probes, which connect the nanowire bundle to the electronic measurement unit [4]. The voltage ΔV, measured at the tips, is proportional to the temperature difference ΔT as follows:

$$\Delta V = \alpha_{m,Ch} \Delta T = \left(\alpha_m - \alpha_{Ch} \right) \Delta T \tag{1}$$

where $\alpha_{m,Ch}$ is the Seebeck coefficient of the tested material (ZnO or CuO) labelled m, with respect to the reference material Ch, i.e., Chromel, while α_m and $\alpha_{Ch} = 28.1$ µV/°C [5] are the absolute Seebeck coefficients of the tested material m and the Chromel, respectively.

For the characterization of the planar thermoelectric generator, the thermoelectric voltage ΔV has been measured using a pair of copper probing tips by means of the above-described experimental setup, as a function of the applied temperature difference ΔT. The thermoelectric voltage ΔV provided by the TEG is proportional to the applied temperature difference ΔT by means of the Seebeck coefficient S of the entire thermoelectric device, as it follows:

$$\Delta V = S\Delta T = N\alpha_{CuO,ZnO}\Delta T = N\left(\alpha_{CuO} - \alpha_{ZnO}\right)\Delta T \qquad (2)$$

where $N = 5$ and $\alpha_{Cuo,ZnO}$ are the number and the Seebeck coefficient of the ZnO–CuO thermocouples composing the thermoelectric device, respectively.

The voltage ΔV has been amplified by means of a low-noise instrumentation amplifier INA111 with a gain of 100 for the characterization of both metal oxide nanowires and planar thermoelectric generator.

4 Experimental Results

Figure 2a shows the applied temperature difference ΔT and the corresponding thermoelectric voltage ΔV measured for a ZnO nanowire sample fabricated at 700 °C as a function of time. From measured data, the generated voltage can be plotted versus the temperature difference, as shown in Fig. 2b.

The Seebeck coefficients of the fabricated samples, estimated by fitting the experimental data with a linear function as predicted by (1) and deriving the slope of the fitting line, are reported in Table 1. The sign of the measured thermoelectric coefficients is negative (positive) for ZnO (CuO) as expected for n-type (p-type) semiconductors. The experimental results are in agreement with recently reported values of Seebeck coefficient for ZnO nanostructures [6] and CuO thin film [7].

Figure 3a shows the trends of the applied temperature difference ΔT and the thermoelectric voltage ΔV versus time for the thermoelectric generator. The

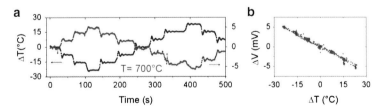

Fig. 2 (a) Trends of the applied temperature difference ΔT and the thermoelectric voltage ΔV versus time and (b) the voltage ΔV versus the applied temperature difference ΔT for the ZnO sample fabricated at 700 °C

Table 1 Measured Seebeck coefficients of the fabricated ZnO and CuO nanowire samples

Material	Deposition temperature (°C)	Seebeck coefficient α_m (mV/°C)
ZnO	700	−0.19
	870	−0.11
	1,070	−0.10
CuO	250	+0.82
	400	+0.43

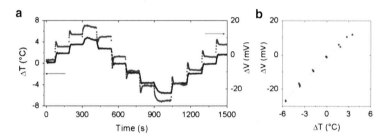

Fig. 3 (**a**) Trends of applied temperature difference ΔT and the voltage ΔV versus the time and (**b**) the generated voltage ΔV versus the applied temperature difference ΔT for the TEG

thermoelectric coefficient of the TEG has been estimated by linear fitting of the experimental data (Fig. 3b) and results of about 4 mV/°C; therefore, a single nano-structured ZnO–CuO thermocouple exhibits a Seebeck coefficient of about 0.8 mV/°C. The experimental data suggest that TEG based on the ZnO–CuO system could be investigated to supply low-power sensors and microsystems.

5   Conclusions

ZnO and CuO nanowire bundles have been deposited by thermal techniques and investigated for energy-harvesting applications by measuring Seebeck coefficient. A planar thermoelectric device based on five ZnO–CuO nanostructured thermo-couples has been fabricated. Experimental data confirm the feasibility of fabricating planar thermoelectric generators for power generation based on ZnO and CuO nanowires deposited via shadow masks.

Acknowledgments Authors gratefully acknowledge partial financial support by the IIT, Project Seed 2009 "Metal oxide NANOwires as efficient high-temperature THERmoelectric Materials (NANOTHER)."

References

1. A. Vomiero, I. Concina, E. Comini, C. Soldano, M. Ferroni, G. Faglia, G. Sberveglieri, One-dimensional nanostructured oxides for thermoelectric applications and excitonic solar cells, Nano Energy, **1** (3), 372–390 (2012).
2. E. Comini, Metal oxide nano-crystals for gas sensing, Analytica Chimica Acta, **568** (1–2), 28–40 (2006).
3. D. Zappa, E. Comini, R. Zamani, J. Arbiol, J.R. Morante, G. Sberveglieri, Copper oxide nanowires prepared by thermal oxidation for chemical sensing, Procedia Engineering, **25**, 753–756 (2011).
4. S. Dalola, G. Faglia, E. Comini, M. Ferroni, C. Soldano, D. Zappa, V. Ferrari, G. Sberveglieri, Seebeck effect in ZnO nanowires for micropower generation, Procedia Engineering **25**, 1481–1484 (2011).
5. S. O. Kasap, Principles of electronic materials and devices, 2nd edition, McGraw Hill, New York (2005).
6. Y. Kinemuchi, M. Mikami, K. Kobayashi, K. Watari, Y. Hotta, Thermoelectric Properties of Nanograined ZnO, Journal of Electronic Materials, **39** (9), 2059–2063 (2009).
7. M Muhibbullah, M. O. Hakim, M.G.M Choudhury, Studies on Seebeck effect in spray deposited CuO thin film on glass substrate, Thin Solid Films, 423 (1), 103–107 (2003).

Piezopolymer Interdigital Transducers for a Structural Health Monitoring System

L. Capineri, A. Bulletti, M. Calzolai, and D. Francesconi

1 PPTs Fabrication Characteristics

The flexible piezopolymer transducers made with thin 100 μm PVDF film were proposed by the authors in previous works [1–3].

The proposed technology uses an array of transducers which can be designed to excite particular types of ultrasonic Lamb waves in laminate materials (metallic or composite). The finger to finger distance of interdigital electrode patterns determines the central wavelength of the transducer (λ) and then the corresponding Lamb waves mode. The piezoelectric polymer film can operate in a range of temperature (−80 to +60 °C) that is broader than that of ceramic transducers. Furthermore, the high mechanical compliance makes it adaptable to curved surfaces. A laser-based microfabrication design process can be tuned for efficient excitation and reception of selected Lamb waves. This property depends on the characteristics of laminate material and defect types. In our case we adopted 8 mm finger distance and four pairs of electrodes (see Fig. 1). Permanently glued on the vessel surface, a series of PZTs (Acellent Technologies, Inc., Sunnyvale, CA) used during previous tests is still present.

The piezopolymer interdigital transducers have been characterized at low temperature (−80 °C) and high temperature (+60 °C) showing no significant change of behavior, confirming that they can operate in the temperature range required for space application. Also levels of driving voltages and excitation burst frequency have been selected depending on the monitored medium. Because of the lower

L. Capineri (✉) • A. Bulletti • M. Calzolai
Department of Information Engineering, University of Florence,
Via S. Marta 3, 50139 Florence, Italy
e-mail: lorenzo.capineri@unifi.it

D. Francesconi
Thales Alenia Space Italia S.p.A, Strada Antica di Collegno 253, 10146 Torino, Italy

C. Di Natale et al. (eds.), *Sensors and Microsystems: Proceedings of the 17th National Conference, Brescia, Italy, 5-7 February 2013*, Lecture Notes in Electrical Engineering 268, DOI 10.1007/978-3-319-00684-0_2, © Springer International Publishing Switzerland 2014

Fig. 1 The fabricated interdigital PPT made with thin 100 μm-thick PVDF film with λ = 8 mm. The two holes (φ = 2 mm) are used for cables connection. Dimensions 26 mm × 37 mm

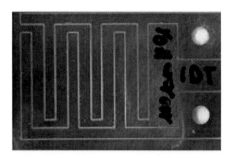

Fig. 2 Interdigital PPT with protective adhesive cover and the matching network designed. Implementation of the Piezopolymer Transducers Network

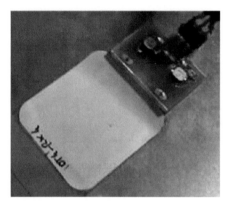

Fig. 3 Programmable acquisition system for active SHM on COPV with interdigital PPTs network

efficiency of PVDF with respect to PZT, higher voltages are necessary to drive the IDT transducers starting from low-voltage (±12 V) integrated driven circuit. Voltages up to 150 Vpp are obtained by designed matching networks (see Fig. 2).

All operations have been carried out with the scope of testing a piezopolymer transducers (PPTs) network applied on the surface of a scaled model of a carbon composite overwrapped pressurized vessel (COPV) to proof its efficiency for the localization of artificially induced damages on the COPV surface (see Fig. 3).

PPTs network has been designed for the wavelength $\lambda = 8$ mm, assembled on matching network designed connected to the receiving electronics. Adhesion of the sensors to the COPV surface is assured by smoothing with sandpaper and then degreasing with isopropyl alcohol the vessel surface. Then 8 PPTs (4 for the transmission Tx and 4 for the detection Rx) are installed on the vessel surface to investigate a portion of the cylindrical surface.

The chosen configuration (4+4 PPTs at 180 mm of distance spread along 180 mm depending on the defect characteristic and dimension) allows the acquisition of data along the paths between all possible pairs of transmitting and receiving transducers with a good signal-noise ratio.

2   Experimental Tests

A developed electronic system is programmed for the selection of multiple couples of pitch and catch PPTs and for the acquisition of relevant data. Front-end electronics has been designed and optimized for high signal-to-noise ratio taking into account the low power space requirements. The system includes also dedicated software that is developed to process off-line signals up to 8×8 different pairs of transducers for detecting both the damaged area and the size of the damage. The developed software interface controls the experiment and evaluates a damage index (DI) both in no damaged and damaged conditions and finally shows a colored scale image connected to the levels of probability of the presence of the damage.

Algorithm development for damaged area detection includes:

- The definition of signal paths from transmitter to the receiver
- The definition of excitation signal parameters
- The definition of artificially induced damages (drops of gel of 3 cm² size)
- The definition of Matlab routine managing the acquired signals of different paths before the DI calculation
- The calculation of DI based on the difference of both shape and amplitude of two acquired signals across the same path in undamaged and damaged conditions

Different artificial defects have been detected successfully by signal processing with the developed laboratory system: defects with area of about 3 cm² made by an ultrasonic gel drop on the surface of COPV have been detected successfully by using two arrays of four transmitters and four receivers placed at a distance of 180 mm. The antisymmetric A_0 mode at 220 kHz has been used to scan the area. The damage index image is shown in Fig. 4 corresponding to 1 for no defect and 0 for strong defect (no signals received).

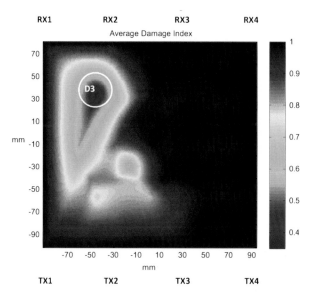

Fig. 4 Image of damage index distribution of the investigation area on COPV with defect D3 with real position ($x = -45°$; $y = +45°$)

References

1. L. Capineri, A. Gallai, L. Masotti, M. Materassi, "Design criteria and manufacturing technology of Piezo-polymer transducer arrays for acoustic guided waves detection", Ultrasonics Symposium, 2002. Proceedings. 2002 IEEE, Volume: 1, 8–11 Oct. 2002, pp. 857–860
2. F. Bellan, A. Bulletti, L. Capineri, L. Masotti, G. G. Yaralioglu, F. Levent Degertekin, B. T. Khuri-Yakub, "A new design and manufacturing process for embedded Lamb waves interdigital transducers based on piezopolymer film", Sensors and Actuators A 123–124 (2005) 379–387
3. F. Bellan, A. Bulletti, L. Capineri, A. Cassigoli, L. Masotti, O. Occhiolini, F. Guasti,, E. Rosi, "Review of non destructive testing techniques for composites materials and new applications ", Proceedings of The 10th Italian Conference on Sensors and Microsystems, Firenze, Italy, 15–17 February, 2005, World Scientific Publ. Pte. Ltd., Singapore Co., pp. 618–623

Noise and Performance of Magnetic Nanosensor Based on Superconducting Quantum Interference Device

C. Granata, R. Russo, E. Esposito, S. Rombetto, and A. Vettoliere

1 Introduction

The nano superconducting quantum interference device (nanoSQUID) is a powerful tool to study the magnetic properties of the nanoparticles at a microscopic level [1–7]. In fact, in the recent years, it has been shown that the magnetic response of nano-objects (nanoparticles, nanobead, small cluster of molecules) can be effectively measured by using a nanoSQUID [8, 9]. Typically, a dc nanoSQUID consists of a square loop having a side length less than 1 μm interrupted by two nanometric niobium constrictions (Dayem bridges) [2–4]. A small capture area improves the magnetic coupling between the SQUID and the nano-object under investigation allowing to obtain a sensitivity as high as few tens of Bohr magnetons per unit bandwidth [2–4]. In this paper, the design, the fabrication, and the characterization of hysteretic nanoSQUIDs based on Dayem nanobridges are reported. The effectiveness of such devices to measure magnetization and relaxation of iron oxide nanoparticles having a mean size ~8 nm demonstrated that our nano-devices are useful tools for nanomagnetism investigation.

2 Sensor Design and Characterization

We have designed and fabricated nanoSQUIDs consisting of a square hole with an inner side length of 0.75 μm, a width of 0.2 μm, and two Dayem bridges 120 nm long and 80 nm wide. We have used a configuration where the bias leads are

C. Granata (✉) • R. Russo • E. Esposito • S. Rombetto • A. Vettoliere
Istituto di Cibernetica "E. Caianiello" del Consiglio Nazionale
delle Ricerche, 80078 Pozzuoli, Napoli, Italy
e-mail: c.granata@cib.na.cnr.it

C. Di Natale et al. (eds.), *Sensors and Microsystems: Proceedings of the 17th National Conference, Brescia, Italy, 5-7 February 2013*, Lecture Notes in Electrical Engineering 268, DOI 10.1007/978-3-319-00684-0_3, © Springer International Publishing Switzerland 2014

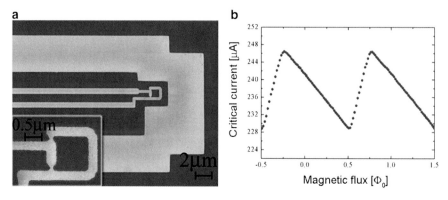

Fig. 1 (**a**) Scanning electron micrograph of a nanoSQUIDs in an asymmetric configuration. The integrated coil allows an effective modulation/calibration of the device. (**b**) Critical currents as a function of the external magnetic flux measured at $T=4.2$ K

asymmetric with respect to the loop which corresponds to an inductance loop asymmetry. As we will show in the experimental results, such a configuration allows us to improve the critical current responsivity. A superconducting integrated coil located very close to the device allows to tune and set the optimal working point of the sensor (Fig. 1a).

The fabrication process is based on electron-beam lithography (EBL) process, which can directly pattern two-dimensional nanostructures on a high-resolution positive-tone polymethyl methacrylate (PMMA) electron-beam resist, spin-coated on a standard silicon wafer, used as substrate [9]. The EBL was used to pattern both the SQUID and an integrated coil into PMMA. A 20-nm-thick Nb film was deposited by dc magnetron sputtering in an ultra high-vacuum system with a base pressure of 10^{-8} mbar. A simple lift-off procedure is employed to obtain the SQUIDs and the related coil. A layer of flowable oxide (Fox-22 DOWCorning) is finally spin coated to obtain a passivation layer. The nanosensors have been characterized in liquid helium ($T=4.2$ K) in a cryoperm magnetic shield using a low noise read-out electronic. In Fig. 2b, the $I-\Phi$ characteristic measured at $T=4.2$ K has been reported. There is an appreciable increase of the curve slope on the steeper side of the characteristic, leading to an increase of the current sensitivity. As evaluated from the figure, the slope ratio between the steepest and the smoothest side is more than a factor three.

The magnetic flux resolution is given by $\Phi_N=\Delta I_{CN}/I\Phi$, where ΔI_{CN} is the minimum critical current difference detectable which depends on the critical current values and $I\Phi=\partial I_c/\partial \Phi_e$ is the magnetic flux to critical current transfer coefficient. For typical I_C value of 100–200 µA, ΔI_{CN} is about 0.01 µA, and considering a maximum $I\Phi$ of about 90 µA/Φ_0 for the asymmetric configuration (Fig. 1b), a magnetic flux resolution of 0.1 mΦ_0 can be estimated.

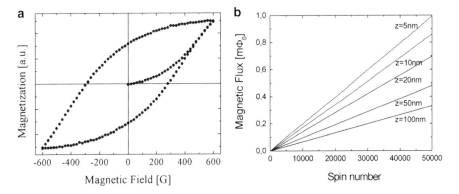

Fig. 2 (**a**) Field dependence of magnetization for 8-nm magnetic nanoparticles (*blue dots*) measured at $T=4.2$ K by using a nanoSQUID. (**b**) Magnetic flux produced by a spin cluster as a function of the spin number, located within the SQUID hole at different distance z from the sensor plane. The overall magnetic flux has been obtained by summing up the flux contributions of each spin

3 Magnetization Measurement

Magnetization measurement relative to magnetite (Fe_3O_4) nanoparticles having a diameter of 8 nm has been performed. The measurement is obtained by measuring the critical current variation due to the change in the magnetic flux threading the SQUID loop. The experimental setup employed to perform these measurements is well described elsewhere [9]. The excitation field is provided by a solenoid coil able to produce a uniform magnetic field in the plane of the nanosensor of about 800 G, while a suitable feedback circuit provides the linearization of the nanoSQUID output. In this configuration, only the magnetic field arising from the nanoparticles will be measured, because the SQUID is sensitive only to the magnetic field normal to the capture area.

The nanoparticles were dispersed in hexane with a concentration of 6 mg/ml. A single drop of the solution of about 10 ml was left to dry on the chip (1 cm²) leaving the nanoparticles stuck on the chip due to Van der Waals force. Supposing a uniform distribution of the solution on the chip, the mass of iron oxide in the active area of the sensor (0.5 µm²) was about 0.3 pg corresponding to 10^4–10^5 nanoparticles [9]. In Fig. 2a, magnetic field dependence of the magnetization relative to Fe_3O_4 nanoparticles for a sweep magnetic field ranging from −600 to 600 G and measured at $T=4.2$ K is reported. The sigmoid shape of the virgin curve at low magnetic field suggests the presence of dipole–dipole interparticle interactions which tend to resist to the magnetization process. The hysteresis indicates ferromagnetic features which are also present for lower field sweeps, suggesting that at $T=4.2$, the magnetic nanoparticles under investigation are well below the blocking temperature. Unfortunately, the available magnetic field did not allow us to observe saturation, so only a minor loop have been recorded. The coercive field of such a loop is about 290 G.

By using the numerical calculations reported in ref. [10], we have computed the magnetic flux produced by a cluster of elementary magnetic moment (Bohr magnetons, $\mu_B = 9.3 \times 10^{-24}$ J/T), located within the SQUID hole at different distance from the sensor plane (Fig. 2b). Since the maximum SQUID output relative to the reported magnetization curve is about 1 Φ_0 and assuming a distance from the nanoparticles and the sensor plane of 10–100 nm, we have approximately estimated from Fig. 2b a value of the total magnetic moment of $(0.6-1.7) \times 10^8 \mu_B$ $(0.6-1.6 \times 10^{-12}$ emu) corresponding to (2–5) emu/g which is a reasonable value since the typical values of the magnetization at the saturation field (few T) reported in the literature are few tens of emu/g.

4   Conclusions

Low noise magnetic nanosensor based on Dayem bridge SQUID has been employed for the measurements of nanoparticles magnetization. A magnetic flux sensitivity of about 10^{-4} Φ_0 has been obtained by using an asymmetric nanoSQUID configuration. The sensor has been used in a hysteretic mode with suitable read-out scheme which has allowed to linearize the SQUID response. The magnetization measurements on iron oxide nanoparticles have shown an evident magnetic hysteresis indicating a blocking temperature well above 4.2 K. These results are very encouraging in view of a wide use of nanoSQUIDs for the nanomagnetism investigations.

References

1. J.P. Cleuziou, W.Wensdorfer, V. Bouchiat, T. Ondarcuchu, M. Monthioux, "Carbon nanotube superconducting quantum interference device" Nature Nanotech. **1**, 53–59 (2006).
2. A.G.P. Troeman, H. Derking, B. Boerger, J. Pleikies, D. Veldhuis, H. Hilgenkamp, "NanoSQUIDs Based on Niobium Constrictions" Nano Lett. **7**, 2152–2156 (2007).
3. C. Granata, E. Esposito, A. Vettoliere, L. Petti, M. Russo, "An integrated superconductive magnetic nanosensor for high-sensitivity nanoscale applications" Nanotechnology, **19**, 275501–275506 (2008).
4. L.Hao, J.C. Macfarlane, J.C. Gallop, D. Cox, J. Beyer, D. Drung, T. Schuring, "Measurement and noise performance of nano-superconducting-quantum interference devices fabricated by focused ion beam" Appl. Phys. Lett. **92**, (192507–102509) (2008).
5. W. Wernsdorfer, "From micro-to-nano-SQUIDs: applications to nanomagnetism" Supercond. Sci. Technol. **22**, 064013 1–13 (2009).
6. C. Granata, A. Vettoliere, R. Russo, E.Esposito, M.Russo, B. Ruggiero," Supercurrent decay in nano-superconducting quantum interference devices for intrinsic magnetic flux resolution" Appl. Phys. Lett. **94**, 062503 1–3 (2009).
7. C. Granata. A. Vettoliere, M. Russo and B. Ruggiero, "Noise theory of dc nano-SQUIDs based on Dayem nanobridges" Phys Rev. **B 84**, 224516 1–7 (2011).

8. L. Hao, C. Aßmann, J. C. Gallop, D. Cox, F. Ruede, O. Kazakova, P. Josephs-Franks, D. Drung, and Th. Schurig, "Detection of single magnetic nanobead with a nano-superconducting quantum interference device", Appl. Phys. Lett. **98**, 092504 1–3 (2011).
9. R. Russo, C. Granata, E. Esposito, D. Peddis, C. Cannas and A. Vettoliere, "Nanoparticle magnetization measurements by a high sensitive nano-superconducting quantum interference device", Appl. Phys. Lett. **101**, 122601 1–3 (2012).
10. C. Granata, A. Vettoliere, P. Walke, C. Nappi and M. Russo, Performance of nano superconducting quantum interference devices for small spin cluster detection", J. Appl. Phys. **106**, 023925 1–5 (2009).

Nonintrusive IR Sensor for Real-Time Measurements of Gas Turbine Inlet Temperature

E. Golinelli, S. Musazzi, U. Perini, and F. Barberis

1 Introduction

Turbine inlet temperature (TIT) is a critical parameter of gas turbine systems influencing both material and coating lifetime of turbines as well as their efficiency. For this reason, online TIT monitoring systems are highly desirable since they would allow both to foresee with higher reliability the residual life of the most critical hot parts of the turbine and, at the same time, to maximize the gas temperature (although it should remain within the limits for materials and coatings operation) so as to increase the gas turbine overall efficiency.

Unfortunately, at present, there are no available methods to accurately measure this parameter on operating plants. For these reasons, we have developed a temperature sensor based on spectroscopic photometric measurements of the infrared radiation emitted in a selected wavelength band by the CO_2 molecules of the combustion gases. The sensor has been designed to provide online and in real-time measurements of the average gas temperature on a plant section transversal to the gas flow. The developed prototype is mechanically robust and thermally resistant to withstand typical operating conditions of industrial gas turbines. It has been tested both at laboratory level in controlled stable conditions and on a full-scale combustor test bed simulating the behavior of real gas turbine machine.

2 Principle of Operation

The measuring technique is a sort of emission-absorption pyrometry like the one utilized in nonreacting gas mixtures with negligible reflectivity [1]. The measured quantity is the grey body spectral irradiance $H(\lambda,T)$ defined by the product of the

E. Golinelli • S. Musazzi (✉) • U. Perini • F. Barberis
Ricerca sul Sistema Energetico—RSE, Via Rubattino 54, 20134 Milano, Italy
e-mail: sergio.musazzi@rse-web.it

C. Di Natale et al. (eds.), *Sensors and Microsystems: Proceedings of the 17th National Conference, Brescia, Italy, 5-7 February 2013*, Lecture Notes in Electrical Engineering 268, DOI 10.1007/978-3-319-00684-0_4, © Springer International Publishing Switzerland 2014

blackbody spectral irradiance $W_B(\lambda,T)$ (over 2π solid angle) and the grey body absorption $Abs(\lambda,T)$, the blackbody spectral irradiance being defined by the Planck's law:

$$W_B(\lambda,T) \times \Delta\lambda = \frac{1}{\lambda^5} \times \frac{C_1}{\exp\left(\dfrac{C_2}{\lambda T}\right) - 1} \times \Delta\lambda \left[W/m^2\right]$$
(1)

where $C_1 = 3.74 \times 10^{-16}$ [J m^2/s], $C_2 = 1.44 \times 10^{-2}$ [m K], and $\Delta\lambda$ is the selected bandwidth. Under the hypothesis that the combustion gas could be considered as a blackbody (i.e., $Abs(\lambda,T) = 1$), the only dependence on temperature is simply given by the blackbody irradiance $W_B(\lambda,T)$. As a consequence, the gas temperature can be determined from an irradiance measurement. To achieve a blackbody condition, irradiance measurements have to be carried out in a very narrow spectral range where the carbon dioxide molecules present in the hot gases strongly absorb the IR radiation. In this way, the detected signal is proportional to the gas temperature and follows its temporal behavior. Actually, since the radiation signal also depends on geometrical and optical parameters that can hardly be theoretically evaluated, it turns out that it is very difficult to directly get the measurement of the absolute temperature. To avoid an external independent reference measurement, we have:

1. To calibrate the system at laboratory level where the dependence on measurement parameters (e.g., transmission of the optical components in the working spectral range, sensor response as a function of the wavelength, optical coupling efficiencies) can separately be evaluated
2. To determine (via Hitran-based simulation programs) the dependence of the measured signal on the gas temperature profile in the test region

3   The Measuring System

The measuring system is made by three units: an optical probe, a detection unit (connected via an optical fiber to the optical probe), and a data acquisition and analysis unit. The optical probe is cooled by water and is purged with air to keep the optical window clean. A ZnSe lens collects the emitted radiation onto the input face of a Hg halide optical fiber. In the detection unit, the radiation emerging from the fiber is first collimated, then filtered by means of an IR interference filter, and finally focused onto the PbSe photoconductive detector (provided with an active cooling system that controls the sensor temperature). The signal out from the detector is first amplified and then fed to the acquisition board. Data acquisition and analysis are carried out via a PC-based analysis unit. All the system is controlled by means of a LabVIEW dedicated software, which extracts the average signal values and records the detected data.

4   Laboratory Tests

The measuring system has been tested by means of a laboratory setup. To reproduce the operating conditions of a gas turbine combustor (in terms of temperature, pressure, and CO_2 concentration), we have manufactured a vertical-axis metallic cell that can be filled with the desired gas mixture and heated up to about 1,000 °C. To let the inner generated optical radiation escape from the hot region and be revealed, the cell is provided with a properly cooled optical access (a CaF_2 optical window transparent to IR radiation).

The heating process is obtained by inserting the cell into a three-section programmable cylindrical vertical furnace. The vertical temperature profile inside the cell is monitored by four thermocouples in contact with the gas. A 45°copper mirror, positioned at the bottom of the cell in correspondence to its optical access, deviates the outcoming IR radiation to the collecting lens of the optical probe.

Different tests have been performed on the cell filled with a mixture of 4% CO_2 in N_2 atmosphere at 15 bars during controlled temperature variations of the furnace. It turned out that the IR signal exhibits good correlation with the theoretical temperature dependence as calculated by the Planck law. As an example we show in Fig. 1 the comparison between the calculated and the measured signal as a function of the total internal gas pressures (i.e., the inner total pressure has been gradually decreased by maintaining constant the ratio between the CO_2 and the N_2 partial pressures).

The calculated signal S (blue line) has been obtained by means of a simplified absorption/emission model based on the test cell temperature profile. That is, the test cell has been divided in three separate staked layers whose average

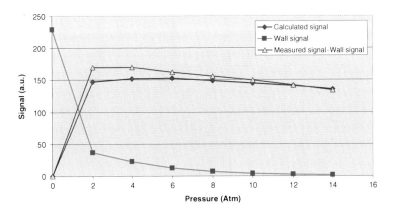

Fig. 1 Comparison between the calculated and the measured signal at different total internal gas pressures

temperatures are known (since measured by the thermocouples) so that the resulting signal can be calculated according to the following relationship:

$$S = K \left[W_{B1} \left(1 - Tr_1\right) + W_{B2} \left(1 - Tr_2\right) Tr_1 + W_{B3} \left(1 - Tr_3\right) \times Tr_1 \times Tr_2 \right] \Delta\lambda$$

where W_{Bn} is the blackbody irradiance of the layer n provided by the Plank law (1), Tr_n is the transmission of the layer n calculated by means of the Hitran program, and $(1 - Tr_n) = Abs_n$ is the corresponding absorption value (that for a blackbody is identical to the emission value).

The measured signal (red line) has been obtained by subtracting to the detected signal the signal recorded with the empty cell (which is solely due to the contribution of the opposite cell wall) multiplied by the calculated transmission of the gas layer. As it can be noticed, the agreement between the two curves is quite good and confirms the foreseen theoretical behavior stating that the signal generated by the opposite wall reduces as the total CO_2 content increases.

5   Conclusions

Laboratory tests have successfully been concluded, and the experimental activity will continue on a full-scale combustor test bed in Italy.

Acknowledgments This work has been financed by the Research Fund for the Italian Electrical System under the Contract Agreement between RSE (formerly known as ERSE) and the Ministry of Economic Development—General Directorate for Nuclear Energy, Renewable Energy, and Energy Efficiency stipulated on July 29, 2009, in compliance with the Decree of March 19, 2009.

Reference

1. G.A. Hornbeck, Optical methods of temperature measurement. Appl. Opt. Vol.5, No.2, 179 (1966).

Thermally Actuated Microfluidic System for Polymerase Chain Reaction Applications

D. Caputo, G. De Cesare, A. De Pastina, P. Romano, R. Scipinotti, N. Stasio, and A. Nascetti

1 Introduction

Lab-on-Chip (LOC) systems, in which functionalities as handling, treatment, and detection of biological solutions are downsized and integrated, represent one of the most attractive device for biomolecular and chemical analysis [1–4]. In particular, implementation of the polymerase chain reaction (PCR) in LOC devices has received a lot of attention [5–7] for applications such as DNA finger printing, genomic cloning, and genotyping for disease diagnosis. Indeed, its miniaturization results in reduced consumption of samples/reagents, shorter analysis times, and higher sensitivity and portability. Different solutions in terms of materials for the substrate and for the microfluidics have been proposed.

In this paper the authors present the design, fabrication, and characterization of a LOC system for DNA amplification based on a PolyDiMethylSiloxane (PDMS) microfluidic structure with integrated thermo-actuated valves and indium tin oxide (ITO) heaters. The use of these materials coupled with their deposition on a microscope glass presents features of easy fabrication, low cost, transparency, and biocompatibility.

2 LOC Design

The structure of the proposed LOC system (see Fig. 1) comprises:

1. A microfluidic network, made in PDMS, that includes a rectangular channel for inlet and outlet, a chamber for the PCR, and two thermo-actuated valves.

D. Caputo (✉) • G. De Cesare • A. De Pastina • P. Romano
R. Scipinotti • N. Stasio • A. Nascetti
Department of Information, Electronic and Telecommunication Engineering,
Sapienza University of Rome, Via Eudossiana 18, 00184 Rome, Italy
e-mail: caputo@die.uniroma1.it

C. Di Natale et al. (eds.), *Sensors and Microsystems: Proceedings of the 17th National Conference, Brescia, Italy, 5-7 February 2013*, Lecture Notes in Electrical Engineering 268, DOI 10.1007/978-3-319-00684-0_5, © Springer International Publishing Switzerland 2014

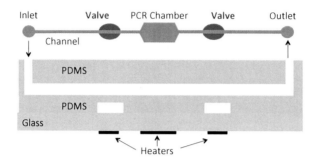

Fig. 1 Top view and cross section of the proposed LOC system

2. Three thin film heaters. Two of them, located at the inlet and outlet positions of the chamber, actuate the valves and allow the chamber isolation, while the last one, positioned below the chamber, is dedicated to the PCR thermal cycle.
3. A glass substrate hosting the microfluidic network on one side and the heaters on the opposite side.

The solution to be amplified is injected in the microfluidic channel through the inlet. As the channel and the PCR chamber are filled, the two heaters below the valves are actuated to avoid sample losses, due to the pressure developed in the reaction chamber during the thermal cycles. Indeed, applying power to the heaters, the air contained in the valve reservoir (rectangular white spaces over the heaters in Fig. 1) heats up and increases its volume creating a pressure that pushes up the PDMS membrane into the channel. As the valves are closed, the PCR cycle may begin by turning on the heater below the PCR chamber.

3 PCR Chamber Design

As reported in Fig. 1, the shape of the PCR chamber is rhomboidal-like in order to avoid air bubble formation [6] during chamber filling. The chamber size is $8 \times 4 \times 0.05$ mm^3 providing 1.6 µl solution volume for DNA amplification.

4 Heater Design

In order to achieve a uniform temperature distribution inside the PCR chamber, the geometry of the PCR chamber heater and the distance between the three heaters have been optimized using the software COMSOL Multiphysics. We found that selecting a chirp geometry for the PCR chamber heater [8], a serpentine geometry for the valve heaters, and 6 mm distance between the valve heaters and the PCR

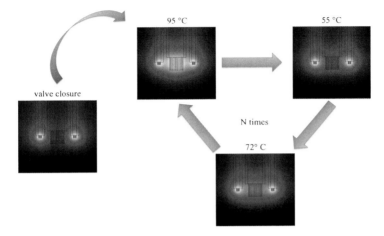

95 °C   55 °C

valve closure

N times

72° C

Fig. 2 Temperature distribution over the heaters during the simulated PCR cycles

chamber, the temperature uniformity in the PCR chamber is better than 3 % for the three temperature steps, satisfying the PCR technique requirements. A simulated temperature distribution during a PCR cycle along the microfluidic network is reported in Fig. 2.

5 Valve Design

The valves have a cylindrical shape with diameter and height equal to 500 μm and 50 μm, respectively. These sizes have been chosen, as a result of finite element simulations, to ensure that the deformation of the valve membranes closes the microfluidic channel when the heaters are turned on.

6 LOC Fabrication and Testing

Device fabrication begins with the deposition of the ITO thin film heaters by magnetron sputtering on one side of a $5 \times 5\text{cm}^2$ ultrasonically cleaned glass substrate. The film thickness is 200 nm, the same utilized in the simulations.

The PDMS structure was obtained by using soft lithography techniques. Two 50-μm-thick SU-8 molds were fabricated and patterned by photolithography: the first for the thermo-actuated valves and the second one for the microfluidic channel and the PCR chamber. PDMS was spun on the first mold in order to obtain a 100-μm-thick layer and a 50-μm-thick PDMS membrane over the SU-8 cylinders. PDMS was also poured onto the second mold to achieve a 3-mm-thick structure. The two PDMS structures were merged using the partial curing method [9], and

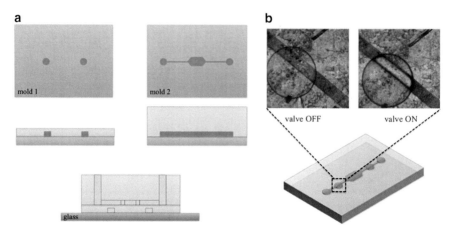

Fig. 3 (**a**) Schematic representation SU-8 molds and PDMS device fabrication (channel cross section: 500 μm×50 μm, PCR chamber volume: 1 μl); (**b**) demonstration of thermal valve actuation

subsequently the overall microfluidic device was bonded to the heaters on the glass substrate by oxygen plasma (Fig. 3a).

The PDMS microfluidic channel was filled with a mix of water and a blue dye in order to easily monitor the fluid position during the experiment. The valve-heater temperature has been raised slowly, and the behavior of the valve has been monitored under the microscope. Valves activation began around 60 °C, and complete closure was observed around to 100 °C as shown in Fig. 3b, without any loss of liquid from the chamber.

7 Conclusions

This paper has presented the design and fabrication of a microfluidic network integrated with thin film heaters for implementation of the PCR technique in Lab-on-Chip systems. The heaters have been designed to achieve temperature uniformity inside the PCR chamber better than 3 % at each temperature of the DNA amplification cycle. The microfluidic network, made in PDMS, includes two thermally actuated valves, whose successful operation in avoiding loss of samples from the PDMS chamber has been demonstrated.

References

1. D. Erickson, L. Dongqing. Integrated microfluidic devices. Analytica Chimica Acta, **507** (1), 11–26 (2004)
2. A.G. Crevillén, M. Hervás, M.A. López, M.C. González, A. Escarpa. Real sample analysis on microfluidic devices. Talanta, **74** (3), 342–357 (2007)

3. D. Janasek, J. Franzke, A. Manz. Scaling and the design of miniaturized chemical-analysis systems. Nature **442** (7101) 374–380 (2006)
4. D. Caputo, M. Ceccarelli, G. de Cesare, A. Nascetti, R. Scipinotti. Lab-on-glass system for DNA analysis using thin and thick film technologies. in Materials Research Society Symposia Proceedings, **1191**, 53–58 (2009)
5. M.A. Northrup, R.F. Hills, P. Landre, S. Lehew, D. Hadley, R. Watson. A MEMS-based DNA analysis system. in Tranducer '95, Eighth International Conference on Solid State Sens Actuators, Stockholm, Sweden. ISBN:9 1-630-3473-5, 764–767 (1995)
6. N.C. Cady, S. Stelick, M.V. Kunnavakkam, C.A. Batt. Real-time PCR detection of Listeria monocytogenes using an integrated microfluidics platform. Sens. Actuators B Chem. **107**, 332 – 341 (2005)
7. Z.Q Niu, W.Y. Chen, S.Y. Shao, X.Y. Jia, W.P. Zhang. DNA amplification on a PDMS–glass hybrid microchip. J. Micromech. and Microeng., **16**, (2), 425–433 (2006).
8. D. Caputo, G. de Cesare, A. Nascetti, and R. Scipinotti. a-Si:H temperature sensor integrated in a thin film heater. Phys. Status Solidi A, **207** (3), 708–711 (2010).
9. M.A. Eddings, M.A. Johnson, B.K. Gale. Determining the optimal PDMS-PDMS bonding technique for microfluidic devices. J Micromech Microengineering, **18**, (6), 06700 (2008).

Microfluidic Sensor for Noncontact Detection of Cell Flow in a Microchannel

M. Demori, V. Ferrari, P. Poesio, R. Pedrazzani, N. Steimberg, and G. Mazzoleni

1 Introduction

Characterization and counting of cells in biological samples is a relevant task in an increasing number of research activities such as in the biomedical, food, agriculture, and environmental fields. The cells can be seen as flow discontinuities with respect to a support fluid in which they are transported. Among the various techniques, cell analyses based on impedance measurements have been widely used in systems based on microfluidic devices [1].

In this work a microfluidic sensor for the detection of cells flowing in a micro-channel is presented. The sensor consists of a PDMS (PolyDiMethylSiloxane) layer with two microreservoirs connected by a microchannel and sensing electrodes formed on a PCB (printed circuit board). A noncontact measurement is ensured by an insulator layer, placed between the electrodes and the fluid, in order to avoid the galvanic contact. The insulator layer allows to avoid the issues associated to double-layer capacitances [2] and degradation of electrodes, offering advantages in terms of measurement repeatability and lifetime of the device.

M. Demori (✉) • V. Ferrari
Department of Information Engineering, University of Brescia,
Via Branze 38, 25123 Brescia, Italy
e-mail: marco.demori@ing.unibs.it

P. Poesio • R. Pedrazzani
Department of Mechanical and Industrial Engineering,
University of Brescia, Via Branze 38, 25123 Brescia, Italy

N. Steimberg • G. Mazzoleni
Department of Clinical and Experimental Sciences, University of Brescia,
Viale Europa 11, 25123 Brescia, Italy

C. Di Natale et al. (eds.), *Sensors and Microsystems: Proceedings of the 17th National Conference, Brescia, Italy, 5-7 February 2013*, Lecture Notes in Electrical Engineering 268, DOI 10.1007/978-3-319-00684-0_6, © Springer International Publishing Switzerland 2014

2 Sensor Description

The sensor consists of a PDMS layer with two microreservoirs connected by a microchannel. The device is fabricated with a hybrid low-cost technique combining the PDMS soft lithography and the PCB milling. As shown in Fig. 1, a rectangular microchannel 100 μm × 50 μm is used as the connection of two microreservoirs obtained in a PDMS layer. The bottom sides of the microreservoirs are respectively faced to two electrode plates formed on a PCB. An insulator layer is placed between the fluid and the electrodes to avoid galvanic contact.

The suspending medium containing the cells is injected in one of the microreservoirs, flowed through the microchannel, and finally ejected from the other microreservoir. In this way the complete system composed by the microchannel and microreservoirs is filled with the suspending medium, and the cells flow through the microchannel. The flowing cells cause changes in the conductivity of the narrow path formed by the fluid in the microchannel, causing variations in the impedance

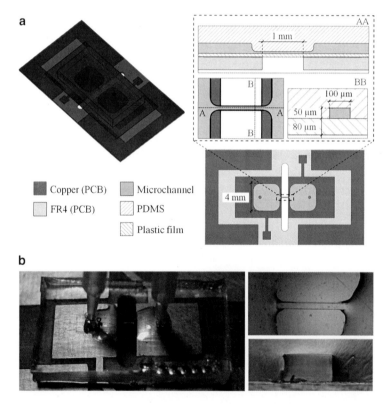

Fig. 1 (a) Sensor layout: 3D sketch, top view, and magnification of the microchannel region with transversal and axial cross sections; (b) picture of the PDMS layer on the PCB and microscope pictures of the microchannel top view and cross section

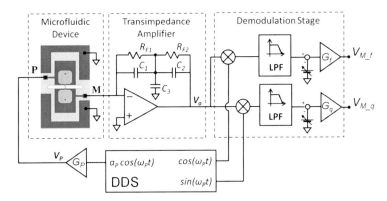

Fig. 2 Block diagram of the electronic interface connected to the microfluidic sensor. The capacitance and conductance sensitivities have been set to $\partial V_{M\_f}/\partial C \approx 0.3$ V/fF and $\partial V_{M\_q}/\partial G \approx 0.1$ V/nS, respectively

between the sensing electrodes. In this way, a Coulter counter configuration is obtained yet exploiting a contactless detection technique [3].

As shown in Fig. 2, the device is connected to a tailored electronic interface, based on a high-sensitivity transimpedance amplifier and a DDS (direct digital synthesis) device [4], to measure the sensor impedance in terms of effective capacitance and conductance variations.

3   Experimental Results

In the experimental tests, a DMEM (Dulbecco's modified eagle medium) solution containing mammal fibroblasts has been flowed into the microchannel between the two microreservoirs. A flow rate of 5 nL/s has been adopted which corresponds to a flow velocity in the microchannel of about 1 cm/s. Since both the plastic layer and the PDMS are transparent, the flow in the microchannel can be visually observed by a microscope through the slit in the PCB.

Based on the results obtained from a modeling of the electrical behavior of the adopted cells [1] in the suspending medium, the electronic interface has been set to operate at an appropriate excitation frequency $f_P = 400$ kHz. As shown in Fig. 3a, b, the system detects the flow of both clusters and single cells with variations in the signals associated to the effective capacitance and conductance between electrodes. The flow of the cell cluster is detected with spikes on the output signals well beyond the noise level. Variations of the effective capacitance and conductance of about 0.9 fF and 1 nS, respectively, have been obtained for a cluster of 5–6 cells with an average diameter of 15 μm each. On the other hand, the flow of a single cell with a diameter of about 20 μm generates spikes, slightly higher than the noise

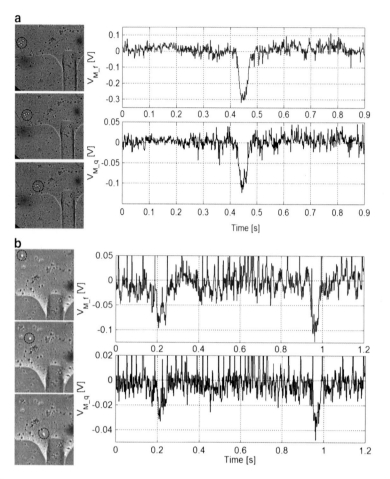

Fig. 3 (**a**) Detection of the flow of a cluster of 5–6 cells with a diameter of about 15 μm each; (**b**) detection of the flow of a sequence of two cells with a diameter of about 20 μm each

level, associated to variations of about 0.2 fF and 0.2 nS. In both cases, the duration of the signal spikes of about 0.1 s is in good agreement with the transit time of the flowing cells.

4   Conclusions

A microfluidic sensor for the detection of the flow of cells in a microchannel is proposed. The sensing electrodes are insulated from the fluid in the microchannel, thereby preventing sample contamination and the degradation of the electrodes. In the experimental tests, the sensor has demonstrated the ability to detect the flow of both cell clusters and single cells. These preliminary results are promising for biological measurements such as counting and sizing of cells in different matrices, by using a low-cost and timesaving portable solution.

References

1. T. Sun, H. Morgan. Single-cell microfluidic impedance cytometry: a review. Microfluid Nanofluid **8**, 423–443 (2010)
2. S. Zheng, M. Liu, Y. Tai. Micro Coulter counters with platinum black electroplated electrodes for human blood cell sensing. Biomedical Microdevices **10** (2), 221–231 (2008)
3. P. Kubán, P.C. Hauser. Ten years of axial capacitively coupled contactless conductivity detection for CZE-a review. Electrophoresis **1**, 176–188 (2009)
4. M Ferrari, V. Ferrari, D. Marioli. Interface circuit for multiple-harmonic analysis on quartz resonator sensors to investigate on liquid solution microdroplets. Sensors and Actuators. B **146** (2), 489–494 (2010)

A Comparison Between Fresh and Thermally Aged Polyaniline Prepared by Different Approaches: On the Conductivity Under High Pressure

C. Della Pina, M. Rossi, and E. Falletta

1 Introduction

Since their discovery to date, electrically conductive organic polymers (ECOPs) have been extensively studied [1–3]. Among them polyaniline (PANI) is unique for its doping/dedoping mechanism, easy preparation, low cost and air stability. Although polyaniline can exist in different oxidation states, only the emeraldine salt (ES, half-oxidated and half-protonated form, Fig. 1) shows electrical conductivity.

The conductivity of ES depends on numerous factors, such as molecular weight, morphology, size, crystallinity and kind of dopants. Some of these parameters can be monitored during the reaction, choosing properly the reaction conditions (i.e. medium, temperature, kind of oxidant and acid), or after the reaction by specific treatments (i.e. change of acid dopant, degree of crystallinity) [4, 5].

The conductivity of ECOPs is the sum of two contributions: the ability of the charge carriers to move along the polymer backbone and the ability of the charge carriers to hop between the polymer chains. This second contribution becomes particularly important when the material is subjected to a force or a pressure. Accordingly, we report on the influence of high pressure, pressing time and thermal ageing of powdered samples of polyaniline doped with a high-boiling inorganic acid (H_2SO_4). In particular we compare the results obtained employing PANI prepared following two synthetic methods: a classical approach (PANI1) and a "green" approach (PANI2).

C.D. Pina • M. Rossi • E. Falletta (✉)
Dipartimento di Chimica, Università degli Studi di Milano,
Via Golgi, 19, 20133 Milano, Italy
e-mail: ermelinda.falletta@unimi.it

C. Di Natale et al. (eds.), *Sensors and Microsystems: Proceedings of the 17th National Conference, Brescia, Italy, 5-7 February 2013*, Lecture Notes in Electrical Engineering 268, DOI 10.1007/978-3-319-00684-0_7, © Springer International Publishing Switzerland 2014

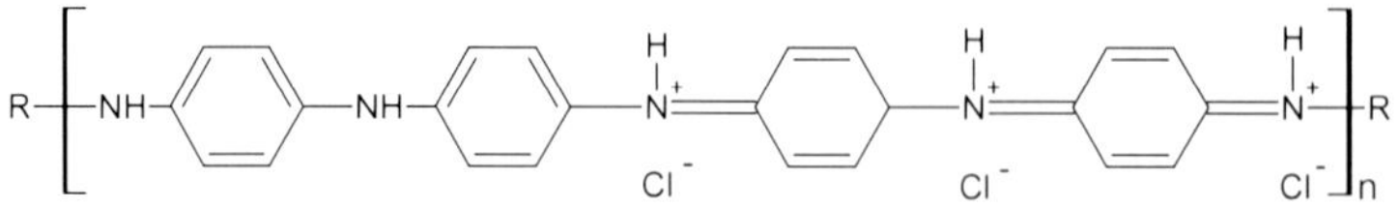

Fig. 1 Molecular structure of emeraldine salt

2 Experimental

2.1 Preparation of PANI1

Aniline was dissolved in an aqueous solution of hydrochloric acid and cooled at 4 °C. Then an aqueous solution of $(NH_4)_2S_2O_8$ (APS) was added dropwise. After 6 h, a dark green solid was collected by filtration, washed several times by water and acetone and dried in air [6]. The green powder, spectroscopically characterized and identified as emeraldine salt, was dedoped by ammonia solution to produce emeraldine (EB1) and redoped with H_2SO_4 (aniline/$H_2SO_4 = 0.5$, molar ratio) thus obtaining PANI1.

2.2 Preparation of PANI2

N-(4-Aminophenyl)aniline (AD, aniline dimer) was dissolved in an aqueous solution of hydrochloric acid; H_2O_2 (H_2O_2/AD = 5, molar ratio) was added as the oxidant and Fe^{3+} as the catalyst (AD/Fe^{3+} = 1,000, molar ratio). After 24 h, a dark green powder was collected by filtration, washed several times by water and acetone and dried in air [7]. The precipitate was spectroscopically characterized and identified as ES and treated as reported above (Sect. 2.1) to obtain EB2 and then PANI2.

3 Results and Discussion

200 mg of PANI1 and PANI2 were pressed to form pellets that were tested at room temperature and after thermal treatment (2 h at 50, 100, 150 and 200 °C). A press connected to a multimeter was employed to measure the conductivity of the pellets at different pressures (0–750 MPa). Figure 2 shows the experimental setup employed.

After a first conditioning cycle, four processes of pressure loading and unloading were repeated for each sample before and after thermal treatment. Figure 3 reports the mean value of all the measurements for each sample.

PANI1 and PANI2 showed a markedly different behaviour during the pressure loading and unloading processes at various temperatures. In fact, although PANI1 displayed overall higher conductivity, smaller hysteresis and better repeatability

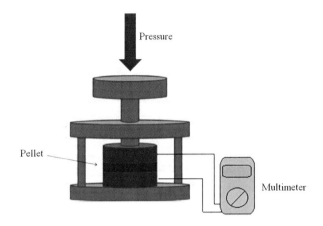

Fig. 2 Experimental setup for the measure of conductivity under different pressure

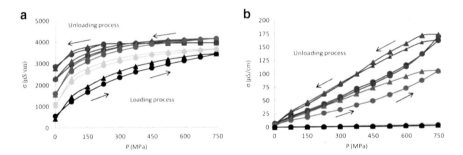

Fig. 3 Conductivity (μS/cm) as a function of pressure (MPa) during the loading and unloading cycles for A) PANI1, B) PANI2 (loading cycles = O, unloading cycles = Δ; *blue line* = room temperature, *red line* = 50 °C, *green line* = 100 °C, *yellow line* = 150 °C, *black line* = 200 °C)

Table 1 Tests results for PANI1 and PANI2

	PANI1					PANI2				
T (°C)	RT	50	100	150	200	RT	50	100	150	200
Max $\Delta\sigma$ (%)	3.4	41.4	57.8	249.9	181.5	688.7	795.7	712.3	1,010.7	1,179.1
Drift time (%)	0.0	0.2	0.4	0.3	0.3	0.5	0.3	0.3	0.8	1.2
RSD (%)	0.5	1.9	3.5	4.1	4.9	8.7	4.6	5.0	9.1	5.7

RT = room temperature, max $\Delta\sigma$ (%) = maximum percentage variation of conductivity, drift time (%) = measured under a pressure of 372 MPa for 10 min, RSD (%) = relative standard deviation

than PANI2, it resulted to be less pressure-sensitive as the lower percentage variation of conductivity by pressure loading shows (Table 1).

However, PANI1 displayed a more regular conductivity decay with the thermal treatment than PANI2, whose conductivity drastically dropped while increasing temperature (Fig. 3). In order to explain the two different behaviours, XRDP analyses were performed for investigating the degree of crystallinity of the samples (Fig. 4).

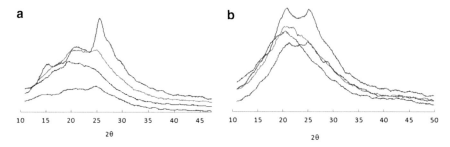

Fig. 4 XRPD (X-ray powder diffraction) patterns of PANI1 (**a**) and PANI2 (**b**) at different temperatures: *blue line* = room temperature, *red line* = 100 °C, *green line* = 150 °C, *black line* = 200 °C

The gradual crystallinity decay for both the materials, PANI1 and PANI2, was very similar, and it started passing from room temperature to 100 °C. As a consequence, the marked drop in conductivity detected for PANI2 after the thermal treatment at 100 °C cannot be ascribed to this reason.

Moreover, the thermogravimetric analyses (TGA data not reported herein) showed that, after exposure to a temperature higher than 100 °C, PANI1 and PANI2 underwent mass losses due to the dopant as well as to structural changes as observed by Peikertová and co-workers [8]. In particular, PANI2 showed higher mass losses than PANI1, likely owing to PANI2 lower molecular weight and its consequent minor stability.

4 Conclusions

In summary, we have investigated the influence of the thermal treatment on the conductivity of polyaniline pellets prepared following two different synthetic ways leading to PANI1 and PANI2. Accordingly, although PANI1 displayed higher conductivity, smaller hysteresis and better repeatability, however, it resulted to be less pressure-sensitive as showed by the lower percentage variation of conductivity by pressure. Both the materials underwent a conductivity drop while increasing temperature from 25 to 200 °C. Regarding the thermal treatment of the samples, TGA analyses of PANI2 presented a higher mass loss after $T = 100$ °C with respect to PANI1, thus highlighting a major thermal instability.

References

1. A. H. El-Shazly, A. A: Wazzan. Using Polypyrrole Coating for Improving the Corrosion Resistance of Steel Buried in Corrosive Mediums. Int. J. Electrochem. Sci. **7**, 1946–1957 (2012)
2. S. K. Dhawan, N. Singh, D. Rodrigues. Electromagnetic shielding behaviour of conducting polyaniline composites. Sci. Technol. Adv. Mater. **4**(2), 105–113 (2003)

3. H. Bai, G. Shi. Gas Sensor Based on Conducting Polymers. Sensors. **7**, 267–307 (2007)
4. J. Joo, Z. Oblakowski, G. Du, J. P. Pouget, E. J. Oh, J. M. Weisinger, Y. Min, A. G. MacDiarmid, A. J. Epstein. Microwave dielectric response of mesoscopic metallic regions and the intrinsic metallic state of polyaniline. Phys. Rev. B **49** (4), 2977–2980 (1994)
5. R. Pelster, G. Nimtz, B. Weissling. Fully protonated polyaniline: Hopping transport on a mesoscopic scale. Phys. Rev. B **49** (18), 12718–12723 (1994)
6. J. C. Chang, A. G. MacDiarmid. Polyaniline: Protonic acid doping of the emeraldine form to the metallic regime. Synth. Met. **13**, 193–205 (1986)
7. Z. Chen, C. Della Pina, E. Falletta, M. Rossi. A "green" route to conducting polyaniline by copper catalysis. J. Catal. **267**, 93–96 (2009)
8. P. Peikertová, V. Matějka, L. Kulhánková, L. Neuwirthová, J. Tokarský, P. Čapková. Thin polyaniline: study of thermal degradation. NanoCon 2011, Sept 21–23 (2011)

Thermal Sensor for Fire Localization

M. Norgia, A. Magnani, and A. Pesatori

1 Introduction

The principles of fire detection and prevention in closed environments have been very important. To prevent the disastrous effects a fire can cause, the fire protection sprinkler system [1, 2] was invented in the last century. The installation of this system is relatively inexpensive and easy; however, there are places where its installation is not possible, such as highway tunnels. For utilization in such environment, a localization system based on fire temperature sensor thermopile matrix [3], which could replace most of the complex and difficult installations of fire safety systems, has been designed and prototyped [3, 4]. The Cool Eye™ thermopile sensor manufactured by Excelitas was chosen to fulfill this task. This sensor is composed of a matrix of 4×4 thermopile sensors with integrated ASIC, E2PROM, and a microcontroller, capable of measuring the temperature within a view angle of $\pm 20°$. Figure 1 displays the response of the sensor. This array will provide only the position of the fire onset; its reliability will be assured by using different fire detection systems like flame and smoke sensors.

2 System Description

Two sensors are utilized for fire localization automated system. These sensors are placed in parallel position at the same height and on the same axis, as shown in Fig. 2. The entire system consistency will be assured by the cooperation of different fire detection systems like smoke and flame sensors. The center of gravity of the

M. Norgia • A. Magnani • A. Pesatori (✉)
Dipartimento di Elettronica Informazione e Bioingegneria,
Politecnico di Milano, Via Ponzio 34/5, Milano, Italy
e-mail: pesatori@elet.polimi.it

C. Di Natale et al. (eds.), *Sensors and Microsystems. Proceedings of the 17th National Conference, Brescia, Italy, 5-7 February 2013*, Lecture Notes in Electrical Engineering 268, DOI 10.1007/978-3-319-00684-0_8, © Springer International Publishing Switzerland 2014

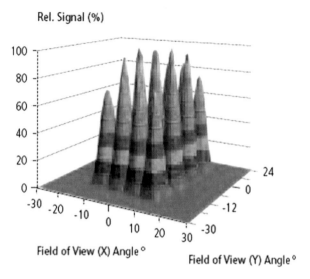

Fig. 1 Thermopile array response due to an infrared source at different field of view angles

onset of fire is calculated through an algorithm based on the triangulation [5, 6] of the two thermopile matrix images; then the three-dimensional "fire coordinates" are sent to an automated fire extinguisher which will take care of directing the beam in that location and extinguishing the fire. The sensor's angle of view is limited to about 20°, which is sufficient for an application in a tunnel, providing a field of view of several tens of meters. To estimate with sufficient accuracy, the distance of the fire onset is necessary to calculate the center of gravity of the image on the two arrays of sensors, with resolution far superior to the individual pixels.

In order to do that, an algorithm for calculating two-dimensional center of gravity was developed, by attributing to each pixel a weight proportional to the optical power measured. To obtain a reliable measure, it is necessary to subtract first the signal generated by the environment where the system is placed. In this specific case, the fifth value in amplitude is subtracted to all the pixels, and all negative results are set to zero. In this way only the four higher pixels give nonzero values: assuming that the image generated by fire cannot cover an area greater than the sensor one. After some experimental tests, it was seen that the image of warm body has a size smaller than a single pixel, with a shape comparable to a Gaussian curve as the one shown in Fig. 3a (compared to the pixels, indicated as gray rectangles). Applying a centroid algorithm with an illumination of this kind, a strongly nonlinear response curve is obtained, very close to the discretization of a pixel as shown in Fig. 3a.

To achieve the required resolution, well below of a single pixel, two solutions have been performed. The first one was to physically modify the sensor, getting out of focus the array of thermopiles, thus enlarging the size of the spot on the sensor and consequently the accuracy of measurement. This technique has produced good results, but has the drawback of requiring a commercial sensor modification,

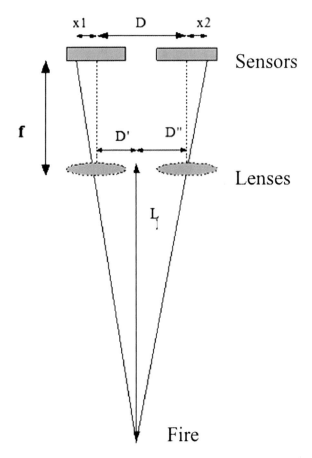

Fig. 2 Setup of fire localization system

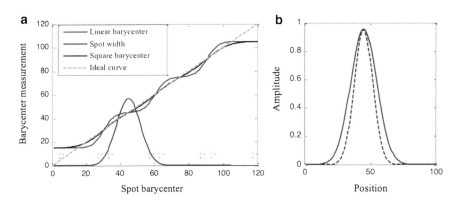

Fig. 3 (**a**) Sensor response. (**b**) Comparison between Gaussian curve (*dotted line*) and its square root

which then leads to a considerable increase in production costs: since it would be necessary to add a second lens or, equivalently, modify the support of the lens of the silicon sensor.

Inspired by the first, the second solution consists of inducing a dummy enlargement of the light spot on the sensor through a nonlinear compression of the measured values.

The function considered is the exponentiation to a power less than 1. As an example, Fig. 3b shows the comparison between a Gaussian curve and its square root, both normalized to the same maximum value.

Figure 3a also shows the extent of the center of gravity obtained by applying compression to the square root of the weight of the individual pixels. As can be seen, the linearity is considerably improved, enabling a resolution of about 1/20 of pixels. The choice of the compression function is dictated by two conflicting requirements: the higher the nonlinear compression (e.g., exponent 0.2), the more linear the sensor's theoretical response becomes, but the compression enhances the noise and the noise floor too. Experimentally the best results were obtained with an exponent equal to 0.5; the square root yields enough spot enlargements, without losing too much signal-to-noise ratio.

Once the single sensor has been characterized, which has shown an angular resolution of about 0.2°, a prototype of the measuring instrument has been made. The two sensors are placed at a distance $b = 1$ m (triangulator baseline), and the focal length of their silicone lenses is 5.5 mm. The data from the sensors are acquired by a microcontroller and transmitted through USB to a personal computer. The performance of this first prototype is promising: at a distance of 10 m, it produces a resolution of about 0.5 m, which worsens quadratically with the distance of the hot target.

3 Conclusions

An automated system for fire onset detection and localization was successfully designed and realized. The system is based on two sensors realized by a thermopile array and a nonlinear elaboration of the 4×4 pixel image. Its performance and very low cost are adequate for fire location inside a highway tunnel or in a house room.

Acknowledgments The authors wish to thank Excelitas Technologies and Caccialanza S.p.A for the collaboration.

References

1. R. Gagnon, Design of Water-Based Fire Protection Systems, Delmar Cengage Learning, Clifton Park, NY; 1996.
2. Mark Bromann, The Design and Layout of Fire Sprinkler Systems, CRC, Boca Raton, FL; 2nd edition April 5, 2001

3. R.F. Aird, "Detection and Alarming of Early Appearance of Fire Using CCTV Cameras", Nuclear Engineering International, Fire & Safety '97 Conference, 1997.
4. A. Pesatori, M. Norgia, C. Svelto, "Automated Vision System for Rapid Fire Onset Detection", I2MTC 2009, pp. 163–166.
5. S. Wang, M. Berentsen, T. Kaiser, "Signal processing algorithms for fire localization using temperature sensor arrays", Fire Safety Journal, **40** (8), pp. 689–697, (2005).
6. R. Gonzalez, R. Woods, Digital Image Processing, Prentice Hall, Upper Saddle River, NJ, 2002.

Localization, Recognition, and Classification of a Superficial Seismic Source in an Inhomogeneous Mean

F. Lo Castro, M. De Luca, and S. Iarossi

1 Introduction

Localization of the source of a signal coming from a superficial seismic source is more complex than the localization of a source into a homogeneous medium [1, 7].

In fact, wave propagation in an inhomogeneous medium, such as the earth ground, is affected by reflections and nonlinearity due to the elastic properties of the terrain and its subjacent layers that modify the shape of the source signal, introducing new attributes noticeable in time and in frequency domain. In addition, prolonged rainfalls and sun exposure cause changes in the properties of the medium itself.

To take account of these phenomena and quantify them, a shaker has been used to characterize the ground.

Understanding the impacts of the variability in soil density, moisture content, and ground grain size in signal transmission may allow further refinement of the developed algorithms for localization [2] and recognition [3] of the event.

The goal of the recognition and classification algorithms is to distinguish signals such as the ground impacts of a device from other surface activities as walker footsteps or the rolling wheels of vehicles [4–6].

2 Experimental Setup and Signal Processing

We conducted a series of seismic experiment in a terrain composed of calcarenites. At the center of the area under testing, four 3-axial geophones are placed in cross constellation with an axis length of 15 m (Fig. 2a).

F.L. Castro (✉) • M. De Luca • S. Iarossi
IDASC-CNR, Area della Ricerca di Roma Tor Vergata,
Via Fosso del Cavaliere 100, Roma 00133, Italy
e-mail: fabio.locastro@idasc.cnr.it

C. Di Natale et al. (eds.), *Sensors and Microsystems: Proceedings of the 17th National Conference, Brescia, Italy, 5-7 February 2013*, Lecture Notes in Electrical Engineering 268, DOI 10.1007/978-3-319-00684-0_9, © Springer International Publishing Switzerland 2014

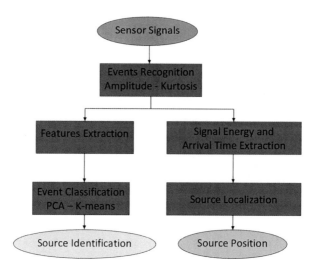

Fig. 1 Flowchart of the localization, recognition, and classification method

A pulse excitation is applied via a shaker to the earth surface in order to probe the terrain. In the experiment, as the seismic source, a person vertically strikes the ground in different locations both with a heavy pickaxe (for far location, more than 15 m from the center of sensor constellation) and with a weight dropped from a height of 1.5 m (in case of near location). A series of ten or three bangs has been made using respectively the pickaxe and the weight dropped for each selected location.

Signals have been acquired and processed by a seismic apparatus (16 channels, 10 kHz sample rate, LabVIEW controlled) developed ad hoc for this experiment.

2.1 Signal Processing

The first step in signal analysis and processing is event recognition [8] of the recorded trace and its time-frequency decomposition [9]. The elliptical and rectilinear polarization of the signal has been used to separate respectively the Rayleigh and the Love superficial waves [10, 11] generated by the bangs. At the end, the DOA (Direction of Arrival) and the RSSR (Received Signal Strength Ratio) techniques have been used for the localization of the seismic source [12, 13].

Moreover, sensor signals have been used for event recognition and classification using PCA and the K-means algorithms (Fig. 1).

As input of the PCA, the features extracted from the impulse received by the geophones were signal width, envelope rise and fall times, angle of decay, signal strength, kurtosis, and the amplitude of the frequency bins centered at 1, 2, 4, 8, 16, 32, 63, and 125 Hz.

3 Results

As shown in Table 1 and in Fig. 2a, localization error changes point-by-point in relation both to the inhomogeneity of the terrain and to the distance from the center of the seismic antenna. Increasing the repetition number of bangs in the same location reduces the localization error around 15 % both in the distance from the center of the antenna and in the angle of arrival.

In this paper, event recognition has been limited to distinguishing the bangs from the pickaxe from other sources such as footsteps, vehicle transit, standard weight impact, and other undefined events.

Table 1 Results of localization

Location	1	2	3
True source location X coordinate (m)	21.5	15.8	−3.4
True source location Y coordinate (m)	−19.1	24.3	−0.4
True θ angle (°)[a]	−42	+57	−173
True distance (m)[a]	28.0	29.0	3.4
Estimated source location X coordinate (m)	21.6	18.7	−1.0
Estimated source location Y coordinate (m)	−16.7	23.3	−0.9
Distance (m)[a]	27.3	29.9	1.4
Uncertainty (m)[b]	±9.8	±10.6	±2.1
Estimated θ angle (°)[a]	−38.0	46.9	−132.6
Uncertainty θ angle (°)[b]	±19.8	±19.5	±56.3

[a]From center of seismic antenna
[b]Expanded uncertainty at 95 % confidence level

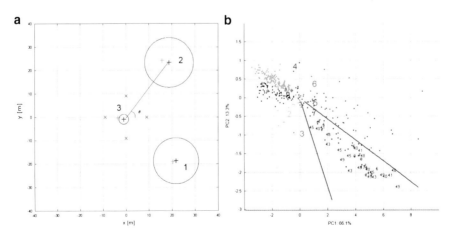

Fig. 2 (a) Localization: X (*blue*) sensors, + (*green*) true source location, + (*red*) estimated source location; (b) PCA and cluster classification of 372 different events; the *numbered dots* are the events of interest (pickaxe bangs)

The identification rate has been around 50 % (true positive and negative) for a pickaxe event with around 10 % of false positive and around 90 % of false positive (Fig. 2b).

4 Conclusions

A whole system for seismic wave acquisition and processing that has permitted the first approach for seismic source localization, recognition, and classification has been developed.

Future development will be focused on the accuracy of single-shot localization, source localization, and feature extraction in order to permit a better event classification.

Acknowledgments We gratefully acknowledge the assistance and the encouragement of Prof. A. D'Amico (Department of Electronic Engineering, Tor Vergata University, Rome, Italy) during the preparation of this work and the technical support given by Mr. Luca Imperatori and Mr. Massimo Di Menno Di Bucchianico (CNR IDASC).

References

1. R.L. Field, R. J. Vaccaro, C. S. Ramalingam, D. W. Tufts. Least-squares time-delay estimation for transient signals in a multipath environment. J. Acoust. Soc. Am. 92 (1), 210–218 (1992)
2. H.C. Schau and A.Z. Robinson. Passive Source Localization Employing Intersecting Spherical Surfaces From Time-Of-Arrival Difference. Ieee Transactions On Acoustics, Speech, And Signal Processing, Vol. ASSP-35, No. 8, 1223–1225 (1987)
3. Luigi Fabbris. Statistica Multivariata: analisi esplorativa dei dati. McGraw-Hill, Milano 1997
4. Michael S. Richman, Douglas S. Deafrick, Robert J. Nation, Scott L. Whitney. Personnel Tracking Using Seismic Sensors. Proceedings of SPIE, 4393, 14–21 (2001)
5. G. Succi, G. Prado, R. Gampert, T. Pedersen, and H. Dhalival. Footstep Detection and Tracking. Proceedings of SPIE, 4393, 22–29 (2001)
6. G. Succi, G. Prado, R. Gampert, T. Pedersen and H. Dhaliwal. Problems in Seismic Detection and Tracking. Unattended Ground Sensor Technologies and Applications II, Edward M. Carapezza, Todd M. Hintz, Editors, Proceedings of SPIE Vol. 4040, 165–174 (2000)
7. Stafsudd J., Asgari S., Ali A. M., Chen C. E., Hudson R. E., Lorenzelli F., Yao K., Taciroglu E. Analysis, Implementation, and Application of Acoustic and Seismic Arrays. ACTA Automatica Sinica Vol. 32, No. 6, 929–937 (2006)
8. P. Wagenaars, P.A.A.F. Waouters, P.C.J.M. van der Wielen, E.F. Steennis. Accurate Estimation of the Time of Arrival of Partial Discharge Pulses in Cable Systems in Service. IEEE Transactions on Dielectrics and Electrical insulation Vol. 15, No. 4, 1190–1199, (2008)
9. Roohollah Askari, R.J. Ferguson. Dispersion And The Dissipative Characteristics Of Surface Waves In The Generalized S-transform Domain. CREWES Research Report – Vol. 22 (2010)
10. Andy Jurkevics, Polarization Analysis of Three-Component Array Data. Bulletin of Seismological Society of America, Vol. 78 No. 5, 1725–1743 (1988)
11. R. Lynn Kirlin, J. Nabelek, G. Lin. Triaxial Array Separation of Rayleigh and Love Waves. IEEE Proceedings of ASILOMAR-29, 722–725 (1996)

12. K.W. Cheung, H.C. So, W.K. Ma, Y.T. Chan. A Constrained Least Squares Approach to Mobile Positioning: Algorithms and Optimality. EURASIP Journal on Applied signal Processing Vol. 2006 Article ID. 20858 doi: 10.1155/ASP/2006/20858
13. P. Bergamo, S. Asgari, H. Wang, D. Maniezzo, L. Yip, R.E. Hudson, K. Yao, D. Estrin. Collaborative Sensor Networking Towards Real-Time Acoustical Beamforming in Free-Space and Limited Reverberance. IEEE Transaction on Mobile Computing Vol. 3, No. 3, 211–224 (2004)

Design and Fabrication of a Compact *p–v* Probe for Acoustic Impedance Measurement

M. Piotto, A.N. Longhitano, P. Bruschi, M. Buiat, G. Sacchi, and D. Stanzial

1 Introduction

The acoustic specific impedance is defined as the ratio between the sound pressure, *p*, and the acoustic particle velocity, *v*, at a given point. Its measurement is useful to characterize the response of passive elements, like wind instruments, absorbing materials, and human ear, to harmonic excitation. Different methods have been proposed to measure the acoustic impedance [1]. The most straightforward way consists in detecting directly the pressure and particle velocity in a point, simultaneously, by means of a *p–v* probe. The pressure is conveniently measured by means of a microphone, while the direct measurement of acoustic particle velocity is usually more difficult due to the lack of compact sensors with a good resolution. For this reason, the particle velocity has been usually derived from a pressure gradient measurement. Since the development of silicon micromachining technology, the fabrication of miniaturized sensors for the particle velocity detection has become feasible [2, 3].

In this work we proposed a novel compact *p–v* probe for the measurement of the acoustic impedance. The pressure is measured with a miniaturized preamplified microphone, while a silicon micromachined device, similar to that described in [3], has been fabricated for the measurement of the particle velocity. Differently from the device proposed in [2], the velocity sensor has been fabricated using a technique

M. Piotto (✉)
IEIIT—Pisa, CNR, via G. Caruso 16, 56122 Pisa, Italy
e-mail: massimo.piotto@cnr.it

A.N. Longhitano • P. Bruschi
Dipartimento di Ingegneria dell'Informazione, University of Pisa,
via Caruso 16, 56122 Pisa, Italy

M. Buiat • G. Sacchi • D. Stanzial
CNR-IDAS "Corbino" c/o Department of Physics and Earth Sciences,
University of Ferrara, v. Saragat 1, 44122, Ferrara, Italy

C. Di Natale et al. (eds.), *Sensors and Microsystems: Proceedings of the 17th National Conference, Brescia, Italy, 5-7 February 2013*, Lecture Notes in Electrical Engineering 268, DOI 10.1007/978-3-319-00684-0_10, © Springer International Publishing Switzerland 2014

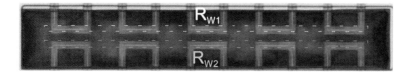

Fig. 1 Optical micrograph of the sensing structure after the silicon removal

compatible with the fabrication processes of standard integrated microelectronic circuits. This paves the way for the integration on the same chip of both the transducer and the read-out circuits, forming the so-called smart sensor [4].

2   Velocity Sensor Description and Fabrication

The velocity sensor is based on the modulation of the heat exchange between two heaters caused by the local velocity of the air. The heaters are two silicided *n-polysilicon* wires placed over suspended silicon dioxide membranes. The wires are heated by an electrical current, and their temperature coefficient of resistance (TCR) is used to convert the heater temperature oscillations caused by the acoustic wave into voltage oscillations.

The device has been designed with the BCD6s process of STMicroelectronics, and a simple post-processing technique has been applied to thermally insulate the sensing structure from the substrate. Details about the fabrication process are reported in [3].

Briefly, dielectric layers have been selectively removed by means of a photolithographic step followed by a standard carbon tetrafluoride (CF_4) plasma RIE (reactive ion etching). Then, the bare silicon has been anisotropically etched with a TMAH solution. Fig. 1 shows an optical micrograph of the device after the silicon removal: the two wires, R_{W1} and R_{W2}, have been divided into five sections in order to increase the device robustness. Each section is placed over a U-shaped dielectric membrane suspended over the cavity etched into the silicon substrate.

3   *p–v* Probe Design and Fabrication

The *p–v* probe is made up of the velocity sensor, a commercial microphone, and an electronic interface to bias the devices and read the signals.

As far as the velocity sensor is concerned, the designed electronic circuit is shown in Fig. 2. The two integrated sensing wires, R_{W1} and R_{W2}, form a Wheatstone bridge with two constant resistors, R_{B1} and R_{B2}. The capacitors C_H and resistors R_H form a fully differential high-pass filter introduced to prevent DC voltages, deriving from unavoidable bridge unbalance, from saturating the amplifier.

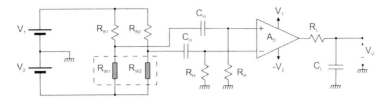

Fig. 2 Circuit used to bias the velocity sensor and filter and amplify the signal

Fig. 3 Photo of the fabricated *p–v* probe

The roll-off frequency of the $C_H R_H$ filters has been set to 10 Hz. The signal has been amplified by a 60 dB gain, low noise instrumentation amplifier (AD8421). The capacitor C_L and resistor R_L form a low-pass filter with a roll-off frequency of 30 kHz to reduce noise in the output signal.

The electronic interface has been fabricated on a double side PCB with surface mount components, and it has been assembled together with the sensor chip and a miniature Knowles® microphone. The PCB has been inserted into a hollow cylinder which is used for the measurement of the acoustic immittance of the ear. The inner diameter of the cylinder is 11 mm. Figure 3 shows the front side of the PCB with the micromachined chip, the amplifier, and the filters. The microphone and the voltage regulators for the power supply have been placed on the back side.

4   Preliminary Characterization

A preliminary test of the *p–v* probe has been performed by means of the standing wave tube technique. A loudspeaker, driven by a signal generator, was fit to one end of a plastic tube of 5 cm diameter and 1.2 m length. The other end of the tube was rigidly closed. The responses of the velocity sensor and the microphone have been acquired with an oscilloscope, and the results are shown in Fig. 4. As expected, the two signals are 90° out of phase.

Fig. 4 Response of the *p–v* probe measured inside a standing wave tube

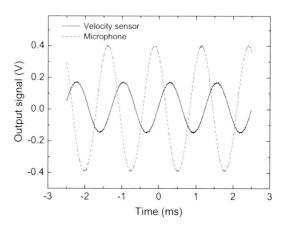

5   Conclusions

A compact *p–v* probe has been designed and fabricated. The device is made up of a commercial miniature microphone, a micromachined velocity sensor, and a read-out electronic interface fabricated on a PCB with surface mount components. Preliminary test inside a standing wave tube confirms the correct operation of the proposed device. Future work will be devoted to a further miniaturization of the probe integrating the velocity sensor and the read-out electronics into a single chip.

Acknowledgments The authors wish to thank (1) the STMicroelectronics R&D group of Cornaredo (MI) for fabricating the chip and (2) DELTATECH srl and Sogliano al Rubicone municipality for supporting this research within the SIHT—Sogliano Industrial High Technology—Project

References

1. J.-P. Dalmont. Acoustic impedance measurement, part I: a review. Journal of Sound and Vibration, **243,** 427–439 (2001)
2. H.-E. de Bree, P. Leussink, T. Korthorst, H. Jansen, T. S. J. Lammerink, M. Elwenspoek. The μ-flown: a novel device for measuring acoustic flows. Sensors and Actuators A: Physical, **54,** 552–557 (1996)
3. P. Bruschi, F. Butti, M. Piotto. CMOS Compatible Acoustic Particle Velocity Sensors. Proc. of IEEE Sensors 2011, 28–31 Oct. 2011, Limerick (Ireland), 1405–1408 (2011)
4. M. Piotto, M. Dei, F. Butti, G. Pennelli, P. Bruschi. Smart flow sensor with on-chip CMOS interface performing offset and pressure effect compensation. IEEE Sensors Journal, **12,** 3309–3317 (2012)

Functional Comparison of Acoustic Admittance Measurements with a CMOS-Compatible *p–v* Microprobe and a Reference One

D. Stanzial, M. Buiat, G. Sacchi, P. Bruschi, and M. Piotto

1 Functional Description of a *p–v* Probe

In acoustics the ratio *p–v* between the sound pressure signal and the particle velocity one defines the impedance, while the sound intensity is given by their product *p–v*. A *p–v* probe, thanks to a velocimetric sensor and a miniaturized preamplified microphone (Fig. 1a, b), can measure directly and simultaneously both these quantities and then indirectly the sound impedance and intensity. The velocity sensor, used here, is based on a silicon micromachined device, which is made up of two silicided *n*-polysilicon wires, each divided into five sections placed over suspended dielectric membranes (see [1] for details). Heat exchange between the wires, heated by an electrical current, is modulated by the air particle flow, producing temperature oscillations, which are transformed into voltage oscillations by the wire temperature coefficient of resistance (TCR).

2 Experimental Setup

The experimental setup for the comparison test is shown in Fig. 2 and has been realized inside a 100 m³ room. It consists of a dodecahedral acoustic source and a tripod supporting the microprobes under comparison (a): the probes have been

D. Stanzial • M. Buiat • G. Sacchi
CNR-IDAS "Corbino" c/o Dept. of Physics and Earth Sciences,
University of Ferrara, v. Saragat 1, Ferrara, Italy

P. Bruschi
Dip. di Ingegneria dell'Informazione, University of Pisa,
v. G. Caruso 16, 56122 Pisa, Italy

M. Piotto (✉)
IEIIT—Pisa, CNR, v. G. Caruso 16, 56122 Pisa, Italy
e-mail: massimo.piottoa@cnr.it

C. Di Natale et al. (eds.), *Sensors and Microsystems: Proceedings of the 17th National Conference, Brescia, Italy, 5-7 February 2013*, Lecture Notes in Electrical Engineering 268, DOI 10.1007/978-3-319-00684-0_11, © Springer International Publishing Switzerland 2014

Fig. 1 Details of the SIHT #1.0 CMOS-compatible $p–v$ microprobe. From left to right: (**a**) velocimetric sensor; (**b**) miniature Knowles® microphone; (**c**) the assembled $p–v$ microprobe

Fig. 2 The experimental setup for the functional comparison test. From left to right: (**a**) dodeca-hedral acoustic source and the tripod; (**b**) the SIHT#1.0 (*right*) and the Microflown® (*left*) $p–v$ probes; (c) the same probes in reverse configuration

located at the jamb of an open window exposed to high-background noise, stressing the robustness of the collected experimental data, and exposed to a logarithmic sine sweep. Two configurations have been tested: the first one with the SIHT#1.0 probe on the left and the Microflown® one on the right (Fig. 2b) and the other with a reverse orientation (Fig. 2c), in order to test the invariance of their admittance responses.

3 Measurement Results

Since, as demonstrated in [2], the essential feature of any $p–v$ probe is its admit-tance response, a direct comparison of the pressure and velocity signal-to-noise ratios measured by the CMOS-compatible prototype probe (model SIHT#1.0), with the ones measured by a reference Microflown® sound intensity probe, is here reported (see Figs. 3 and 4). The comparison was performed in the 100–1,200 Hz frequency range of interest for the target application, in order to check the possibil-ity of using the prototype $p–v$ probe as a functional device for specific acoustic admittance measurements. The obtained results—which are invariant for the probe reversal test illustrated in Fig. 2—show that, despite the low signal-to-noise ratio of the prototype v-sensor (SNR = 20 dB), the correct calibration of the SIHT#1.0 $p–v$ probe prototype in the range of interest of the functional device is anyway feasible (Fig. 5).

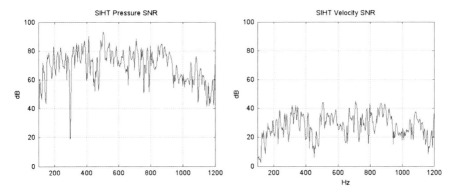

Fig. 3 Signal-to-noise ratio (SNR) of pressure (*left*) and velocity (*right*) measured with the SIHT#1.0 prototype *p–v* probe

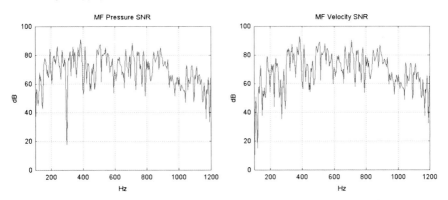

Fig. 4 SNR of pressure (*left*) and velocity (*right*) measured with the reference Microflown® *p–v* probe

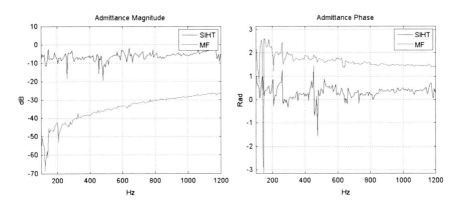

Fig. 5 Admittance amplitude (*left*) and phase (*right*) obtained from rough signals with the prototype probe (*blue*) and the Microflown® one (*red*). The comparison simply shows the feasibility of applying a calibration process to the prototype probe in the range 100–1,200 Hz

4 Conclusions

A comparison between a CMOS-compatible prototype of p–v probe and a reference one has been presented.

Since, as demonstrated in [2], the essential feature of any p–v probe is its admittance response, a direct comparison of the pressure and velocity signal-to-noise ratios measured by the CMOS-compatible prototype probe (model SIHT#1.0), with the ones measured by a reference Microflown® sound intensity probe, has been here reported.

The results—which are invariant for the probe reversal test – show that, despite the low signal-to-noise ratio of the prototype v-sensor (SNR = 20 dB), a comparison calibration of the SIHT#1.0 p–v probe prototype in the range of interest of the functional device is anyway feasible.

Acknowledgments The authors wish to thank DELTATECH and the Municipality of Sogliano al Rubicone (FC) for supporting this research under the "Sogliano Industrial High Technology" (SIHT) project.

References

1. P. Bruschi, M. Piotto. CMOS Compatible Acoustic Particle Velocity Sensors. IEEE Sensors Proceedings, 1405–1408 (2011)
2. D. Stanzial, G. Sacchi, G. Schiffrer. Calibration of pressure–velocity probes using a progressive plane wave reference field and comparison with nominal calibration filters. J. Acoust. Soc. Am **129**, 3745–3755 (2011)

Design and Characterization of a Micro-opto-mechanical Displacement Sensor

E. Schena, M. Cidda, D. Accoto, M. Francomano,
G. Pennazza, E. Guglielmelli, and S. Silvestri

1 Introduction

The growth of interest around micro-fabricated and fiber optic-based sensors (FOS) can be explained in terms of both *economic* factors, such as the cost reduction of micro-fabrication techniques and of key optical components, and *technological* factors, including the increase of components quality and the improvement of metrological characteristics. In the last decades, a big research effort has been devoted in the design of FOS for monitoring parameters of physiological interests [1–3], and it is expected that such sensors will have a central role in several application fields, including medicine and the life sciences.

When a micro-optical element, such as a diffraction grating, is invested by a light source, diffraction patterns are observed on a screen placed beyond it. This phenomenon determines a characteristic light intensity distribution. Several authors investigated the application of diffraction gratings for the measurement of various physical parameters, including force [4] and displacement/strain [5].

The working principle of the proposed transducer is based on the light intensity modulation caused by the relative displacement between two overlapped gratings. A fiber optic is used to transport the laser radiation investing the two micro-diffraction gratings, thus solving issues related to the alignment of optical components as well as to electromagnetic interferences.

E. Schena • M. Cidda • E. Guglielmelli • S. Silvestri
Measurements and Instrumentation Lab, Università Campus
Bio-Medico di Roma, Via Alvaro del Portillo, 21-00128 Rome, Italy

D. Accoto (✉) • M. Francomano
Biomedical Robotics and Biomicrosystems Lab, Università Campus
Bio-Medico di Roma, Via Alvaro del Portillo, 21-00128 Rome, Italy
e-mail: d.accoto@unicampus.it

G. Pennazza
Laboratory for Electronics for Sensor Systems, Università Campus
Bio-Medico di Roma, Via Alvaro del Portillo, 21-00128 Rome, Italy

C. Di Natale et al. (eds.), *Sensors and Microsystems: Proceedings of the 17th National Conference, Brescia, Italy, 5-7 February 2013*, Lecture Notes in Electrical Engineering 268, DOI 10.1007/978-3-319-00684-0_12, © Springer International Publishing Switzerland 2014

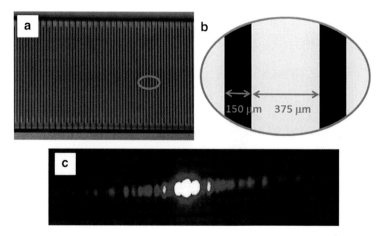

Fig. 1 (a) Snapshot of the microgratings, (b) detail of the microgeometry, (c) diffraction pattern

The aim of this work is to experimentally provide a proof of concept of the novel transduction strategy, also estimating the influence of grating geometry on sensitivity and range of measurement.

2 Sensor Description

The working principle is based on the light modulation induced by the relative displacement of two micro-fabricated diffraction gratings, interposed to a laser source and a fiber tip. Light modulation is monitored by a photodetector, connected to the distal extremity of the fiber in order to avoid misalignment of optical components. Therefore, light intensity can be considered an indirect measurement of the displacement.

The two micro-fabricated gratings were micro-fabricated by lift-off, according to the following steps: (1) photolithography of the negative photoresist MA-N490; (2) deposition of the Pt film (45 nm thick) on a Pyrex substrate by sputtering; and (3) removal of the residual photoresist and platinum in excess.

The gratings, shown in Fig. 1a, b, have a period of 525 μm, each Pt stripe being 150 μm wide and 375 μm apart from the next one (71.4 % duty cycle).

3 Experimental Setup and Results

Experimental trials have been carried out to characterize the sensor. Light emitted by a laser source ($\lambda = 632.8$ nm, Fig. 2a) perpendicularly hits the two overlapped gratings (Fig. 2b): the first is fixed to the optical table; relative displacements are

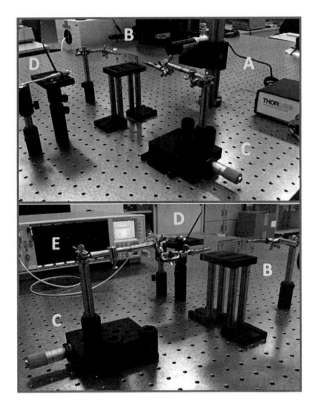

Fig. 2 Experimental setup adopted to perform the static calibration of the sensor: (**a**) laser source, (**b**) micro-diffraction gratings, (**c**) micrometer drives, (**d**) distal extremity of the fiber optic, (**e**) photodetector

applied to the second one by a differential micrometer drive (150–811 ST, Thorlabs, Fig. 2c) perpendicularly to the grating directions. An optical fiber (Fig. 2d) transports the laser radiation transmitted to a photodetector (AQ2200-211, Yokogawa, Fig. 2e).

During the static calibration, five sets of measurements were performed by applying displacement up to 525 μm, in steps of 10 μm. All experimental data are reported as mean±expanded uncertainty (Fig. 3). The uncertainty was calculated using a student reference distribution, with a level of confidence of 95 % and four degrees of freedom.

Figure 3 shows that the transducer has a high and constant sensitivity in the ranges from 30 to 140 μm and from 360 to 490 μm. On the other hand, when the displacement ranges from 140 to 360 μm, the sensitivity decreases, due to the duty cycle (i.e., 71.4 %) of the micro-fabricated gratings. The sensor showed an accuracy better than 4 % in the whole range of measurements.

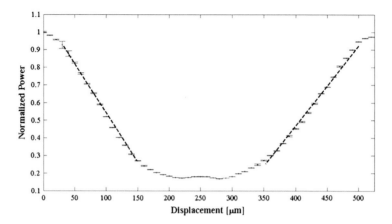

Fig. 3 Experimental data: normalized power vs. displacement; linearity ranges are also reported

4 Discussion and Conclusions

The proposed sensor allows to discriminate displacement lower than 10 μm, using a cost-effective micro-fabrication process. It shows a good linear behavior in two wide ranges of displacement, with high sensitivity and good accuracy (lower than 4 % in the whole range of calibration).

The results also provided useful information to drive the design of the grating geometry. In particular—in order to avoid a marked decrease of the sensitivity, it is necessary to realize gratings with duty cycle of 50 %—the lower the size of the grating period, the higher is the sensitivity; on the other hand, the range of measurement decreases. The use of fiber optics allows to overcome typical alignment issues between the optical components and makes the sensor immune to electromagnetic interferences.

As a step towards the development of fMRI haptic technologies, future work will include the embedding of the sensor in compliant series elastic actuators [6, 7], possibly adapted to incorporate chemical prime movers [8] not generating parasitic electromagnetic fields.

References

1. P. Saccomandi, E. Schena, S. Silvestri. A novel target type low pressure-drop bidirectional optoelectronic air flow sensor for infant artificial ventilation: Measurement principle and static calibration. Rev. Sci. Instrum. **82** (2),024301 (2011).
2. E. Schena, P. Saccomandi, S. Silvestri. A high sensitivity fiber optic macro-bend based gas flow rate transducer for low flow rates: Theory, working principle, and static calibration. Rev. Sci. Instrum. **84** (2), 024301 (2013).

3. S. Silvestri, E. Schena. Micromachined flow sensors in biomedical applications. Micromachines, **3**(2), 2012, pp. 225–243.
4. E. Schena, P. Saccomandi, M. Mastrapasqua, S. Silvestri. An optical fiber based flow transducer for infant ventilation: measurement principle and calibration. In: Medical Measurements and Applications Proceedings (MeMeA), 2011 IEEE International Workshop on. IEEE, 2011. p. 311–315.
5. M. Moscato, E. Schena, P. Saccomandi, M. Francomano, D. Accoto, S. Silvestri. A micromachined intensity-modulated fiber optic sensor for strain measurements: working principle and static calibration. In: IEEE International Conference on Engineering in Medicine and Biology Society Proceedings (EMBC), 2012, pp. 5790–5793.
6. N.L. Tagliamonte, F. Sergi, D. Accoto, G. Carpino, E. Guglielmelli. Double actuation architectures for rendering variable impedance in compliant robots: a review. Mechatronics, **22**(8), 2012, pp. 1187–1203.
7. N.L. Tagliamonte, F. Sergi, G. Carpino, D. Accoto, E. Guglielmelli. Design of a variable impedance differential actuator for wearable robotics applications. In IEEE/RSJ International Conference on Intelligent Robots and Systems (IROS), 2010, pp. 2639–2644.
8. F. Vitale, D. Accoto, L. Turchetti, S. Indini, M.C. Annesini, E. Guglielmelli. Low-temperature H2O2-powered actuators for biorobotics: Thermodynamic and kinetic analysis. In IEEE International Conference on Robotics and Automation (ICRA). 2010.

Multisensor Acoustical Systems: Calibration and Related Problems

S. Ameduri, O. Petrella, V. Quaranta, G. Betta, and M. Laracca

1 Introduction

In the last decades, multi-acoustic sensor systems have done well because of their possible uses in different fields. These systems, also called acoustic antennas, consist of a set of N microphones distributed in a certain way in space, following linear, planar, or three-dimensional geometries.

Acoustic antennas have a wide range of applications in different fields. In automotive, they are used to highlight noise propagation path; in the multimedia sector, these sensors allow to localize a speaker without portable microphones. The civil safety and military fields benefit from these systems as well: gunshot detection in city areas [1], fire prevention in wooded zones [2, 3], and soldier protection from enemy attacks [4] are just some of their possible applications. Even in the aerospace field, there are interesting applications: among others, one recalls the monitoring of air traffic zones. This application was investigated by the authors of the GUARDIAN National Project, aimed at developing innovative acoustic antennas (see Fig. 1a) able to detect approaching aircrafts and to track their acoustic signature [5]. In this paper, starting from their past experience in sensor realization, characterization, and calibration [6–12], the authors propose a calibration procedure for acoustic antennas in order to improve the uncertainty in the estimation of acoustic source position.

S. Ameduri • O. Petrella • V. Quaranta
Italian Aerospace Research Center, Via Maiorise, Capua, Italy

G. Betta • M. Laracca (✉)
Department of Electrical & Information Engineering,
University of Cassino, Via G. Di Biasio, Cassino, Italy
e-mail: m.laracca@unicas.it

C. Di Natale et al. (eds.), *Sensors and Microsystems: Proceedings of the 17th National Conference, Brescia, Italy, 5-7 February 2013*, Lecture Notes in Electrical Engineering 268, DOI 10.1007/978-3-319-00684-0_13, © Springer International Publishing Switzerland 2014

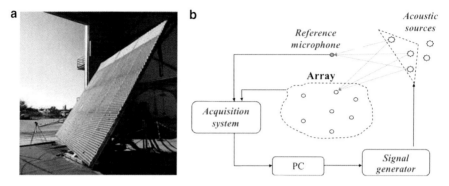

Fig. 1 (**a**) GUARDIAN acoustic antenna; (**b**) architecture of the proposed calibration system

2 Acoustic Antennas: Working Principles and Calibration Problems

An acoustic antenna detects the position of the acoustic source by suitable process-ing of the signals sensed by different microphones. In particular, the signal gener-ated by an acoustic source arrives to the array with a delay depending on the sound speed and on the distance of a single sensor from the source.

The measure of these delays plays a critical role. In fact, besides the intrinsic mechanical and electrical features of a single microphone, its position within the array also contributes to the measurement of signal delay with respect to a reference microphone. All these aspects cause a delay variation of the reception, that is, to receive the signal in a position different from the geometric location, namely, "acoustic" location [13].

The purpose of calibration is to accurately estimate the acoustic position of microphones, a fundamental input for signal processing performed in locating the acoustic source. In the work at hand, an innovative calibration strategy is illustrated, dedicated to the acoustic multisensor systems.

3 The Proposed Calibration Idea

The proposed solution to detect the acoustic position of the microphones is based on a triangulation technique using a certain amount of acoustic sources, located in known points, and evaluating the time delay among the signals sensed by each microphone with respect to a reference one.

Figure 1b shows the architecture of the calibration procedure. It is composed of a set of sources fed by a signal generator which creates the reference acoustic signals. The data acquisition system acquires the signals transduced by the microphones. Finally, a personal computer, by means of a suitable measurement software, manages the calibration system and estimates the acoustic position of the microphones.

A suitable numerical model was developed to estimate the acoustic location of the microphones and the related time delay signals.

This model, based on the solution of the spherical wave equation, provides an estimate of the acoustic pressure on sensor location, produced by an assigned source distribution. The delay between each microphone signal and a reference one is computed by means of a cross correlation function. The result provided by this function strongly depends on two parameters: the sample rate and the phase shift between the compared signals. The first is linked to measurement accuracy; the second has to be lower than the signal period to avoid indetermination related to the cross correlation definition. Moving from the estimated phase delays for each microphone, the triangulation technique is used. In practice the location of each microphone is estimated considering all the possible source triplets. By applying the standard deviation (SD) function, the variation from the average measures was evaluated, this way having an estimate of the accuracy [14]. Through the aforedescribed model, the effects on the SD of layout parameters like source number and location, sample rate, and signal frequency were then analyzed. This study proved the advantages obtainable through a suitable compromise between the number of sources (i.e., the measurement efforts) and the sample frequency (i.e., the acquisition system performance).

Also, the effects due to source location from the array were investigated by pointing out an optimal distance minimizing the SD. In this regard, the negative effects due both to the small delay difference (comparable with the sample dimension and occurring at large distances) and to a wide dispersion of the delay (occurring at small distances) were observed. An additional study was finally carried out aiming at enabling the characterization of large-sized arrays. The calibration of this kind of arrays is made difficult by the fact that it is often impossible to simultaneously satisfy the time delay condition for all the sensors inside the limited domain of an anechoic room. It makes sense, therefore, to carry out an optimization process aiming at finding sources and reference microphone positions satisfying, for the widest number of microphones, the time delay condition.

In the present work, the specific antennas (Fig. 1) developed in the GUARDIAN Project were considered.

A first optimization process, based on genetic algorithm logic, was carried out this way, identifying a first source-microphone distribution with respect to the time delay condition for the 70 % of the microphones. Another optimization was then performed to find out a new layout covering the remaining sensors. The precious information given by the theoretical model (i.e., the influence of source number and location and of sample frequency on SD) allowed to define a dedicated setup that will be exploited to estimate the accuracy of the process.

4 Conclusions

A calibration method to allow the calibration of acoustic antennas is proposed in this paper. The measurement accuracy depends on several parameters, among others, the sampling frequency, the number of sources, the extension of their distribution, and the distance from the array. After the definition of the calibration procedure, an optimization phase was made in a suitable simulation environment in order to obtain precious information on the effects of the mentioned parameters on achievable accuracy for the calibration of even large-sized arrays.

References

1. ShotSpotter Gunshot Location System® (GLS) - http://www.shotspotter.com
2. EU-FIRE Innovative optoelectronic and acoustic sensing technologies for large scale forest fire long term monitoring - www.eufire.org
3. Claus Blaabjerg, Karim Haddad, Wookeun Song, Ignazio Dimino, Vincenzo Quaranta, Alessandro Gemelli, Natalia Corsi, D. X. Viegas, L. P. Pita, "Detecting and Localizing Forest Fires from Emitted Noise" (VI International Conference on Forest Fire Research, 15–18 November 2010, Coimbra, Portugal)
4. Hostile Artillery Locating System (HALO), Selex Galileo, http://www.finmeccanica.com/Corporate/EN/Corporate/Settori/Elettronica_per_la_Difesa_e_Sicurezza/Prodotti/HALO_Selex_Galileo/index.sdo
5. V. Quaranta, S. Ameduri, D. Donisi, M. Bonamente, An in-air passive acoustic surveillance system for air traffic control - GUARDIAN project, ESAV 2011 September 12–14, 2011 - Island Of Capri, Italy.
6. G. Betta, L. Ferrigno, M. Laracca, "Calibration and adjustment of an eddy current based multi-sensor probe for non-destructive testing", Proceeding of IEEE Sensors for Industry Conference – SICON/02, Houston, Texas, USA, pp. 120–124, November 2002.
7. A. Bernieri, G. Betta, L. Ferrigno, M. Laracca, "An automated self-calibrated instrument for non destructive testing on conductive materials", IEEE Transactions on Instrumentation and Measurement, Vol. 53, No. 4, pp. 955–962, August 2004.
8. G. Betta, D. Capriglione, M. Laracca, "Reproducibility and uncertainty of conducted emission measurements for adjustable-speed electrical power drive characterization", IEEE Transactions on Instrumentation and Measurement, Vol. 55, No. 5, pp. 1502–1508, October 2006.
9. A. Bernieri, L. Ferrigno, M. Laracca, A. Tamburrino, "Improving GMR Magnetometer Sensor Uncertainty by Implementing an Automatic Procedure for Calibration and Adjustment", Proceeding of XIX IEEE Instrumentation and Measurement Technology Conference-IMTC/07, Warsaw, Poland, pp. 1–6, May 2007.
10. D. Amicone, A. Bernieri, G. Betta, L. Ferrigno, M. Laracca, "On the Remote Calibration of Electrical Energy Meters", Proceeding of 16th IMEKO TC4 Symposium, Florence, Italy, September 2008.
11. G. Betta, S. Esposito, M. Laracca, M. Pansini, "A novel Sol-Gel-based sensor for humidity detection", Proceeding of 16th IMEKO TC4 Symposium, Florence, Italy, September 2008.
12. S. Esposito, A. Setaro, P. Maddalena, A. Aronne, P. Pernice, M. Laracca, "Synthesis of Cobalt Doped Silica Thin Film for Low Temperature Optical Gas Sensor", Journal of Sol-Gel Science and Technology, Springer, Vol. 60, No. 3, pp. 388–394, May 2011.
13. O. Petrella, V. Quaranta, S. Ameduri, G. Betta, "Preliminary operations for calibrating a phased microphone arrays for air traffic monitoring", ICSV19, July 8–12, 2012, Vilnius, Lithuania
14. O. Petrella, V. Quaranta, S. Ameduri, G. Betta, "Acoustic antenna calibration accuracy: a parametric investigation" Inter-noise 2012, 19–22 August, New York.

Automatic System to Measure the Impedance of Piezoelectric Actuators Used in Ultrasonic Scalpels

A. Bulletti, L. Capineri, and D. Floridia

1 Introduction

The project may be divided into two sections: in the first section is an active volt-amperometric circuit that provides voltage and current of the load impedance when excited with a programmable sinusoidal generator; in the second section, a LabVIEW program has been developed for acquiring from an external oscilloscope or a data acquisition board the amplitude and the phase of the voltage and current sinusoidal signals. When the system is connected to a digital oscilloscope, automatic measurements can be read directly by the Ethernet interface.

Different tests are based on the equivalent model of the actuator which is based on the equivalent circuit of the quartz introduced by Mason [1].

2 Measuring Facility

In the developed software interface, the implementation in the use of simple instructions in automatic and repetitive modes of ultrasonic scalpels is also considered.

The realized system is composed of four items as shown in Fig. 1: a digital oscilloscope for automatic measurements, a programmable signal generator, and a personal computer with LabVIEW software to program and communicate with the external laboratory instruments via USB or Ethernet. The USB port is also used to

A. Bulletti (✉) • L. Capineri
Department of Information Engineering, University of Florence,
Via S. Marta 3, Florence 50139, Italy
e-mail: andrea.bulletti@gmail.com

D. Floridia
Elettronica Bio Medicale s.r.l., Via F. Bettini 13, Foligno 06034, Italy

C. Di Natale et al. (eds.), *Sensors and Microsystems: Proceedings of the 17th National Conference, Brescia, Italy, 5-7 February 2013*, Lecture Notes in Electrical Engineering 268, DOI 10.1007/978-3-319-00684-0_14, © Springer International Publishing Switzerland 2014

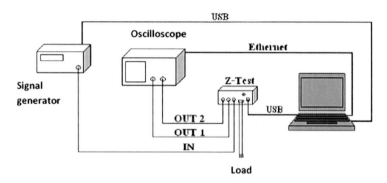

Fig. 1 Block scheme of the realized system

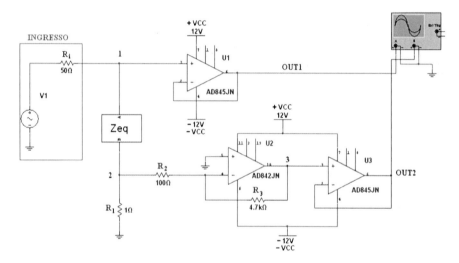

Fig. 2 Schematic electronic circuit of the Z-test impedance tester

provide a power supply to the impedance tester (Z-test) designed. This impedance tester is able to carry out volt-ammeter measurements and is powered directly by the USB port. The schematic electronic circuit of the Z-test is shown in Fig. 2.

The purpose of the Z-test instrument is to determine the values of the electrical impedance Z (magnitude and phase) in the range of frequencies useful for diagnostics of the actuator with a sinusoidal excitation:

$$\vec{Z} = \vec{V} / \vec{I} = |\vec{Z}| e^{j\varphi} \tag{1}$$

The software that manages the entire system has been realized with LabVIEW 8.5 and performs a frequency scanning in a fixed interval with a scanning step fixed by the operator.

Based on the results of many tests carried out on a reference actuator, it has been possible to limit the range of frequencies of interest in the interval from

Fig. 3 The realized Z-test PCB prototype

55.3 to 56 kHz; this frequency range includes the parallel resonance (f_p) and series resonance (f_s) in a narrow range of values estimated at 30 Hz.

Preliminary measurements have been carried out on some ultrasonic scalpels with the serial number and the status description reported below:

- D9E43K035: good ultrasonic scalpel, widely used
- F9G02M075: good ultrasonic scalpel, used
- D9DE5A063: good ultrasonic scalpel, used
- D9DX05057: poor ultrasonic scalpel, inefficient
- D9DV62064: new ultrasonic scalpel

The realized Z-test PCB prototype used for repeated test measurements of the module and the phase response of new or used ultrasonic actuators is shown in Fig. 3.

The most important result, however, was on the standard deviation between the parallel resonance frequency and series resonance frequency which is repeatable in each test with variations of 1–2 Hz. In addition to the repeatability of the measurements with consistent numerical results, we observed a small systematic error due to circuit and parasitic errors, and this error is corrected by the software (Fig. 4).

3   Conclusions

The analysis of database and viewgrams of impedance demonstrates the efficiency of the scalpels during the life cycle. The standard deviation between the parallel and series resonance frequencies is repeatable in each test with variations within 2 Hz.

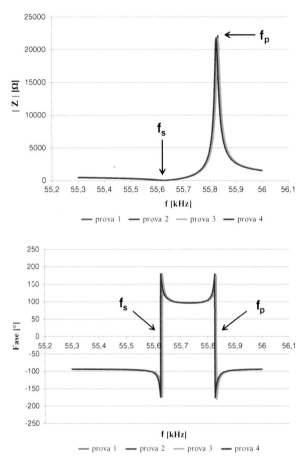

Fig. 4 Phase and module response of the electrical impedance on four repeated tests

Reference

1. W.P. Mason, Electromechanical transducers and wave filters, D. Van Nostrand Co., New York, 1948

Part II
Chemical Sensors

Amperometric Determination of Strong Oxidising Species Through Titanium Electrode Systems

F. Terzi, B. Zanfrognini, J. Pelliciari, L. Pigani, C. Zanardi, and R. Seeber

1 Introduction

Strong oxidising species are commodity chemicals employed in a number of applications, from cellulose pulp to textile bleaching, treatment of drinking and wastewaters and synthesis of inorganics and organics [1]. As a result, strong oxidants need to be determined in many different matrices over a wide concentration range, from less than 1 mM to more than 10 % w/w. The protocols for the quantitative evaluation of these chemicals usually require off-line or at-line approaches, such as titrations, which are often the source of inconveniences.

Electroanalysis may overcome these drawbacks, allowing the measurement system to work on-line and even in-line. A number of electrode systems have been developed for the determination of strong oxidants. In most cases the electrode material is Au, Pt or C; however, they suffer from fouling of the surface and heavy overvoltages affecting the cathodic reduction. Furthermore, most of the electrode systems potentially suitable for the determination of strong oxidants at high concentration in industrial environments require a membrane. Unfortunately, the use of a membrane suffers from serious drawbacks, e.g. the necessity to work under specific flux conditions of the solution, limited stability over time of the membrane itself and narrow operative temperature range.

The present contribution aims at proposing an unusual electrode material for determination of strong oxidants, namely, Ti. Ti has been rarely employed for determining dissolved species by amperometric techniques [2]. In most cases, the studies have been limited to determination of analytes at low concentration values, while examples of the use of titanium electrodes for the analysis of concentrated solutions are still scarce. As to strong oxidants, the interaction between hydrogen peroxide

F. Terzi (✉) · B. Zanfrognini · J. Pelliciari · L. Pigani · C. Zanardi · R. Seeber
Department of Chemical and Geological Sciences, University of Modena
and Reggio Emilia, Via G. Campi 183, Modena 41125, Italy
e-mail: fabio.terzi@unimore.it

C. Di Natale et al. (eds.), *Sensors and Microsystems: Proceedings of the 17th National Conference, Brescia, Italy, 5-7 February 2013*, Lecture Notes in Electrical Engineering 268, DOI 10.1007/978-3-319-00684-0_15, © Springer International Publishing Switzerland 2014

and titanium, e.g. the corrosion processes, has been studied [2], while literature reports on other oxidants are rare.

We report here the cathodic responses of Ti electrodes to different oxidants, namely, H_2O_2 and HClO, in a wide concentration range (1–500 mM) at two pH values (4 and 7). The results have been fruitfully employed to the development of analytical procedures for the determination of the different oxidising species at high concentration values. A statistical treatment of the electrochemical responses suggests that repeatability and reproducibility are quite good.

2   Experimental Section

All reagents were from Sigma. H_2O_2 was from Carlo Erba (40 w/v). Samples of detergents were from Officina Naturae (Italy): a general-purpose detergent for hard surfaces has been tested (trade name: Detersivo Universale).

The electrochemical measurements were performed with an Autolab PGSTAT12 (Eco Chemie) potentiostat/galvanostat, in a single-compartment three-electrode cell, at room temperature. 2 mm diameter Ti disks (grade 1) were used as working electrodes. They were polished subsequently with 1, 0.3 and 0.05 µm Al_2O_3 powder and then rinsed with Millipore water. An aqueous saturated calomel electrode, KCl sat. (Amel), was the reference electrode; all the potential values given are referred to it. A GC rod was the auxiliary electrode.

The solutions for the electrochemical studies were 1 M phosphate and citrate buffers at pH 7 and 4 for H_2O_2 and at pH 7 for HClO. pH values and concentration range of H_2O_2 and HClO have been selected considering the relevant most diffused industrial applications [1]. Voltammetric traces have been recorded starting from 0.0 V at 50 mV s^{-1} potential scan rate; in correspondence to the fourth scan, a steady state was reached, and the response could be used for analytical purposes. Random additions have been employed for tracing the calibration curves. All the experiments have been carried out in solutions in equilibrium with the atmosphere, i.e. in the presence of dissolved O_2, in order to mimic the conditions typical of the applicative contexts. The concentration of H_2O_2 both in buffered and in detergent samples has been determined through a standard titration procedure, namely, ASTM D2180 test.

3   Results and Discussion

The voltammetric responses of Ti electrode at the highest concentrations explored are reported in Fig. 1. Above these concentrations the solution pH is not faithfully buffered. In the case of H_2O_2, two current peaks, located at ca. −1.1 and −1.6 V, are evident in the forward potential scan at pH = 7; at pH = 4 they are shifted toward less negative potentials and the peak with Ep ca. −1.0 V appears as a shoulder. In the case of HClO, only one peak (Ep ca. −1.0 V) is present at pH = 7.

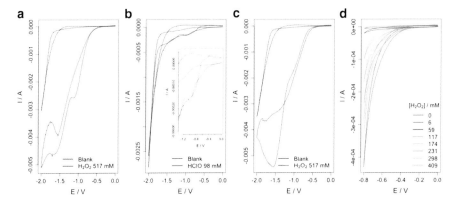

Fig. 1 Voltammetric curves recorded on H_2O_2 (**a**, **c**) and HClO (**b**) aqueous solutions at pH 7 (**a**, **b**) and 4 (**c**)—fourth scans. Representative voltammetric traces recorded in the presence of different H_2O_2 concentrations at pH 7—fourth scans (**d**)

Calibration curves have been constructed by recording voltammetric curves from 0.0 V over a wide potential range at pH = 4 and 7, finally choosing the current values at the −0.8 V. This potential has been adopted because it represents the best compromise between sensitivity and reproducibility: the sensitivity [3] decreases at decreasing the polarisation potential, e.g. 0.09, 0.54 and 1.30×10^{-6} A mM^{-1} for H_2O_2, pH = 7, at −0.6 V, −0.8 V and −0.9 V, respectively, and the confidence interval of the prediction at the midrange concentration is significantly higher in the case of −0.6 and −0.9 V, e.g. 44, 16 and 30 mM for H_2O_2, pH = 7, at −0.6 V, −0.8 V and −0.9 V, respectively, which accounts for higher random errors. Moreover, the best reproducibility between different sets of measurements has been obtained by cycling the potential from 0.0 to −0.8 V (50 scans, 50 mV s^{-1} scan rate) in the presence of the maximum H_2O_2 or HClO concentration explored.

The experimental data are best fitted by a parabolic curve; the complexity of the underlying mechanism implies that an explanation to this behaviour is speculative. The electrode systems and the relevant responses result are also well reproducible, as deduced, by use of t-tests, by the lack of significant differences between the calibration plots obtained with three electrodes on the highest and lowest sensitivity values and the relevant standard deviations ($p > 95$ %).

The system and experimental conditions were not specifically developed for the determination of particularly low H_2O_2 or HClO concentrations. However, calibration plots over a range of low concentrations, much narrower with respect to the previous case (0–20 mM for H_2O_2 and 0–5 mM for HClO), have been drawn to estimate the potentiality of the system in similar situations. The results show that the limit of detection (LOD), computed on the calibration plot over a low concentration range, is below ca. 2 mM for both species at the different pH values. By computing LOD as three times the standard deviation of the blank, as is common practice in the literature, it results below ca. 0.01 mM.

Table 1 Main figures of merits of the different calibration plots

	Sensitivity ($\times 10^6$ A mM^{-1})	Confidence interval of prediction in correspondence of midrange concentration value (mM, $p=0.95$)
H$_2$O$_2$ pH 7 (0–400 mM)	0.54	16
HClO pH 7 (0–100 mM)	0.59	4
H$_2$O$_2$ pH 7 (0–400 mM)	0.40	13
Detergent sample (31–1,100 mM)	0.015	25
Detergent sample (0.013–126 mM)	0.25	3

As a sound example of the antifouling properties of Ti electrode material, the H$_2$O$_2$ content in detergent samples was determined. The formulation of detergents is very complex [4]; in our case, it consists of cocoglycerides and cocamidopropyl betaines (4 % w/w) and ethanol (8 % v/v) in aqueous solvent. It is to notice that it is impossible to remove H$_2$O$_2$ from the matrix without altering its composition, due to the presence of volatile compounds. A conventional calibration, consisting of addition of different, known amounts of H$_2$O$_2$ to the background, is hence impractical. That is why the starting solution consisted of a detergent sample containing 31.0 mM H$_2$O$_2$ and was obtained by dilution of the pure detergent solution with demineralised water (10 ml, pH = 5.3). Concentration of H$_2$O$_2$ in this solution was computed by titration, according to the cited standard procedure; then, calibration plots were constructed, adding known aliquots of standard H$_2$O$_2$ solution (40 % w/v), dissolved in small volumes (50 µl each), not to alter solution composition. Repeated tests led to well-reproducible calibration curves in a concentration range from 31 mM to 1 M using a parabolic regression equation. When applied for the determination of an unknown sample, the H$_2$O$_2$ concentration can be determined with good accuracy ($p > 95$ %).

In particular, the specific case of detergents requires that solutions for the different applications are obtained by dilution of the commercial product. By such a procedure, an H$_2$O$_2$ concentration range from ca. 126 to 0.013 mM was explored. A parabolic function still results to regress at best the experimental points. These tests also show good repeatability and reproducibility, showing that the analytical determination is possible in-line, at any stages of the detergent production and use (Table 1).

4 Conclusions

Ti electrodes represent an interesting amperometric probe for in-line determination of strong oxidants over a wide concentration range. In particular, repeatable and reproducible voltammetric traces are obtained in buffered solution containing H$_2$O$_2$ and HClO. The resistance to fouling in complex matrices has been checked on detergent samples.

References

1. C.W. Jones, Applications of Hydrogen Peroxide and Derivatives (RSC Publishing, London 1999)
2. T. Clark, D.C. Johnson. Activation of Titanium Electrodes for Hydrogen Peroxide in Alkaline Media. Electroanal. **9** (4), 273–278 (1997)
3. M. Reichenbacher, J.W. Einax, Challenges in Analytical Quality Assurance (Springer, Heidelberg, 2011)
4. K.-Y. Lai, Liquid detergent, 2nd edn. (CRC Press, Boca Raton, 2005)

Oligopeptides-Based Gas Sensing for Food Quality Control

D. Pizzoni, P. Pittia, M. Del Carlo, D. Compagnone, and C. Di Natale

1 Introduction

Food aromas and flavours are mainly dependent on the original volatile molecules composition and product matrix but are also strictly influenced by the production and modification occurring during food processing and packaging. The monitoring of the aromatic patterns released by food during production and processing is an important task to ensure quality of the final product.

In the last years, electronic nose has proved to be a very useful tool for food and aromas analysis [1].

In literature, many works can be found reporting the use of peptides as sensing elements in gas sensors array [2, 3].

In this work, a new approach in the realization of peptide-based electronic nose with application to real food samples is presented.

Peptide was linked to gold nanoparticles (GNPs) by self-assembling, and sensors were then modified by drop casting of a peptide-GNP suspension.

D. Pizzoni (✉) • P. Pittia • M. Del Carlo • D. Compagnone
Department of Food Science, University of Teramo, Via C.R. Lerici 1,
Mosciano S.A. (TE) 64023, Italy
e-mail: dpizzoni@unite.it

C. Di Natale
Department of Electronic Engineering, University of Tor Vergata,
Via del Politecnico 1, Rome 00133, Italy

C. Di Natale et al. (eds.), *Sensors and Microsystems: Proceedings of the 17th National Conference, Brescia, Italy, 5-7 February 2013*, Lecture Notes in Electrical Engineering 268, DOI 10.1007/978-3-319-00684-0_16, © Springer International Publishing Switzerland 2014

2   Materials and Methods

2.1   Materials

Reagents were all purchased from Sigma-Aldrich Italia (Milan, Italy). 20 MHz quartz crystal microbalances (QCMs) were from Elba Tech (Isola d'Elba, LI, Italy).

Gummy candies samples were provided by GELCO Srl (Castelnuovo Vomano, TE, Italy).

2.2   GNP Synthesis and Deposition on QCM

GNPs were synthesized by sodium borohydride reduction method [4]. Briefly, 100 mL of a 10^{-4} M aqueous solution of tetrachloroauric acid was reduced by 0.01 g of NaBH$_4$ at room temperature. A ruby-red gold hydrosol, containing GNPs of 2 nm average diameter, was formed after the reaction.

GNPs were then capped by self-assembly incubating 10^{-4} M aqueous solution of coated thiolated compounds at room temperature for 4 h.

GNPs were coated with seven different peptides (or single amino acid). The seven peptides were GNP-Cys-Gly, GNP-glutathione (GNP-GSH), GNP-Cys, GNP-γ-Glu-Cys, GNP-Cys-Arg-Gln-Val-Phe (GNP-P1), GNP-Cys-Ile-His-Asn-Pro (GNO-P3) and GNP-Cys-Ile-Gln-Pro-Val (GNP-P4) (the last three penta-peptides were synthesized by F-moc chemistry on solid phase using Wang resin).

20 MHz QCM sensors were modified by drop casting of 50 mL of the GNP suspension on each side of the crystal and let to dry in desiccator at room temperature.

2.3   Measurement Set-Up

The electronic nose (model TEN 2011) was developed by Tor Vergata Sensors Group (University of Tor Vergata, Rome) and allowed allocation of up to 8 QCM sensors in the same measuring chamber.

N$_2$ was used as carrier gas and a home-made tubing system with three-way stopcocks allowed the flow to pass through the sample headspace or going directly to the measuring chamber.

In a typical experiment the sample (10 mL of pure solvent or about 500 mg of candies) was introduced in a 100 mL glass laboratory bottle. Three-way stopcocks were open for 2 min to completely remove the air in the bottle (preopening). The stopcocks were then closed for 10 min to equilibrate the headspace at controlled temperature

Table 1 Sensors ΔF responses (in Hz) to ethyl acetate and acetonitrile (measurement condition as in Sect. 2.3)

Sensor	Ethyl acetate	Acetonitrile
GNP-GSH	33	0
GNP-TA	165	44
GNP-P3	89	55
GNP-P4	80	43
GNP-Cys	89	18
GNP	95	3
GNP-Cys-Gly	65	37
GNP-P1	58	10

(25 °C for solvents or 40 °C for candies samples) in stationary conditions. Finally stopcocks were open to carry the head space to the measuring chamber.

The frequency shift (ΔF) was calculated as the difference between the average of the last ten measurements before injecting the vapour sample (baseline) and the average of the last ten values before closing the stopcocks (after equilibration of the signal) for a cleaning cycle carried out with N_2.

3 Results and Discussion

A first approach in the use of GNP-peptide-based sensors was carried out by our group, on pure solvents and aroma solutions, with a slightly different sensors array based only on glutathione and its amino acids and dipeptides (data not shown, see [5]).

After verifying the feasibility of a GNP-peptide-based sensors array, new peptides were synthesized in order to demonstrate that different amino acid sequences (even if with the same length) lead to different responses for the sensors. So, three penta-peptides were chosen on the basis of their different hydrophilic behaviours, isoelectric point and charge at pH 7 that were supposed to guarantee different behaviours in sensing responses and, therefore, a better discrimination.

Table 1 shows ΔF responses for all sensors to ethyl acetate and acetonitrile

For the compounds tested, as supposed, the three pentapeptides gave different relative responses ranging from 10 to 82 %. For ethyl acetate the highest ΔF was given by GNP-TA, while for acetonitrile by GNP-P3. GNP-GSH gave the lowest signal for both compounds, and GNP sensor gave the second lowest ΔF for acetonitrile and the second highest for ethyl acetate.

After demonstrating good performance with pure compounds, electronic nose was challenged with real food sample (gummy candies).

The GNP-peptide-based sensors array was able to discriminate between different candy structures after a PCA analysis of data.

All samples were formulated with strawberry aroma, but three different compounds (gelatin, pectin and gum arabic) were used as structuring agents.

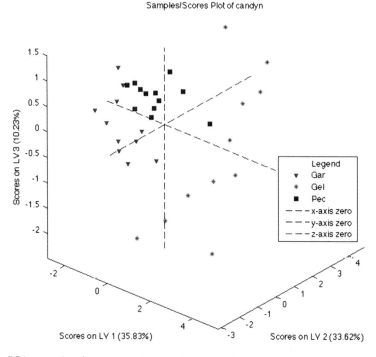

Fig. 1 PCA score plot of gummy candies samples with different structuring agent (gelatin = *green*, pectin = *blue*, gum arabic = *red*)

ΔF data were normalized according to (1) in order to reduce the effect of concentration:

$$\Delta F_i^{'} = \frac{\Delta F_i}{\sum_{j=1}^{8}\Delta F_j} \tag{1}$$

PCA score plot (Fig. 1) showed a clear separation between the samples prepared with different structuring agents demonstrating the ability of the electronic nose to discriminate between different aroma patterns from real matrix.

4 Conclusions

A new gas sensors array, based on GNP-peptides, has been developed.

Pure compounds and real food sample headspace have been analysed with good results in terms of discriminating ability.

Considering the simplicity of this approach and the data obtained on real food samples, GNP-peptide-based gas sensing has, in our opinion, a great potential to be used for food quality and process control.

Taking also into account the large number of possible combination, new peptides, computationally designed on specific purpose, could be realized and tested as future development.

References

1. M. Santonico, P. Pittia, G. Pennazza, E. Martinelli, M. Barnabei, R. Paoloesse, A. D'Amico, D. Compagnone, C. Di Natale. Study of the aroma of artificially flavoured custards by chemical sensor array fingerprinting. Sens Actuators B Chem **133**, 345 (2008).
2. T. Okada, Y. Yamamoto, H. Miyachi, I. Karube, H. Muramatsu. Application of peptide probe for evaluating affinity properties of proteins using quartz crystal microbalance. Biosens Bioelectron, **22** (7), 1480–1486 (2007)
3. S. Sankaran, S. Panigrahi, S. Mallik. Olfactory receptor based piezoelectric biosensors for detection of alcohols related to food safety applications. Sens Actuators B Chem **155**(1), 8–18 (2011)
4. S. Aryal, B.K.C. Remant, N. Dharmaraj, N. Bhattarai, C.H. Kim, H.Y. Kim. Spectroscopic identification of S-Au interaction in cysteine capped gold nanoparticles. Spectrochim Acta A Mol Biomol Spectrosc, **63**(1), 160–163 (2006).
5. D. Compagnone, G.C. Fusella, M. Del Carlo, P. Pittia, E. Martinelli, L. Tortora, R. Paolesse, C. Di Natale. Gold nanoparticles-peptide based gas sensor arrays for the detection of food aromas. Biosens Bioelectron **42**, 618–25 (2013)

Dissolved Oxygen Sensor Based on Reduced Graphene Oxide

S.G. Leonardi, D. Aloisio, M. Latino, N. Donato, and G. Neri

1 Introduction

Oxygen is one of the most important chemical species present in the environment and it is the basis of life of most living species. Oxygen is also involved in several chemical and biochemical processes that are part of the ecosystem. Its concentration level in the gaseous atmosphere as well as in water is then extremely important.

In this work, the attention has been focused on monitoring of dissolved oxygen (DO) in water. The determination of DO has been a subject of considerable interest in the last decades, due to its importance in environmental monitoring, industrial safety, energy conversion, waste management, medical applications, aquaculture, and many other biochemical processes [1]. Classical DO measurement was performed by Winkler titration method. Recently, the development of chemical sensors has replaced this technique. The two most common technologies available for dissolved oxygen sensing are the optical and electrochemical ones [2]. Clark-type sensor has been the first electrochemical device employed to monitor DO, and variations of this one are still widely used [1]. Commercial Clark sensors have high accuracy and sensitivity; however, they are often not very compact and relatively expensive due to the complexity of manufacture and the use of noble metals as electrodes. Actually, many of these problems can be overcome by using screen-printing techniques and more performance electrocatalytic materials [3]. Carbon nanostructures such as CNTs and graphene have attracted considerable interest in the field of electrochemical sensors [4, 5]. Due to their larger surface area, fast electron transfer and excellent support to metal catalyst and protein immobilization, carbon nanostructures promise electrocatalytic ability when used as electrodes modifier in electrochemical reactions.

S.G. Leonardi (✉) • D. Aloisio • M. Latino • N. Donato • G. Neri
Department of Electronic Engineering, Chemistry and Industrial Engineering,
University of Messina, Messina, Italy
e-mail: leonardis@unime.it

C. Di Natale et al. (eds.), *Sensors and Microsystems: Proceedings of the 17th National Conference, Brescia, Italy, 5-7 February 2013*, Lecture Notes in Electrical Engineering 268, DOI 10.1007/978-3-319-00684-0_17, © Springer International Publishing Switzerland 2014

In order to fabricate a simple DO sensor, in this work RGO has been used to modify the electrode surface of a commercial screen-printed device. So, no noble metal has been used in the formulation of the sensing layer with obvious advantages in saving costs.

2 Experimental

2.1 Graphene Synthesis

RGO was prepared by microwave-assisted chemical reduction method starting from graphene oxide (GO) according to what was already reported in a previous paper [6]. GO was prepared from graphite powder (<20 μm, synthetic, Aldrich) by a modified Hummers method. GO was added in benzyl alcohol and sonicated until the homogenous dispersion in solvent. The suspension was heated under microwave irradiation at 185 °C for 10 min, giving the RGO sample.

2.2 Sensor Fabrication

Sensor was fabricated by using a commercial screen-printed electrode device. It consists of a flexible plastic support above which are deposited two silver electrical contacts. The two electrodes, with function of working and counter/reference electrodes, are made of carbon and Ag/AgCl, respectively. Their surfaces are 1.8 mm^2 and 6.2 mm^2, respectively. The working electrode was modified by depositing above it a suspension of RGO dispersed in 5 % Nafion solution and dried at room temperature till the solvent evaporation.

2.3 Characterization

Transmission electron microscopy (TEM) images were obtained on a JEOL EM-2010 at an accelerating voltage of 200 kV. The XPS experiments were performed in an UHV multipurpose surface analysis system (Sigma Probe, ThermoFisher Scientific, UK).

Electrochemical measurements were carried out in a two-electrode mode using a homemade, portable, USB power electronic potentiostat. A dedicated software allows to operate in both chronoamperometric and voltammetric methods. The DO measurements were carried out in 0.1 M KCl solution. DO level was controlled saturating the electrolyte solution by bubbling N_2/O_2 gas mixtures.

CV was performed in the potential range 0 to −2 V at scan rate of 100 mV/s in both nitrogen- and oxygen-saturated solution.

Chronoamperometry experiments were carried out at working electrode potential of −1.35 V. Sensor signal was registered during the bubbling of oxygen into solution. To obtain a homogeneous oxygen concentration, solution was kept under magnetic stirring.

3   Results and Discussion

3.1   Graphene Characterization

XPS and TEM analyses of materials investigated were performed in order to evaluate the nature of the surface functional groups and the morphological characteristics. Figure 1a shows the XPS analysis of the as-prepared RGO and GO precursor. XPS spectra indicate a high contribution due to C–O bond for the GO, while O/C ratio decreases notably in RGO sample. This means that benzyl alcohol under microwave treatment acts as a reducing agent leading to the partial reduction of GO to reduced graphene oxide. Figure 1b reports a TEM micrograph showing the morphological characteristic of exfoliated RGO synthesized.

3.2   Electrochemical Characterization

CV was used to investigate the electrochemical behavior of RGO-modified screen-printed commercial sensor. Figure 2 shows the CV cycles carried out in both

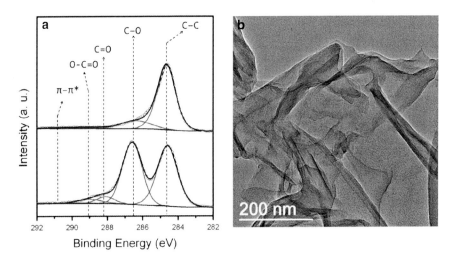

Fig. 1 (**a**) XPS spectra of RGO (*top*) and GO (*bottom*), (**b**) TEM image of RGO

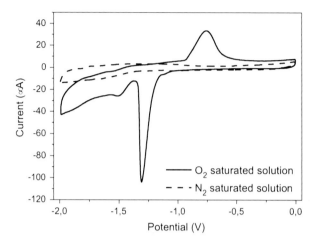

Fig. 2 CV of RGO-modified sensor electrode in N_2- and O_2-saturated 0.1 M KCl solution

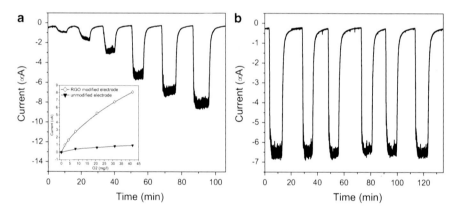

Fig. 3 (**a**) Dynamic response and calibration curve (*inset*), (**b**) response reproducibility

deaerated and oxygen-saturated 0.1 M KCl solution, in the potential range 0 to −2 V and at scan rate of 100 mV/s. Preliminarily, it has been observed that no significant electrochemical process occurred when the sensor with unmodified electrodes was tested under these conditions. Instead, an intense reduction peak has been registered by the RGO-modified electrode at −1.35 V, with onset potential higher than −0.7 V, when oxygen was present in solution.

Figure 3a, b shows the dynamic response of the sensor to increasing concentrations of DO in water and the response reproducibility, respectively. The sensor follows quickly the diffusion of oxygen in solution and completely recovers the baseline signal when the solution is deaerated. A high reproducibility of the response has been also observed. Calibration curves obtained by plotting the measured current

Table 1 Sensor response in two different sample solutions

DO concentration 6.8 mg/l	Current (µA)
0.1 M KCl	2.45
Seawater	2.35

values to different molar concentration of DO, for both RGO-modified electrode and unmodified sensors, are shown in the inset of Fig. 3a. The average sensitivity was calculated to be 6.3 µA/mM with resolution of about 2 mg/l. Instead, much lower sensitivity was shown by the sensor with unmodified electrodes.

In order to assess the possibility to use the fabricated sensor in real applications, it was tested in measuring of the DO level in a seawater sample. The currents obtained by the sensor operated at −1.35 V in 0.1 M KCl solution and in seawater, in the presence of the same molar concentration of DO, are shown in Table 1. Sensor response variation was lower than 5 %. Furthermore, it has demonstrated that the high concentrations of ions present in the seawater do not interfere with the measurement.

4 Conclusion

RGO, prepared by a chemical reduction method from graphene oxide, was used as material for the electrocatalytic reduction of oxygen. Such material was then used for the fabrication of a simple DO electrochemical sensor modifying the working electrode surface of a commercial screen-printed device. The sensor showed excellent sensitivity and reproducibility. Furthermore, tests carried out in seawater samples have shown the possibility of use in real applications.

References

1. M. L. Hichman. The Measurement of Dissolved Oxygen. (John Wiley, New York, 1978), 255
2. The Dissolved Oxygen Handbook. (YSI Incorporated, 2009). www.ysi.com/weknowdo
3. I. Shitanda, S. Mori, M. Itagaki. Screen-printed Dissolved Oxygen Sensor Based on Cerium Oxide-Supported Silver Catalyst and Polydimethylsiloxane Film. Anal. Sci. 27 (10), 1049–1052 (2011)
4. A. Ahammad, J. Lee, M. Rahman. Electrochemical Sensors Based on Carbon Nanotubes. Sensors 9 (4), 2289–2319 (2009)
5. Y. Shao, J. Wang, H. Wu, J. Liu, Ilhan A. Aksay, Y. Lin. Graphene Based Electrochemical Sensors and Biosensors: A Review. Electroanalysis 22, 1027–1036 (2010)
6. P. A. Russo, N. Donato, S. G. Leonardi, S. Baek, D. E. Conte, G. Neri, N. Pinna. Room-Temperature Hydrogen Sensing with Heteronanostructures Based on Reduced Graphene Oxide and Tin Oxide. Angew. Chem. Int. Ed. 51, 11053–11057 (2012)

COST Action TD1105 on New Sensing Technologies for Air-Pollution Control and Environmental Sustainability: Overview in Europe and New Trends

Michele Penza

1 Introduction

This international networking, coordinated by ENEA (Italy), includes over 75 big institutions from 27 COST countries (EU zone: *Belgium, Bulgaria, Czech Republic, Denmark, Finland, France, Germany, Greece, Hungary, Iceland, Ireland, Israel, Italy, Latvia, the former Yugoslav Republic of Macedonia, Netherlands, Norway, Poland, Portugal, Romania, Serbia, Slovenia, Spain, Sweden, Switzerland, Turkey, United Kingdom*) and 5 non-COST countries (extra-Europe: *Australia, Canada, China, Russia, United States*) to create a S&T critical mass in the environmental issues.

This COST Action [1, 2] (see logo in Fig. 1) focuses on a new detection paradigm based on sensing technologies at low cost for air-quality control (AQC) and sets up an interdisciplinary top-level coordinated network to define innovative approaches in sensor nanomaterials, gas sensors, devices, wireless sensor systems, distributed computing, methods, models, standards, and protocols for environmental sustainability within the European Research Area (ERA).

The state of the art showed that research on innovative sensing technologies for AQC based on advanced chemical sensors and sensor systems at low cost, including functional materials and nanotechnologies for eco-sustainability applications, the outdoor/indoor environment control, olfactometry, air-quality modeling, chemical weather forecasting, and related standardization methods, is performed already at the international level but still needs serious efforts for coordination to boost new sensing paradigms for research and innovation. Only a close multidisciplinary

M. Penza (✉)
Technical Unit for Materials Technologies—Brindisi Research Center,
ENEA—Italian National Agency for New Technologies, Energy and Sustainable
Economic Development, PO Box 51 Br-4, Brindisi 72100, Italy
e-mail: michele.penza@enea.it

C. Di Natale et al. (eds.), *Sensors and Microsystems: Proceedings of the 17th National Conference, Brescia, Italy, 5-7 February 2013*, Lecture Notes in Electrical Engineering 268, DOI 10.1007/978-3-319-00684-0_18, © Springer International Publishing Switzerland 2014

Fig. 1 COST office, ESSEM Domain, and COST Action TD1105 EuNetAir logo

cooperation will ensure cleaner air in Europe and reduced negative effects on human health for future generations in smart cities, efficient management of green buildings at low CO_2 emissions, and sustainable economic development.

2 Objectives

The aim of the Action is to create a cooperative network to explore new sensing technologies for low-cost air-pollution control through field studies and laboratory experiments to transfer the results into preventive real-time control practices and global sustainability for monitoring climate changes and outdoor/indoor energy efficiency. Establishment of such a European network, involving non-COST key experts, will enable EU to develop world capabilities in urban sensor technology based on cost-effective nanomaterials and contribute to form a critical mass of researchers suitable for cooperation in science and technology, including training and education, to coordinate outstanding R&D and promote innovation towards industry and support policy makers. Main objectives of Action are listed, but not limited to:

- To establish a top-level Pan-European multidisciplinary R&D platform on new sensing paradigm for AQC contributing to sustainable development, green economy, and social welfare
- To create collaborative research teams in the ERA on the new sensing technologies for AQC in an integrated approach to avoid fragmentation of the research efforts
- To train early-stage researchers (ESR) and new young scientists in the field for supporting competitiveness of European industry by qualified human potential
- To promote gender balance and involvement of ESR in AQC
- To disseminate R&D results on AQC towards industry community and policy makers as well as general public and high schools

The Workplan is organized in four complementary Working Groups (WGs), each devoted to a progressive development of synthesis, characterization, fabrication,

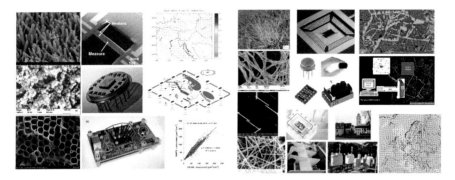

Fig. 2 Selected R&D technological products developed by some partners (academia, research institutes, agencies, industry) involved in the COST Action TD1105 EuNetAir. Courtesy from EuNetAir partnership

integration, prototyping, proof of concepts, modeling, measurements, methods, standards, tests, and application aspects. The four WGs with the specific objectives are:

- *WG1*: Sensor materials and nanotechnology
- *WG2*: Sensors, devices, and sensor systems for AQC
- *WG3*: Environmental measurements and air-pollution modeling
- *WG4*: Protocols and standardization methods

This Action is involved in the study of the sensor nanomaterials and nanotechnologies exhibiting unique properties in terms of chemical and thermal stability, high sensitivity, and selectivity. Nanosize effects of functional materials will be explored for integration in the gas sensors at low power consumption. Furthermore, specific nanostructures with tailored sensing properties will be developed for gas sensors and sensor systems with advanced functionalities.

Selected high-quality research products and innovative technologies developed by the partnership of COST Action TD1105 are shown in Fig. 2.

3 Scientific Program

This COST Action focuses on the new sensing technologies for low-cost AQC. This is of strategic importance for the development of European industrial capability in environmental sustainability and green economy and to protect the public health. In the past, noticeable scientific efforts have been realized to develop functional materials and devices for AQC, air-pollution modeling, and chemical forecasting, but developed nanosensors for cost-effective AQC have not yet commercialized.

The national research projects in this field have to be coordinated in this Action aiming to share road maps and common purposes. In fact, the national efforts, coherently with the research projects worldwide, are considered more promising for the technological applications in a global scenario at an international level.

The Action will coordinate different research tasks to achieve the objectives described in previous section. The Workplan is organized in four complementary Working Groups (WGs), each devoted to a progressive development of synthesis, characterization, fabrication, integration, prototyping, proof of concepts, modeling, measurements, methods, standards, tests, and practical applications in real-world scenario. Calibration tests of the air-quality sensors in laboratory and field tests by experimental campaigns are planned as key issues for quality control and quality assurance in order to evaluate the chemical sensors performance for environmental monitoring.

Enhanced microsensors and nanosensors with functional nanomaterials will be studied in sensor networks at distributed deployment with high spatial and temporal resolution and wireless communications of data. Improved air-pollution models based on real-time datasets from deployed sensor networks will provide high-resolution mapping of air pollutants to provide accurate chemical weather forecasting and air-pollution modeling. Long-term experimental campaigns of portable sensor systems by environmental measurements in field are expected to assess the huge potential of the new sensing technologies for cost-effective AQC.

In addition to the mentioned four WGs, several Special Interests Groups (SIGs) devoted to specific issues with interdisciplinary contents will be set up to pursue planned objectives involving research, academia, and industry people regarding gender balance and including early-stage researchers in the Action agenda.

4   Conclusions

This COST Action TD1105 seems to be the best approach to coordinate, streamline, integrate, and harmonize the interaction between material scientists, environmental modelers, chemists, computing engineers, sensor manufacturers, end users, and stakeholders for a wide international community to address environmental issues and to establish European leadership on air-quality control technologies to support green economy and competitiveness of the European industry.

Acknowledgments This COST Action TD1105 *EuNetAir* has been approved by the COST Committee of Senior Officials (CSO) at its 183rd meeting on 1 December 2011 at Brussels. Memorandum of Understanding (MoU) oc-2011-1-9706 has been signed by 27 COST countries' governments at the date on May 2013 in the COST Domain *Earth System Science and Environmental Management (ESSEM)—Trans-domain Action*. The proposer/chair is highly indebted to all partners for excellent scientific interests and expertise to support the Action.

References

1. Action Memorandum of Understanding and Action Fact Sheet: http://www.cost.eu/domains_actions/essem/Actions/TD1105
2. Action website: http://www.cost.eunetair.it

A Smart System to Detect Volatile Organic Compounds Produced by Hydrocarbons on Seawater

A. Tonacci, D. Corda, G. Tartarisco, G. Pioggia, and C. Domenici

1 Hydrocarbons and Environment

The detection of hydrocarbons is a key issue for environmental monitoring and marine pollution monitoring in particular [1, 2]. Although this analysis is not often easy and can carry on many problems, it is extremely important to address efforts towards the realization of a smart system for this purpose, aiming to find possible presence of oil spills and/or hydrocarbons leakage on the sea surface.

To the best of our knowledge, few system were realized with good performances, low cost, high portability, and capability of detection and discrimination between different compounds, and to overtake this problem, a smart solution is proposed in this work.

In this paper, we describe the sensor system realized within the European project ARGOMARINE and employed for the detection of oil and/or hydrocarbons spills on the sea surface.

2 Materials and Methods

The first step for the realization of the smart system was the choice of a proper battery of sensors for this purpose. Our choice was in favor of piD sensors, whose driving force was represented by photoionization. These commercial sensors are distributed by Baseline-Mocon, Inc., and we chose to employ three different categories of piDs (namely, Black, Bronze, and Silver piDs), with different sensitivity to volatile organic compounds (VOCs).

A. Tonacci (✉) • D. Corda • G. Tartarisco • G. Pioggia • C. Domenici
Institute of Clinical Physiology, National Research Council, Pisa, Italy
e-mail: alessandro.tonacci@ifc.cnr.it

C. Di Natale et al. (eds.), *Sensors and Microsystems: Proceedings of the 17th National Conference, Brescia, Italy, 5-7 February 2013*, Lecture Notes in Electrical Engineering 268, DOI 10.1007/978-3-319-00684-0_19, © Springer International Publishing Switzerland 2014

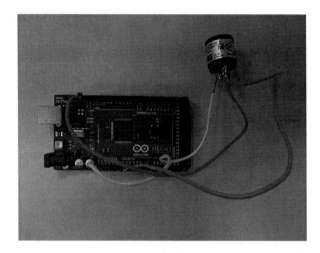

Fig. 1 A "Silver" piD sensor connected to the Arduino Mega 2560 electronic board

Fig. 2 AUV equipped with sensor system floating on seawater

Once the sensors were chosen, we realized a cylindrical flow chamber where the sensors were located. The flow chamber was composed by an air inlet, an air outlet, and six radial lodging for up to six sensors, properly placed upon up to six sockets, in order to receive the same amount of air flow [3]. The material employed for the realization of the flow chamber was PEEK (polyether-ether-ketone), a very light organic thermoplastic polymer with good mechanical and chemical characteristics, having a lower density with respect to metals or other polymeric materials like Teflon and allowing, in this way, to consistently reduce the weight of the complete system.

A control and acquisition electronics was chosen, using an Arduino™ Mega 2560 as electronic board, which represents a low-cost, easy-to-use, smart option for this purpose (Fig. 1).

An integrated structure composed by the flow chamber with the array of sensors, a sampling system with pumps and valves for air inlet and outlet, a humidity sensor, and an air aspiration cone, together with the control electronics, was placed into a floating Autonomous Underwater Vehicle (AUV), shown in Fig. 2, and into a moored buoy for a dynamic and static analysis of the state of pollution of seawater.

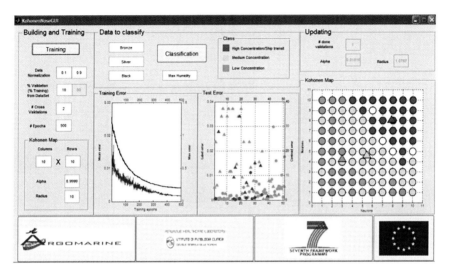

Fig. 3 Kohonen Self-Organizing Map designed for the classification of the VOCs detected

For the detection of pollutants on the sea surface, an artificial neural network (ANN) of the type Kohonen Self-Organizing Map (KSOM, Fig. 3) was designed and properly trained, by employing a dataset obtained by preliminary tests performed, within the ARGOMARINE project activities, in the La Spezia Gulf and in Elba Island, part of the protected area of the National Park of Tuscan Archipelago.

3 Results

The first analysis performed aimed to divide the detected stimuli into three different categories of warning, in order to be able to alert authorities by triggering alarms concerning the pollution state of the sea.

The results obtained in this analysis were encouraging, with a percentage of 73.9 % of correct discrimination of the stimuli into the three given classes.

The second analysis aimed to classify the detected stimuli into four classes, depending on the nature of the stimuli detected. To perform this analysis, we used as training set some data obtained in the test sessions mentioned above, employing four types of hydrocarbons representing the most frequent pollutants due to leakages on seawater. These substances were gasoline, diesel fuel, kerosene, and crude oil.

With the employment of these four substances, the capability of the KSOM to discriminate between these four classes was equal to 63.3 %, a satisfying result considering the difficult task requested, the not easy conditions of analysis and the characteristics of low cost and high portability of the system.

The system described gave reliable results in detecting VOCs produced by the four kinds of hydrocarbons mentioned above, all of them detectable at around 100 ppm with "Silver" piD, 1,000 ppm with "Bronze" piD, and up to 5,000 ppm with "Black" piD.

4   Conclusions

In this work, a smart system for the detection of volatile organic compounds produced by the presence of hydrocarbons on the sea surface was described. The system gave satisfying results in terms of correct discrimination between odorous compounds and could probably form the basis for future low-cost systems capable to detect the presence of VOCs for environmental but also for biomedical purposes.

Future research should probably deal with the increase of the sensor system performances, in order to allow the realization of a portable system able to better discriminate between different substances. Moreover, it could be useful to increase the reliability of the sensors' signal in particular environments, such as places with high humidity ratio, being the output signal span not negligible when relative humidity (RH) is over 70 %. This fact could represent a limitation for the employment of such sensors in marine environment and/or very humid places, but it was overtaken in this work with the employment of a humidity sensor activating pumps and valves for humidity expulsion.

Acknowledgments This work has been realized within, and partially funded by, the FP7 European Project ARGOMARINE—Automatic Oil spill Recognition and Geopositioning integrated in a Marine Monitoring Network (Grant Agreement: SCP8-GA-2009-234096-ARGOMARINE), coordinated by the National Park of Tuscan Archipelago.

References

1. M.J. Kachel, Particularly Sensitive Sea Areas: The IMO's Role in Protecting Vulnerable Marine Areas, Series: Hamburg Studies on Maritime Affairs **13** (Springer, Berlin, 2008).
2. I.N.E. Onwurah, V.N. Ogugua, N.B. Onyike, A.E. Ochonogor, O.F. Otitoju. Crude oil spills in the environment, effects and some innovative clean-up biotechnologies. Int. J. Environ. Res. **1** (4), 307–320 (2007).
3. F. Di Francesco, M. Falcitelli, L. Marano, G. Pioggia. A radially symmetric measurement chamber for electronic noses. Sens. Actuator B-Chem. **105** (2), 295–303 (2005).

Electrical and Morphological Characterization of TiO$_2$ Electrospun Nanofibers

S. Capone, C. De Pascali, A. Taurino, M. Catalano, P. Siciliano,
E. Zampetti, and A. Macagnano

1 Introduction

Electrospinning (ES) came into spotlight as simple, low-cost, single-step bottom-up, and versatile technique for fabricating one-dimensional (1D) nanofibers, exceptionally long in length, and uniform in diameter from various materials, e.g., polymers, metal oxides, and composites, in different configurations and assemblies. This is highly beneficial for a lot of functional devices, as energy conversion and storage devices; nanofilters for environmental, chemical, or biological protection; bioaffinity membranes; tissue engineering scaffolds and drug delivery; photocatalysts for air decontamination; and sensors to detect chemicals [1].

In particular, in recent years, many research papers on chemical gas sensors based on semiconductor metal oxides (SMOs) nanofibers fabricated by ES were published. This is a consequence of the unique combination of the intrinsic gas-sensing properties of some SMOs and the advantages offered by ES technique, as easiness, flexibility in depositing multilayered of nanofibers on any kind of substrates grounded to collecting electrode, and a large area-to-volume ratio, which is an attractive parameter for gas-sensing applications. Nanocrystalline SnO$_2$, WO$_3$, ZnO, and TiO$_2$ electrospun nanofibers showed excellent gas-sensing properties in terms of sensitivity, reversibility, response, and recovery time [2–5]. The composition of the ES solution, the process conditions, as well as the crystallization conditions strongly influences the thermal, chemical, and physical properties of SMOs electrospun nanofibers.

S. Capone (✉) • C. De Pascali • A. Taurino • M. Catalano • P. Siciliano
Institute of Microelectronics and Microsystems (CNR-IMM),
via Monteroni, Lecce 73100, Italy
e-mail: simona.capone@le.imm.cnr.it

E. Zampetti • A. Macagnano
Institute of Microelectronics and Microsystems (CNR-IMM),
Via del Fosso del Cavaliere 100, Roma 00133, Italy

C. Di Natale et al. (eds.), *Sensors and Microsystems: Proceedings of the 17th National Conference, Brescia, Italy, 5-7 February 2013*, Lecture Notes in Electrical Engineering 268, DOI 10.1007/978-3-319-00684-0_20, © Springer International Publishing Switzerland 2014

In this work, we prepared nanocrystalline TiO_2 fibers by electrospinning, and we fabricated chemoresistive gas sensors based on the deposited TiO_2 electrospun nanofibers. We studied the relationship between electrospinning parameters and microstructure and between microstructure and gas-sensing properties. At this aim, electrical-functional characterization by DC-sensing tests and AC impedance spectroscopy in controlled environment at different temperatures and different concentrations of ethanol in dry air were carried out together with morphological and structural characterization by SEM and TEM.

2 Experimental

Electrospinning solution (7.877×10^{-5} M) was prepared by dissolving polyvinylpyrrolidone (PVP, Mn 1,300,000, Aldrich) in anhydrous ethanol (EtOHa, Aldrich) and stirred for 2 h. A 2-ml aliquot of 1:4 (w/v) solution of titanium isopropoxide (TiiP, Aldrich) solved in 1:1 (v/v) mixture of glacial acetic acid (AcAcg) and EtOH was freshly prepared and added dropwise to 2.5 ml PVP solution under stirring, until it resulted in a clear yellow solution, in order to obtain a 1.95 (w/w) TiiP/PVP final ratio. Both mixtures were prepared in a glove box under low humidity rate (<5 %, obtained under continuous N_2 flow rate). The resulting polymer solution was then poured into a 2-ml glass syringe equipped with a 24-gauge stainless steel needle (din=0.0122 in., purchased from Hamilton). Figure 1 showed a scheme and an image of the electrospinning apparatus and an image of the sensor substrates.

The TiO_2 nanofibers were electrospun onto the active area of Pt interdigitated electrodes (50-nm spaced fingers) realized on Si substrates (2 mm $\times$ 2 mm squared) whose reverse side was integrated with a Pt heater. After annealing at 550 °C, the devices were bonded on TO-39 sockets and characterized as gas sensor by both

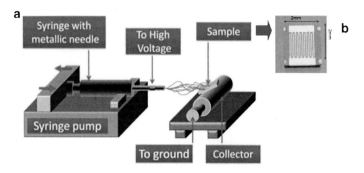

Fig. 1 (**a**) Scheme of the electrospinning apparatus. (**b**) Sensor substrates. Si substrate with Pt electrodes (IDE) and Pt integrated heater

DC-sensing tests and AC impedance spectroscopy in controlled environment at different temperatures and different concentrations of ethanol (100–150–200 ppm ethanol in dry air).

3 Results

SEM morphological analysis showed a highly porous mesh of TiO_2 nanofibers, having an average diameter of 60 nm and a length reaching hundreds of µm, and a quite rough surface resulting from a nanograin structure. Ramification or coalescence between adjacent nanofibers can occur. High-magnification TEM images indicated that the wires are fine-grained with very small nanodomains (<10 nm).

Electron diffraction patterns demonstrated that both samples have the same structure compatible with the brookite phase of TiO_2. No transformation to anatase or rutile phases was observed after gas-sensing tests.

After gas-sensing tests, morphological changes were observed; nanofibers surface appears characterized by the presence of bright contrast particles, compatible with a reorganization of the grain structure inside the wires, probably associated with a higher degree of crystallinity (Fig. 2).

Moreover, samples showed clear signs of thermal stress, such as breaking or shattering of the wires. This effect proves that a long and uninterrupted period during which the samples work at high temperature (450–500 °C) can modify and/or damage the structure of the nanofibers network.

The dense tangle of TiO_2 nanofibers has demonstrated good sensing properties to ethanol vapors (high sensitivity, fast response time, and lower operation temperature compared to TiO_2 films). The best working temperature was 450 °C. Gas-sensing tests (a) in DC and (b) in AC (by electrical impedance spectroscopy at T = 450 °C) are shown in Fig. 3.

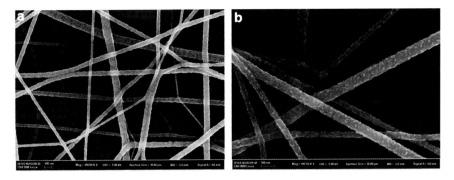

Fig. 2 SEM image of TiO₂ electrospin nanofibers: (**a**) before gas-sensing tests and (**b**) after gas-sensing tests

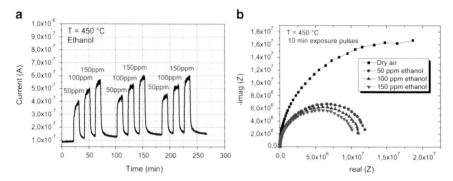

Fig. 3 (**a**) Dynamic response to ethanol in DC polarization; (**b**) Nyquist plot as in AC polarization

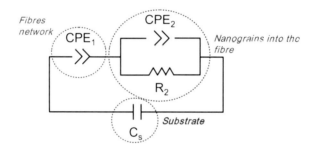

Fig. 4 Equivalent circuit that may be related to contributions from fiber network, nanograins into the fiber and substrate

Impedance spectra of the TiO_2 nanofiber-based devices were fitted by complex nonlinear least-squares method using LEVM software, and a first tentative for an equivalent circuit was made (Fig. 4).

4   Conclusions

Potential application of nanocrystalline TiO_2 fibers prepared by electrospinning in chemoresistive gas sensors was demonstrated. By ES spectra fitting, first tentative of modeling of TiO_2 electrospun nanofibers, describing the main contributes to conduction processes involved during the gas/semiconductor interactions, was provided in order to correlate specific morphological/structural features with electrical and sensing properties. Work is in progress to improve the modeling and validate it by more electrical tests.

References

1. V. Thavasi, G. Singh, S. Ramakrishna, Electrospun nanofibers in energy and environmental applications, Energy & Environ. Sci. **1**, 205–221 (2008)
2. Il-Doo Kim, A. Rothschild, B. Hong Lee, D. Young Kim, S. Mu Jo, H. L. Tuller, Ultrasensitive chemiresistors based on electrospun TiO2 nanofibers, NanoLetters 6 (9), 2009–2013 (2006)
3. J. Moon, J.-A. Park, S.-J. Lee, T. Zyung, Il-Doo Kim, Pd-doped TiO$_2$ nanofiber networks for gas sensor applications, Sensors and Actuators B 149, 301–305 (2010)
4. J. Moon, J.-A. Park, S.-J. Lee, S. C. Lim, T. Zyung, Structure and electrical properties of electrospun ZnO-NiO mixed oxide nanofibers, Current Applied Physics 9, S213–S216 (2009)
5. T.-A. Nguyen, S. Park, J. Beom Kim, G. H. Seong, J. Choo, Y. S. Kim, Polycrystalline tungsten nanofibers for gas-sensing applications, Sensors and Actuators B 160, 549–554 (2011)

A Handy Method for Reproducible and Stable Measurements of VOC at Trace Level in Air

A. Demichelis, G. Sassi, and M.P. Sassi

1 Introduction

The accurate and stable measurements of volatile organic compound (VOC) at trace level (below the ppm) in air are required in different concerns, e.g., to improve the modelling uncertainty in climate change previsions [1], to measure indoor and outdoor air quality [2, 3]. To perform an accurate and stable VOC measurement at trace level, it is necessary to use a repeatable, i.e., precise, and stable gas measurement system calibrated by comparison with accurate VOC standard, e.g., certified VOC cylinders or dynamic generation of VOC mixtures.

The state-of-the-art repeatable and stable gas measurement system is expensive and cumbersome gas chromatographs with different installed detectors. Fast response and cheap measurements are required for real-time continuous monitoring for occupational safety and health [4], GAW stations [1], in the process line of electronic and agro-alimentary industries, for local artworks protection. More handy gas instrumentations, portable and of lower cost, are online gas chromatographs with installed gas detectors and commercial cheap and compact gas detectors. They do not always guarantee the requested repeatability and stability (expanded uncertainty better than 5 % below the ppm). This is commonly due to the impossibility to keep them working in the same condition (after a switch on/off, after some time, with another operator) in order to have a reproducible and stable response. Moreover, online gas chromatographs are not suitable when fast or in situ VOC measurements are needed because of its cost and dimension.

A. Demichelis (✉) • M.P. Sassi
Istituto Nazionale di Ricerca Metrologica INRIM, Torino, Italy
e-mail: a.demichelis@inrim.it

G. Sassi
Istituto Nazionale di Ricerca Metrologica INRIM, Torino, Italy

Dipartimento di Scienza Applicata e Tecnologia DISAT, Politecnico di Torino, Torino, Italy

C. Di Natale et al. (eds.), *Sensors and Microsystems: Proceedings of the 17th National Conference, Brescia, Italy, 5-7 February 2013*, Lecture Notes in Electrical Engineering 268, DOI 10.1007/978-3-319-00684-0_21, © Springer International Publishing Switzerland 2014

The scope of the present work is to project a compact and cheap measuring system based on a commercial photoionization detector PID that performs repeatable and stable VOC concentration measurements.

The system is projected, the sensor is characterized in comparison with its specifications, and the influence parameters to sensor reproducibility and stability are identified and analyzed.

This method should be suitable for reproducible and stable in situ trace VOC measurements in the industrial and hygiene field, where fast and cheap analysis is required.

2 Materials and Methods

A photoionization detector is employed, hereafter referred as PID (Alphasense, model PID-AH). It is a very compact sensor (2 cm height for 2 cm width), suitable for fast and online total VOC measurements. The sensor has been cabled, inserted in a homemade glass chamber, and a sampling and vent line are realized. To analyze the detector response, different concentrations of VOC mixtures (with a 5 % expanded uncertainty) are dynamically generated with the stable tunable INRIM reference VOC generator (a detailed description is reported in [5]). Acetone has been employed as model VOC. The PID response has been analyzed in function of the PID relative gas pressure and massive gas flow in the detector chamber.

All these evidences are tested among different sensors taken from the same production line (named PID-A and PID-X).

3 Results

In Fig. 1, the obtained PID response over time is reported.

At the beginning, self-produced zero-air has been analyzed, after that three different acetone concentrations are dynamically generated and analyzed with a positive relative pressure in the PID chamber (+300 Pa), and finally three different lower acetone concentrations are analyzed with a negative relative pressure (−120 Pa). The absolute gas pressure in the PID chamber has not been regulated.

As a result, the sensor response seems to be linear (as outlines Fig. 2), with a minimum detectable concentration of 1 ppm of acetone and a response time of minutes in the investigated range of 0.25–5 ppm. Some of these evidences are in contrast with the PID specifications of a 40 % higher acetone response, a minimum detectable limit for acetone of 5 ppb, and a response time of seconds. The stability of the sensor signal at 1 ppm acetone mixture was found to be 0.1 % in 1 h. Measurements done in aspiration mode, at negative relative pressure in the sensor chamber, show a signal more stable.

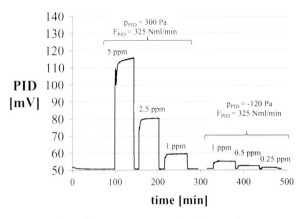

Fig. 1 PID-X response to dynamic acetone concentrations in zero-air

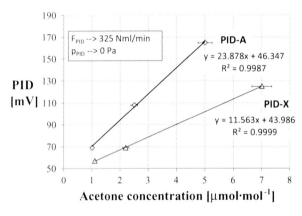

Fig. 2 Analysis of the response of two sensors PID-X and PID-A to reference dynamic acetone mixtures in the ppm range

A significant difference in the PID response is obtained varying the relative gas pressure and flow in the detector chamber, making these parameters crucial influence variables to the signal reproducibility.

In Fig. 2, the linear response behavior of the two different sensors (PID-X and PID-A) in a 1–7 ppm range for acetone mixtures is reported. It can be shown that the two sensors present a linear response with different behaviors; the PID-A response is higher. But in both cases, as reported in Fig. 3, a significant influence to the response of the relative gas pressure regulated in the PID chamber is noticeable, at constant regulated massive gas flow in the PID chamber (325 Nml/min). The higher is the pressure, the higher is the response because the increase of the number of acetone molecules that reach the PID chamber at higher gas pressure increases

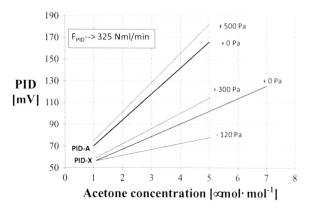

Fig. 3 Behavior of the sensor response to different relative gas pressures in the PID chamber

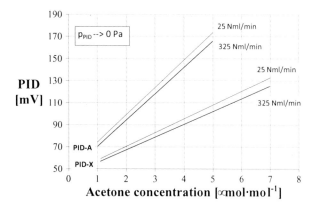

Fig. 4 Behavior of the sensor response to different massive gas flows in the PID chamber

the generated ionic current. This suggests that an accurate control of the relative gas pressure in the PID chamber can contribute to improve the sensor reproducibility and the sensor response.

In Fig. 4, the sensor response with a null relative gas pressure is displayed, at different massive gas flows in the chamber. The massive gas flow gives a low contribution to the measurement reproducibility. In general, the sensor response increases while decreasing the gas flow throughout because of the increase of the permanence time of the acetone molecules in the sensor chamber at lower flow rate that causes an increase in the generated ionic current. This suggests that a very accurate control of the massive gas flow in the PID chamber is not necessary to guarantee a good sensor reproducibility.

4 Conclusions

A compact measurement system for environmental trace gases is projected using a commercial compact PID detector in order to realize a handy method for reproducible and stable measurements of trace gases. The sensor is inserted in a homemade detector chamber equipped for gas sampling and electrical connections. The gas parameters inside the detector chamber (gas pressure and mass flow) are measured with accurate gas instrumentation, and the ambient humidity, not regulated, is 30–70 % RH.

A discrepancy between the measurements and the optimistic sensor specifications in terms of detection limit (1 ppm of acetone has been measured instead of 5 ppb), sensor response (40 % less), and response time (minutes has been measured instead of seconds in diffusion mode) has been found. Significant difference in the sensor linearity has been found among sensors of the same product line. This stresses the necessity to calibrate each sensor before use by comparison with accurate VOC standard.

The absolute gas pressure in the sensor has not been regulated, but a significant influence of the gas relative pressure and gas massive flow rate in the detector chamber to the sensor response and to the signal stability has been found. This maximized the sensor signal obtained, setting a high positive relative pressure and a low massive gas flow. These parameters in fact allow an increase in the generated ionic current at the sensor, in lieu of the higher number of ionizable gas molecules resident for a longer time in front of the lamp.

Regulating these parameters with the commercially available gas instrumentation, it is possible to realize a very reproducible (better than 3 %) and stable (better than 0.1 % in 1 h) gas concentration measurement.

In order to realize a not-expensive portable gas measurement system, these gas parameters can be regulated, installing a stable compact suction pump (preferable not volumetric) with a pre-calibrated geometry of the sampling tube.

Even though a good sensor stability is found at the optimal experimental conditions, a stability better than 0.1 % can be obtained when sampling is performed in aspiration mode.

References

1. S. Penkett, "GAW Report No.171: A WMO/GAW Expert Workshop on Global Long-Term Measurements of Volatile Organic Compounds (VOCs)", WMO Geneva, TD No. 1373 (2007).
2. D. Camuffo et al., "Environmental monitoring in four European museums", Atmos. Env., vol. 35 (2001), pp. S127–S140.
3. D. Helmig et al., "Calibration system and analytical considerations for quantitative sesquiterpene measurements in air", J. Chromatogr. A, vol. 1002 (2003), pp. 193–211.
4. OSHA, 146 Permit-Required Confined Spaces. 1910 Occupational Safety and Health Standards. 29, Information Collection Requirements, (1999), App E.
5. G. Sassi, A. Demichelis and M.P. Sassi "Uncertainty analysis of the diffusion rate in the dynamic generation of VOC mixtures". Meas. Sci. Technol. 22 (2011), 105104 -105111

UV Effect on Indium Oxide Resistive Sensors

S. Trocino, P. Frontera, A. Donato, C. Busacca, P. Antonucci, and G. Neri

1 Introduction

Metal oxide semiconductors have been the most intensely investigated materials in gas sensing area for the past several years due to their unique electric properties. Among them, indium oxide (In_2O_3) is an important n-type III–V semiconductor with a band gap of 3.6 eV, which has been widely used for the detection of both oxidizing (e.g., O_3 and NO_x) and reducing (e.g., CO, H_2S, and H_2) gases [1]. As is well known, the response of gas sensors can be improved by increasing the surface area of the sensing element. Accordingly, gas sensors based on In_2O_3 nanowires (NWs) and nanotubes (NTs) have been extensively studied because of their improved properties in comparison with the traditional bulk materials, such as the lower operating temperature and better response in the detection of gas concentration at trace levels [2, 3]. Being a new, simple, and flexible technique to synthesize NWs and NTs, electrospinning process has gained increased interest in gas sensing area. Furthermore, to overcome some problems during the functioning at room, low operating temperature, such as the poor sensitivity and long recovery time, the illumination of the sensor with UV radiation is the most studied and promising solution [4].

S. Trocino (✉) • P. Frontera • A. Donato • C. Busacca • P. Antonucci
DICEAM, University of Reggio, Calabria, Italy
e-mail: stefano.trocino@unirc.it

G. Neri
DIECII, University of Messina, Messina, Italy

C. Di Natale et al. (eds.), *Sensors and Microsystems: Proceedings of the 17th National Conference, Brescia, Italy, 5-7 February 2013*, Lecture Notes in Electrical Engineering 268, DOI 10.1007/978-3-319-00684-0_22, © Springer International Publishing Switzerland 2014

2   Experimental Details

Polyvinylpyrrolidone (PVP), dimethylformamide (DMF), and indium nitrate ($In(NO_3)_3 \cdot xH_2O$) were analytical grade and were used without further purification. The precursor solution was prepared by dissolving 0.6 g of indium nitrate powder into DMF; after 30 min, 0.76 g of PVP was added into the precursor solution to make the weight ratio of the inorganic to PVP (In/PVP) equal to 0.3. The obtained solution was stirred for 3 h and then loaded into a 20 cc glass syringe for electrospinning. In our experiment, a voltage of 20 kV was applied for electrospinning. An aluminum foil was served as the counter electrode, and the distance between the capillary and the grounded collector was 12 cm. The as-electrospun In_2O_3/PVP composite nanofibers were dried for 12 h at room temperature and then calcined in air at 600 °C for 2 h.

The structure and morphology of the prepared composites were investigated by thermo gravimetric-differential scanning calorimetry (TG/DSC), X-Ray diffraction (XRD), and scanning electron microscopy (SEM) investigations. The as-spun composite nanofibers were subjected to TG/DSC analysis using NETZSCH-STA-409 instrumentation. The analyses were performed with a heating rate of 10 °C/min in static air up to 1,000 °C. XRD measurement was performed using a Philips X-Pert diffractometer equipped with a Ni β-filtered Cu-Kα radiation at 40 kV and 20 mA. Data were collected over a 2θ range of 10–85°, with a step size of 0.05° at a speed of 0.05° per second. Diffraction peak identification was performed on the basis of the JCPDS database of reference compounds. The morphology of the samples was investigated by SEM, using a Philips XL-30-FEG scanning electron microscope at an accelerating voltage of 20 kV.

Sensor devices were fabricated mixing the electrospun annealed fibers with water to obtain a paste and then printing it on alumina substrates (3 mm × 6 mm) supplied with interdigitated Pt electrodes and a Pt resistance as heating element. Sensing tests were carried out in a system which allows to operate at controlled temperature and perform resistance measurements while varying the nitrogen dioxide concentration in the carrier stream, from 250 ppb to 2 ppm, making use of flowmeter controller. Measurements were performed under a total stream of 100 sccm, registering the sensors resistance data in the four-point mode by means of an Agilent 34970A multimeter, checking and adjusting the sensor temperature by the Agilent E3631A. The gas response is defined as $S = R/R_0$, where Ris the resistance of the sensor at different concentrations of NO_2 and R_0 the electrical resistance in synthetic air. UV illumination has been provided by a commercial LED with wavelengths of 350 nm and supplied at 5 V, introduced in test chamber with an angle of 45° with respect to stream.

3   Results and Discussion

In order to understand the processes occurring during the annealing step, TG and DSC analyses of the as-spun composite In_2O_3/PVP nanofibers (not shown) reveal that most of the organic materials PVP and solvents, NO_3^- groups, and the other

Fig. 1 XRD patterns of
electrospun fibers as prepared
(**a**) and after annealing at
600 °C (**b**)

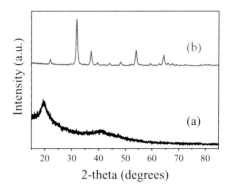

volatiles (H_2O, COx, etc.) were removed below 600 °C. The weight loss occurred in three stages. The initial weight loss could be ascribed to the loss of the surface adsorbed solvent or the residual solvent molecules in the composite fibers. The second weight loss appeared between 200 and 400 °C accompanied by exothermic peaks due to the nitrate decomposition and due to the different steps of polymer degradation, which happen through two degradation mechanisms involving both intra- and intermolecular transfer reactions.

The achievement of pure phase of indium oxide by annealing is confirmed by XRD analysis. The spectra for In_2O_3/PVP composites are reported in Fig. 1. Curve (a) related to the as-spun sample shows the broad peaks typical of long-range three-dimensional molecular disorder of PVP. From XRD spectrum of the annealed sample at firing temperature of 600 °C (curve (b)), it can be noted that this temperature was enough for In_2O_3/PVP composite to promote its complete crystallization to In_2O_3. In fact, the diffraction peaks of composite disappear, indicating that PVP was decomposed and removed out from composite fibers. Then, pure In_2O_3 was obtained after annealing at 600 °C; the d values and relative peak intensities are in agreement with those of *JCPDS* standard card (06-0416) for body-centered cubic In_2O_3. Furthermore, the strong and sharp diffraction peaks indicate a high degree of crystallization.

The SEM characterization of the as-spun fibers is shown in Fig. 2a. The composite materials obtained by sol system exhibit inhomogeneity on large scale. In particular, it can be noted that the presence of many beads anchored to the fibers. The formation of the beads can be attributed to the high surface tension of the PVP/DMF solutions. The composite fibers obtained have nanometric dimensions. The calcined products obtained are nanofibers with a distribution of diameter centered around 90 nm (Fig. 2b); they have a rough surface and hollow inside. The internal pore has a diameter of about 25 nm.

Sensing behavior towards nitrogen dioxide has been then tested both in the dark and under UV light, utilizing an LED with a wavelength of 350 nm. To analyze in detail the effect of UV radiation, sensing test has been performed by using three different procedures: without UV illumination, under continuous UV illumination,

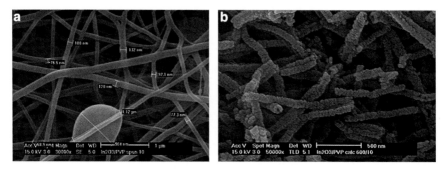

Fig. 2 SEM micrographs of In₂O₃/PVP sample: (**a**) "as-spun" and (**b**) after calcination at 600 °C

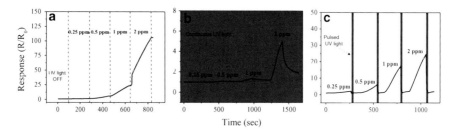

Fig. 3 Sensor response to successive pulses of NO_2 (250 ppb–2 ppm) at different illumination: (**a**) in the dark, (**b**) under continuous UV illumination, and (**c**) under pulsed UV illumination

and under pulsed UV illumination (provided only during the recovery time). As shown in Fig. 3a, in the absence of UV light, the response is high (at 2 ppm about 100 times), but the recovery time is extremely long. Under continuous UV light (Fig. 3b), the sensor is reversible but much less responsive (at 2 ppm about 5 times). Intermittent UV irradiation (Fig. 3c) allows the maintaining of high sensitivity (at 2 ppm about 25 times) along to the fast recovery time of the sensor (at 2 ppm about 30 s).

4 Conclusions

The gas sensing properties of In_2O_3 nanofiber-based sensor to NO_2 under UV illumination and without one were reported. It has been proved that the optimal sensing characteristics (good sensibility and shorter recovery time) of the sensor have been registered when the sample is irradiated with pulsed UV light.

References

1. P.C. Xu, Z. X. Cheng, Q. Y. Pan, J. Q. Xu, Sensors and Actuators B: Chemical **130**, 802-808 (2008)
2. E. Li, Z. Cheng, J. Xu, Q. Pan, W. Yu, Y. Chu, Crystal Growth & Design **9**, 2146-2151 (2009)
3. T. Zhai, X. Fang, M. Liao, X. Xu, H. Zeng, B. Yoshio, D. Golberg, Sensors **8**,6504-6529 (2009)
4. S.P. Chang, K. Y. Chen, ISRN Nanotechnology **2012**, 453517 (2012)

P-Type NiO Thin Films Prepared by Sputtering for Detection of Pollutants

Raj Kumar, C. Baratto, G. Faglia, G. Sberveglieri, E. Bontempi, and L. Borgese

1 Introduction

Since the last five decades, semiconductor oxide materials are playing a big role in sensor industries, to detect harmful gases (CO, O_3, H_2S etc.) in petroleum industry, coal mine, etc. Still there are lots of challenges with MOX sensor like stability for the long time and sensitivity at the low temperature. On the other hand, these materials have great advantage in sensing field and have good detection probability with the target gases like ethanol (EtOH), H_2, CO, O_2 and acetone. During the last five decades, n-type semiconductor-based sensors were popular in the market and research field: metal oxides like ZnO, SnO_2 and TiO_2 were extensively investigated [1].

During the last two decades, p-type semiconductor oxide-based sensors have carved a niche in the sensor market and industries. We observed that p-type semiconductor oxides have good potentials for sensor application. Wisitsoraat et al. tested the NiO sensing layer for ethanol and CO and confirmed the p-type semiconductor oxide [2]. NiO can be a promising oxide for sensor application [4], which is naturally a p-type semiconductor oxide because of the oxygen ions vacancy. We investigated preparation conditions of NiO thin films by sputtering from a NiO target and investigated them by glancing incidence X-ray diffraction (GIXRD) and scanning electron microscopy (SEM) techniques. Gas sensing properties were studied as a function of preparation conditions with volt–amperometric techniques: the gas sensing properties towards CO, ethanol and NO_2 are discussed.

R. Kumar • C. Baratto (✉) • G. Faglia • G. Sberveglieri
Department of Information Engineering, CNR-IDASC SENSOR Laboratory
and University of Brescia, Via Valotti 9, 25133 Brescia, Italy
e-mail: baratto@ing.unibs.it

E. Bontempi • L. Borgese
INSTM and Chemistry for Technologies Laboratory, Department of Mechanical and
Industrial Engineering, University of Brescia, Via Branze 38, 25133 Brescia, Italy

C. Di Natale et al. (eds.), *Sensors and Microsystems: Proceedings of the 17th National Conference, Brescia, Italy, 5-7 February 2013*, Lecture Notes in Electrical Engineering 268, DOI 10.1007/978-3-319-00684-0_23, © Springer International Publishing Switzerland 2014

2   Experimental

NiO thin films were deposited from a NiO target (99.99 % pure, 4″ size from CERAC) by RF reactive magnetron sputtering on different substrates: glass, silicon substrates for GIXRD tests and alumina 2 mm × 2 mm substrates for functional characterization as gas sensor. Deposition temperature of the substrate was varied from RT to 400 °C with step size of 100 °C in high vacuum (10^{-6} m bar) at 100 W powers. Ratio of O_2/Ar was varied from 0, 10, 25 to 50 % for sensing film at 200 °C.

The thickness of the layers was measured after deposition by a step profiler (Alpha step KLA TENCOR): the thickness of deposited films varied as a function of deposition time and sputtering conditions: thickness was 240–550 nm for thin films deposited on alumina for gas sensing tests and 220–580 nm for other tests.

Phase determination was determined using GIXRD (Bruker D8 Advance). GIXRD is performed by means of a Bruker 'D8 Advance' diffractometer equipped with a Göbel mirror and a Cu Kα radiation tube ($\lambda = 0.154$ nm) (Chem4Tech, Brescia, Italy) [3]. The beam was collimated by slits; the cross section of the X-ray beam behind the slit is 600 μm high and 1.5 cm wide. The incidence angle was fixed at 0.5°, to be more sensitive to film structure.

Surface morphology was investigated by field emission SEM (LEO 1525). During SEM measurements, the samples were observed in secondary electron imaging mode by operating the microscope at 5 keV accelerating voltage.

For electrical characterization, Al_2O_3 substrates, 2 mm × 2 mm in size and equipped with interdigitated Pt contacts and backside Pt heater, were used. The volt–amperometric technique at constant bias (1 V direct current) was applied to the sensing film, and the electrical current was measured by a picoammeter Keithley model 486. All characterizations were performed in a stainless steel test chamber (volume 500 cm³), using a constant flux of 300 sccm and keeping the ambient at a temperature of 20 °C, at atmospheric pressure and at 50 % relative humidity (RH).

3   Results and Discussions

GIXRD analysis carried out on thin films deposited at 100 °C for O_2/Ar ratio varying from 0, 10, 25 to 50 % is reported in Fig. 1. Despite of a little shift with respect to the tabulated values, crystalline peaks can be attributed to NiO phase that can be observed in any samples. The peaks shift can be due to a slight different content of oxygen in the NiO cell [4]. The change in the intensities of the peaks may be due to a preferred orientation effect. Preferred orientation (111) and (200) was already observed at 0 % O_2 or low oxygen content [4].

Thin films are deposited on glass insulating substrate (Fig. 2): the low conductance of thin films and glass substrate prevents images from being obtained at higher magnification than 100,000. For thin films deposited in Ar atmosphere, we observed

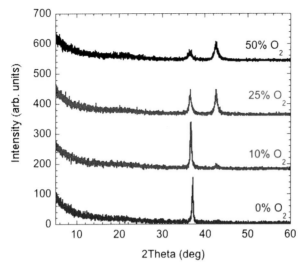

Fig. 1 GIXRD measurements of NiO thin film deposited at increasing O_2/Ar ratio at 100 °C temperature

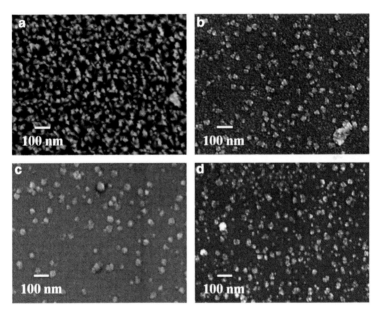

Fig. 2 SEM images of NiO thin films deposited by RF sputtering at 200 °C in 0 % O_2 (**a**), 10 % O_2 (**b**), 25 % O_2 (**c**), 50 % O_2 (**d**)

a dense packing of tetrahedral-shaped particles, as shown in Fig. 2a. In Fig. 2b–d, particles are agglomerated on the surface of the thin film, with approximately rounded shape.

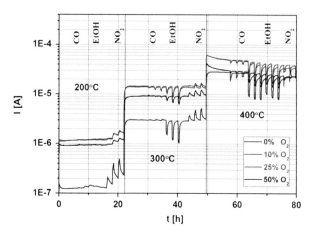

Fig. 3 Dynamic response of NiO thin films towards ethanol (100–300–500 ppm), CO (100–300–500 ppm) and NO$_2$ (1–5–8 ppm) in RH = 50 %

3.1 Sensing Study

Gas sensing performances of NiO thin films were screened for the detection of pollutant gases. The working temperature varied from 200 to 400 °C. Figure 3 reports the dynamic response of NiO thin films to CO, ethanol and NO$_2$. All NiO films showed p-type behaviour in gas sensing as shown in Fig. 3.

At 200 °C, negligible response to CO and ethanol is observed, while NO$_2$ is detected even in the lower concentration. For NO$_2$, we can notice higher response from the 0 % O$_2$ sensor, but the response and recovery times are too high to reach the steady state. At 300 °C, either CO or ethanol can be detected, and for NO$_2$, response and recovery times are much lower, being T$_{rise}$ = 660 s and T$_{fall}$ = 720 s at 300 °C. At 400 °C, NO$_2$ is no more detected and an unclear behaviour is observed.

Figure 4 resumes the relative response data for NO$_2$: the maximum response at 300 °C was recorded for the sensor deposited in inert atmosphere.

The sensor response to 500 ppm of CO: the maximum of the response curve was recorded at 300 °C for 10–25 % O$_2$ sensors and at 500 °C for 0 and 50 % O$_2$ sensors.

The relative response to ethanol (100 ppm) with W.T. (200–500 °C): the maximum relative response can be observed at 400 °C for every sensor, while the maximum response at 400 °C was recorded for sensor deposited in inert atmosphere.

4 Conclusions

Phase formation of NiO was confirmed by GIXRD analysis. P-type nature of NiO thin film was confirmed by sensing results. Sensing film of NiO shows promising results towards NO$_2$; optimum response at 300 °C was observed for sensor deposited

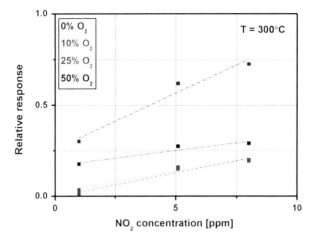

Fig. 4 Sensor response of NO_2 at 300 °C to NO_2 (1–5–8 ppm)

in inert atmosphere. The maximum response to 100 ppm of ethanol was observed at 400 °C working temperature. Low response was recorded towards CO. NiO sensing film deposited in inert atmosphere shows the optimum response. We ascribed this behaviour to higher porosity observed in this case with respect to other thin films, due to tetrahedral-shaped particles on the surface.

Acknowledgments The research leading to these results has received funding from the European Communities 7th Framework Programme under grant agreement NMP3-LA-2010-246334. Financial support of the European Commission is therefore gratefully acknowledged.

References

1. A. Wisitsoraat, A. Tuantranont, E. Comini, G. Sberveglieri, W. Wlodarski, Characterization of n-type and p-type semiconductor gas sensors based on NiO_x doped TiO_2 thin films. Thin Solid Films **517**, 2775–2780 (2009)
2. I. Hotovy, J. Huran, L. Spiess, H. Romanus, S. Caponez, V. Rehacek, A. M. Taurino, D. Donoval1, and P. Siciliano, Au-NiO nanocrystalline thin films for sensor application. Journal of Physics: Conference Series **61**, 435–439 (2007)
3. P. Colombi, P. Zanola, E. Bontempi, R. Roberti, M. Gelfi, E. L. Depero, Journal of Applied Crystallography **39**, 176-179 (2006)
4. I. Hotovy, J. Huran, L. Spiess, R. Capkovic, S. Hascik, Preparation and characterization of NiO thin films for gas sensor application. Vacuum **58**, 300-307 (2000)

Well-Ordered Titania Nanostructures for Gas Sensing

V. Galstyan, E. Comini, A. Ponzoni, N. Poli, G. Faglia, G. Sberveglieri, E. Bontempi, and M. Brisotto

1 Introduction

Solid-state semiconducting films can monitor levels of gaseous species in the atmosphere through changes in the resistivity of the material. These resistivity changes are caused by electronic transfer upon adsorption of gas molecules over the thin film surface. Properly controlling the electronic, morphological, and chemical properties of the material, namely, band-gap, Fermi level position, dispersion of catalyst, size of crystallites, and their network connection, is fundamental to enhance the sensitivity of chemoresistive devices [1]. TiO_2 is one of the most studied metal oxides for fabrication of chemical gas sensors due to its unique electrochemical and electrophysical properties. To improve the properties of nanostructured TiO_2, different shapes have been obtained by different fabrication methods [2, 3]. In this context the electrochemical anodization is a low-cost and facile method for modification of surface structure of TiO_2 [4]. This work demonstrates the pliability of electrochemical anodization to prepare pure and Nb-contained TiO_2 nanotubes (NTs) on different substrates for the development of gas sensing devices.

V. Galstyan (✉) • E. Comini • A. Ponzoni • N. Poli • G. Faglia • G. Sberveglieri
SENSOR Lab, Department of Information Engineering, University of Brescia
and CNR IDASC, Via Valotti 9, 25133 Brescia, Italy
e-mail: vardan.galstyan@ing.unibs.it

E. Bontempi • M. Brisotto
INSTM and Chemistry for Technologies Laboratory, University of Brescia,
Via Branze 28, 25133 Brescia, Italy

C. Di Natale et al. (eds.), *Sensors and Microsystems: Proceedings of the 17th National Conference, Brescia, Italy, 5-7 February 2013*, Lecture Notes in Electrical Engineering 268, DOI 10.1007/978-3-319-00684-0_24, © Springer International Publishing Switzerland 2014

2   Experimental

The NTs of pure and Nb-containing TiO_2 have been obtained by electrochemical anodization of metallic Ti and Nb-Ti deposited on alumina and Kapton HN flexible substrates. The deposition of metallic films has been carried out by means of RF (13.56 MHz) magnetron sputtering. The concentration of Nb versus Ti was about 0.14 at.% [5]. Since the increase in conductance was not satisfactory, we increased further the Nb content in the film to 0.42 at.%. A Pt foil has been used as a counter electrode. Anodization has been carried out in 0.3–0.5 wt% NH_4F and 1–2 mol L^{-1} H_2O-contained glycerol and ethylene glycol by potentiostatic mode at room temperature. As-prepared structures have been crystallized by thermal annealing according to the regimes of our previous investigations [6]. The morphological characterization has been studied by means of atomic force microscopy (AFM), using a Bruker-Thermomicroscope CP-Research, and scanning electron microscopy (SEM), using a LEO 1525 microscope equipped with field emission gun. The functional properties of prepared NTs have been tested towards NO_2.

3   Results and Discussions

3.1   Morphological Characterizations

Figure 1 shows the two-dimensional (2D) AFM scans and the respective sample profiles measured on alumina and Kapton HN substrates. The analysis of the alumina surface is reported in Fig. 1a, b, which shows the substrate surface imaged over a scanned area of 5 μm × 5 μm and a sample profile, respectively.

This substrate shows a very rough morphology, being composed by grains with size in the range of 0.1–1 μm. Differently, Kapton HN shows a quite flat surface and to appreciate its granularity is necessary to scan its surface at the submicron scale, as shown in Fig. 1c, d. In terms of roughness, alumina shows an RMS (root mean squared) roughness of about 100 nm, while Kapton HN shows an RMS lower than 1 nm. SEM observations show that after the anodization in ethylene glycol, as-prepared TiO_2 structures are composed of smooth and well-ordered NT arrays (Fig. 2a). Differently, Ti-Nb films did not show the tubular structure after the anodization process in the ethylene glycol, featuring instead the morphology of a porous film (Fig. 2b). Instead Nb-TiO_2 structures showed highly ordered tubular arrays (Fig. 2c) in glycerol. This behavior can be explained based on the different viscosity of the electrolytes [7]. It is important to achieve equilibrium between oxide growth and local oxide dissolution for the preparation of NTs. In viscous electrolytes fluorine ion mobility is strongly reduced; therefore in glycerol anodization becomes a slower process allowing equilibration between the aforementioned competitive phenomena leading to the formation of NTs.

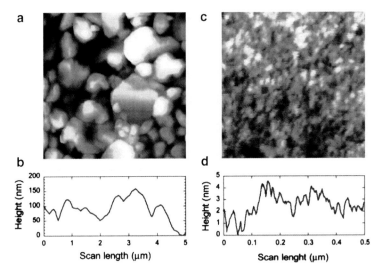

Fig. 1 AFM analysis of Alumina and Kapton substrates: (**a**) 5 μm × 5 μm 2D topography image of the alumina substrate; (**b**) sample profile of the alumina substrate; (**c**) 0.5 μm × 0.5 μm 2D topography image of the Kapton HN substrate; (**d**) sample profile of the Kapton HN substrate

Fig. 2 (**a**) SEM image of TiO$_2$ NTs on alumina, (**b**) SEM image of 0.42 at.% Nb-TiO$_2$ anodized in ethylene glycol, (**c**) SEM image of 0.42 at.% Nb-TiO$_2$ NTs obtained on Kapton HN

As a consequence, the inter-pore regions are also etched, and Nb-TiO$_2$ NT arrays appeared. The inner diameter of pure TiO$_2$ and 0.42 at.% Nb-TiO$_2$ NTs is about 75 and 65 nm, respectively.

3.2 Gas Sensing Characteristics

Gas test characterizations were carried out by varying nitrogen dioxide concentration between 1 and 5 ppm at 40 % RH. Working temperature was varied between 100 and 500 °C. Operating temperatures higher than 400 °C destroy the sensitivity to nitrogen dioxide. Figure 3 illustrates the dynamic response of NTs operated at their optimal temperature (300 °C). Due to the presence of an oxidizing species, sample conductance decreases, a standard behavior of n-type semiconductors. The response towards 1 ppm of NO$_2$, defined for n-type semiconductor as the normalized variation of conductance, is 5 % for the pure TiO$_2$ and 180 % for the

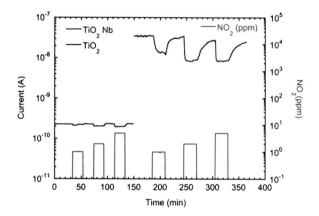

Fig. 3 Dynamical response of 0.42 at.% Nb-TiO₂ and TiO₂ NTs towards square concentration pulses of NO₂ at a working temperature of 300 °C and 40 % RH@20 °C

Nb TiO₂, respectively. The Nb introduction results in an improvement in the sensing performances together with an increase in the air conductance value, even with respect to the results reported in ref. [5] related to 0.14 at.% Nb-TiO₂ NTs. Dynamic response is also affected by operating temperature; response and recovery time improve with increasing temperatures since chemical reaction kinetic at the surface speeds up.

4 Conclusions

TiO₂ and Nb-TiO₂ NTs have been successfully fabricated on stiff and flexible substrates by means of electrochemical anodization method. Nb-TiO₂ NTs (with the higher concentration of Nb) have been prepared when ethylene glycol has been replaced by more viscous glycerol. The introduction of Nb increased both the conductance value of tubes in air and the sensitivity of the material to NO₂.

Acknowledgments The research leading to these results has received funding from the Italian Ministry of Education through project FIRB "Rete Nazionale di Ricerca sulle Nanoscienze ItalNanoNet" (Protocollo: RBPR05JH2P, 2009–2013) and Region Lombardia (Italy) Xnano Project.

References

1. N. Yamazoe and K. Shimanoe. Roles of shape and size of component crystals in semiconductor gas sensors. Journal of the Electrochemical Society, **155** (4), J85-J93 (2008)
2. S. Yamabi, H. Imai. Crystal phase control for titanium dioxide films by direct deposition in aqueous solutions. Chemistry of Materials **14** 609-614 (2002)

3. C. Das, P. Roy, M. Yang, H. Jha, P. Schmuki. Nb doped TiO2 nanotubes for enhanced photo-electrochemical water-splitting. Nanoscale **3** (8), 3094-3096 (2011)
4. V. Galstyan, A. Vomiero, E. Comini, G. Faglia, G. Sberveglieri. TiO_2 nanotubular and nanoporous arrays by electrochemical anodization on different substrates. Rsc Advances. **1** 1038-1044 (2011)
5. V. Galstyan, E. Comini, G. Faglia, A. Vomiero, L. Borgese, E. Bontempi, G. Sberveglieri. Fabrication and investigation of gas sensing properties of Nb-doped TiO_2 nanotubular arrays. Nanotechnology **23** 235706 (2012)
6. V. Galstian, A. Vomiero, I. Concina, A. Braga, M. Brisotto, E. Bontempi, G. Faglia, G. Sberveglieri. Vertically aligned TiO_2 nanotubes on plastic substrates for flexible solar cells. Small **7** 2437-2442 (2011)
7. J.M Macak, P. Schmuki. Nonaqueous Viscous Electrolytes for Growth of Anodic Titania Nanotubes. Electrochimica Acta **52** 1258 (2006)

A Disposable Ammonia Sensor by Low-Cost Ink-Jet Printing Technology

B. Andò, S. Baglio, G. Di Pasquale, C.O. Lombardo, and V. Marletta

1 Introduction

Gas sensing is a wide diffused need in several contexts [1]. Just to give some examples, detection of NO and NO_2 is adopted for monitoring and controlling exhaust from automobiles; CO, CO_2, NO_x, H_2S, and CH_4 detection are required for environmental protection, safety, and process control in homes, industries, agriculture, and mine; and NO detection is mandatory for the estimation of physiological responses of human body.

Several approaches have been used for the realization of gas sensors both in terms of sensing methodologies, technologies, and materials adopted for the realization of functionalized layers. As an example, semiconductor gas sensors based on metal oxides have been used extensively to detect toxic and inflammable gases [2]. A review of mixed potential gas sensor as an alternative to classical gas sensors is given in [3]. Among low-cost technologies, especially related to the sensing layer deposition, CVD, screen printing, and ink-jet printing are the most diffused. Typical examples of sensing materials adopted for the realization of printed gas sensors are PEDOT-PSS (poly 3,4-Ethylenedioxy-Thiophene-PolyStyrene-Sulfonate), PANI (Polyaniline), graphene, and ZnO. The gas sensing properties of graphene synthesized by a chemical vapor deposition (CVD) method are investigated in [4]. The use of sputtered zinc oxide (ZnO) for the realization of CO_2 sensor is discussed in [5].

In recent years wide availability of materials, low-cost production techniques, high speed of prototyping, and the possibility of combinations of different materials and geometries allowed development of novel printed sensors [6]. Screen-printed ZnO thick films on glass substrates for different gases such as H_2S, CO, and ethanol are presented in [7]. Fabrication process of an ammonia gas sensor obtained by ink-jet printing of polyaniline suspension on alumina substrates is treated in [8].

B. Andò (✉) • S. Baglio • G. Di Pasquale • C.O. Lombardo • V. Marletta
D.I.E.E.I., D.I.I., University of Catania, Catania, Italy
e-mail: bruno.ando@dieei.unict.it

C. Di Natale et al. (eds.), *Sensors and Microsystems: Proceedings of the 17th National Conference, Brescia, Italy, 5-7 February 2013*, Lecture Notes in Electrical Engineering 268, DOI 10.1007/978-3-319-00684-0_25, © Springer International Publishing Switzerland 2014

Above sensors are usually developed by hybrid printed technology, exploiting screen printing for electrodes and ink-jet printing for deposition of functional layers. Such approach requires the use of mask-based process and high-cost ink-jet printing solution.

This paper focuses on a novel ammonia gas sensor realized by a full printing technology based on low-cost ink-jet printing and cheap post-processing. The sensor developed is able to detect the presence of low ammonia concentration in the order of few ppm, with a fast time response. The latter is strategic for early warning systems aimed to monitor risk sites (e.g., refrigeration systems where early warning for ammonia leakage is required and workplaces where the time-weighted average allowed concentration of ammonia is 25 ppm). Above already mentioned advantages, main claims of the proposed solution come from the low-cost direct printing technology adopted which allows for the rapid prototyping of complex electrodes, thus enabling specific applications. Such features are dramatic to address specific contexts, where the possibility to develop low-cost and disposable sensing devices is mandatory, such as research laboratories and harsh monitoring scenario (e.g., workplace and refrigeration systems).

2   An Overview of the Printed Sensor and the Characterization System

The sensor prototype is composed of interdigitated electrodes (200 μm finger width and spacing), realized by printing a silver ink (Metalon® JS-015 Water-based Silver Inkjet Ink by Novacentrix™) on a PET (polyethylene terephthalate) substrate through a common desktop printer (Epson Stylus S22). Dimensions of the sensing area of the device are 10 mm by 17 mm. About the conducting polymer, PEDOT/PSS was used (see Fig. 1a) due to good electrochemical, ambient, and thermal stability of its electrical properties. It is commercially available as an aqueous dispersion, CLEVIOS™ PHCV4, by H.C. Starck. The ink, prepared diluting the starting solution in distilled water (1:1, v/v), was deposited over the electrode by a calibrated spreader (12 μm). The conductivity of the film was enhanced by annealing at 80 °C for 50 min. A real view of the sensor prototype is shown in Fig. 1b, while Fig. 1c gives a stack-up schematization of the sensor. The sensing mechanism is based on the dependence of PEDOT/PSS layer resistivity on ammonia concentration.

A dedicated sealed chamber was used for the sake of sensor characterization. Ammonia concentration inside the chamber has been independently measured by the reference-calibrated sensor Figaro-TGS2444. Temperature and relative humidity inside the measurement chamber are measured by the standard temperature sensors LM35 by Texas Instruments and the humidity sensor HIH-4000-001 by Honeywell. The printed sensor and the TGS2444 reference sensor are positioned in the upper section of the chamber, far from the ammonia inlet section, in order to assure a uniform ammonia concentration in the sensing area.

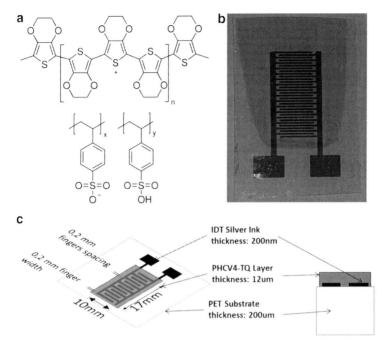

Fig. 1 The developed printed ammonia sensor. (**a**) PEDOT/PSS; (**b**) the real device; (**c**) schematization of the sensor stack-up

A MuIn board is used to acquire signals provided by the reference sensor, to implement the measurement protocol, and to manage the serial communication with the LabVIEW™ measurement environment.

Conditioning circuit of printed sensor is composed by standard bridge configuration. A dummy sensor is used to compensate temperature influence on the device behavior. The bridge circuit is supplied by 0.5 V peak to peak sinusoidal waveform with a frequency of 1 kHz through a buffer circuit. The bridge output signal is conveyed to an instrumentation amplifier INA111. Output signals from conditioning electronics are acquired by a Data Acquisition Board and successively processed by the dedicated LabVIEW™ environment shown in Fig. 2.

3   Results and Conclusions

Experiments have been implemented to observe the sensor response as a function of increasing values of ammonia concentration. To such aim known fraction of ammonia solution of distilled water and ammonia has been spread inside the chamber by using a graduated syringe. At first, low ammonia concentration (in the order of 10 ppm) has been spread into the chamber sequentially. The device response shown in Fig. 3

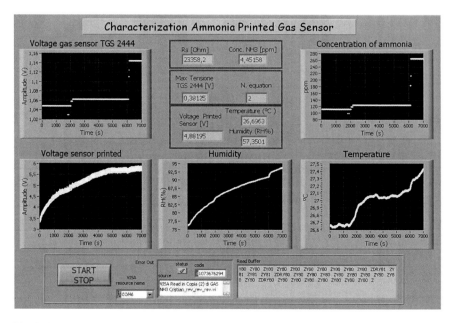

Fig. 2 GUI of the LabVIEW™ Tool

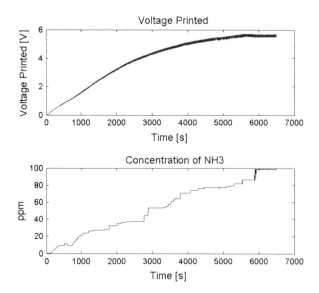

Fig. 3 Time response of the printed sensors

allows to affirm that for such low concentration value the sensor behaves in an integral mode. The steady-state output regime is observed for ammonia concentration starting from 100 ppm. As it can be observed, the sensor is characterized by a very fast time response which could be strategic for early warning applications.

References

1. S. Zhuiykov, N. Miura, Development of zirconia-based potentiometric NOx sensors for automotive and energy industries in the early 21st century: what are the prospects for sensors?, Sens. Actuators B **121**, 639–651 (2007).
2. Y. Takao, K. Miyazaki, et al.: High ammonia sensitive semiconductor gas sensors with double layer structure and interface electrode. J. Electrochem. Sco. **141**, 1028–1034 (1994).
3. A. Morata, J.P. Viricelle, A. Tarancon, et al., Development and characterization of a screen-printed mixed potential gas sensor, Sensors and Actuators B **130**, 561–566 (2008).
4. Madhav Gautam, Ahalapitiya H. Jayatissa, Gas sensing properties of graphene synthesized by chemical vapor deposition, Materials Science and Engineering C **31**, 1405–1411 (2011).
5. P. Samarasekara, N. U. S. Yapa, N. T. R. N. Kumara, M. V. K. Perera, CO2 gas sensitivity of sputtered zinc oxide thin films, Bull. Mater. Sci., Vol. 30, No. 2, pp. 113–116 (2007).
6. M. Mäntysalo, V. Pekkanen, K. Kaija, J. Niittynen, et al., "Capability of Inkjet Technology in Electronics Manufacturing", Elec. Comp. and Tech. Conf. (2009).
7. V. B. Gaikwad, M. K. Deore, P. K. Khanna, D. D. Kajale, S. D. Shinde, D. N. Chavan, G. H. Jain, Studies on Gas Sensing Performance of Pure and Nano- Ag Doped ZnO Thick Film Resistors, Recent Adv. in Sens. Tech. Lecture Notes in Elect. Eng. Vol. **49**, 293-307 (2009).
8. F. Loffredo, G. Burrasca, L. Quercia, D. Della Sala, Gas Sensor Devices Obtained by Ink-jet Printing of Polyaniline Suspensions, Macromol. Symp., 247, 357–363 (2007).

Reproducibility of the Performances of Graphene-Based Gas-Sensitive Chemiresistors

Ettore Massera, Maria Lucia Miglietta, Tiziana Polichetti, Filiberto Ricciardella, and Girolamo Di Francia

1 Introduction

Graphene, a monatomic planar sheet of carbon atoms arranged in honeycomb lattice existing at room temperature, has put itself as the leader of the new discovered materials in every research field.

Thanks to its bi-dimensionality, highest mobility and lowest resistivity values, the sensing properties have been among the most explored ones because this monatomic layer has shown formidable performances, in particular detecting toxic gases [1]. Graphene sensor performances and reproducibility seem to be strongly dependent on material quality, in terms of layer number and mean lateral size of flakes, as well as on device fabrication, including film deposition.

Here, chemiresistive devices based on chemically exfoliated natural graphite, produced from colloidal suspensions, are presented. Varying some parameters in the material preparation, a study on conductance variation of chemiresistors upon exposure to 350 ppb of NO_2 is performed.

E. Massera (✉) • M.L. Miglietta • T. Polichetti • G. Di Francia
ENEA-UTTP-MDB Laboratory for Materials and Devices Basic Research,
1, Piazzale E. Fermi 1, 80055 Portici (Naples), Italy
e-mail: ettore.massera@enea.it

F. Ricciardella
ENEA-UTTP-MDB Laboratory for Materials and Devices Basic Research,
1, Piazzale E. Fermi 1, 80055 Portici (Naples), Italy

Department of Physics, University of Naples «Federico II »,
Via Cinthia, 80126 Naples, Italy

C. Di Natale et al. (eds.), *Sensors and Microsystems: Proceedings of the 17th National Conference, Brescia, Italy, 5-7 February 2013*, Lecture Notes in Electrical Engineering 268, DOI 10.1007/978-3-319-00684-0_26, © Springer International Publishing Switzerland 2014

2 Experimental

Stable colloidal suspensions of graphene were prepared by liquid phase exfoliation of graphite flakes (Sigma-Aldrich, product 332461) in two organic solvents, namely, N-methyl-pyrrolidone (NMP) and N,N-dimethylformamide (DMF), as elsewhere specified [2, 3].

In Table 1 the parameters for the different preparations are listed: solvents, speed, and time of centrifugation.

Morphological analysis, performed by transmission electron microscopy (FEI TECNAI G12 Spirit-Twin) and dynamic light scattering technique (DLS Zetasizer Nano, Malvern Instruments), shows that films consist in a mixture of nanostructured graphite crystals and graphene flakes with lateral size of hundreds of nanometers (Fig. 1).

Chemiresistive devices listed in Table 1 were realized by drop-casting few microliters of the so-obtained material, and their resistances soon after preparation were

Table 1 Recipes utilized for the material preparation

Name	Resistance (kΩ)	Solvent	Centrifugation speed (rpm)	Centrifugation time (min)
Device 1	50,000	NMP	500	45
Device 2	180	NMP	500	45
Device 3	460	NMP	500	45
Device 4	5.14	NMP	500	90
Device 5	73.7	NMP	500	90
Device 6	8.5	NMP	2,500	90
Device 7	4.2	NMP	2,500	90
Device 8	44.6	NMP	13,000	90
Device 9	4.14	NMP	13,000	90
Device 10	173	DMF	13,000	10
Device 11	400	DMF	13,000	10
Device 12	16.6	DMF	13,000	10

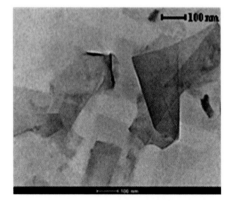

Fig. 1 TEM image of graphene and nanostructured graphite crystals

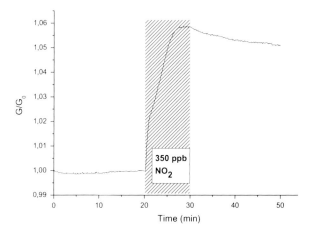

Fig. 2 Typical conductance variation of the chemiresistive device exposed to 350 ppb of NO_2

Table 2 Performances and final benchmark attributed to the devices

Name	Conductance variation (%)	SNR	Final benchmark
Device 1	1.3	42	0
Device 2	0.9	199	2
Device 3	7.7	42	3
Device 4	5.6	1,656	8
Device 5	1.9	134	3
Device 6	0.6	1,646	7
Device 7	1.2	128	3
Device 8	2.2	1,150	6
Device 9	4.1	1,012	6
Device 10	3	197	3
Device 11	11.7	226	6
Device 12	3.2	510	4

recorded. Tests for sensing measurements were performed in a stainless steel chamber (Kenosistec equipment) placed in a thermostatic box (T = 22 °C). The employed protocol consists basically of three steps: a baseline of 20 min in wet carrier gas (N_2 at RH = 50 %), 10 min of exposure to 350 ppb of NO_2, and 20 min of flushing with carrier gas. The typical behavior of device is shown in Fig. 2.

3   Results and Discussion

As it can be observed in Table 1, regardless the recipe adopted, sometimes graphene films deposited from the same batch could not have the same properties, namely, the initial resistance value. In order to obtain information on reproducibility of the graphene-based devices, a benchmark was considered (Table 2), taking into account three device parameters: resistance, conductance variation towards the same NO_2 exposure (calculated as $\Delta G/G_0$ where G_0 represents the conductance value soon before the introduction of the analyte), and signal-to-noise ratio (SNR). For each parameter, the performance was awarded attributing the highest (10) and the lowest mark (0) to the best and to the worst performance, respectively. In particular, referring to the resistance, the huge range of the obtained values required the adoption of the logarithmic scale: the less resistive device (Device 9) was considered to be the best. The final benchmark, representing the global evaluation of the device performances, was attributed weighting differently the parameters: 20 % for initial resistance, 30 % for SNR, and 50 % for the response.

A correlation between the three considered parameters appeared not clearly. As general consideration, albeit the recipe is not so crucial on the device performance, samples prepared using NMP as solvent produced a more homogenous film on the substrate. Starting from the same solvent and same centrifugation speed, higher homogeneity and larger covered device area allowed obtaining chemiresistors with lower resistance. However, devices with a low resistance could have poor and very noisy response, and better homogeneity of the sensing layer did not necessary result in higher response value. On the other hand, large area graphene films determined a good SNR.

4   Conclusions

Considering three different parameters such as device initial resistance, conductance variation, and SNR, we have performed a study on reproducibility of graphene-based chemiresistors upon exposure to an extremely low concentration of NO_2.

We have inferred that it could be possible to improve device performances, e.g., in terms of conductance variation, by increasing the area of transducer covered by the dispersion. Alternatively, a good trade-off between the three considered parameters could be obtained with a large and homogeneous area covered by the sensing layers.

References

1. F. Schedin, A.K. Geim, S. V. Morozov, E.H. Hill, P. Blake, M.I. Katsnelson, K. S. Novoselov. Detection of Individual Gas Molecules Adsorbed on Graphene. Nat. Mater. **6**, 652-655 (2007)
2. M.L. Miglietta, T. Polichetti, E. Massera, I. Nasti, F. Ricciardella, S. Romano, G. Di Francia. Sub-ppm nitrogen dioxide conductometric response at room temperature by graphene flakes based layer. Proc. of Sensors and Microsystems: AISEM 2011, Feb 7-9, Rome (IT), 121-125 (2011)
3. M.L. Miglietta, E. Massera, S. Romano, T. Polichetti, I. Nasti, F. Ricciardella, G. Fattoruso, G. Di Francia. Chemically exfoliated graphene detects NO_2 at the ppb level. Proc. Eng. **25**, 1145-1148 (2011)

Exfoliation of Graphite and Dispersion of Graphene in Solutions of Low-Boiling-Point Solvents for Use in Gas Sensors

Filippo Fedi, Filiberto Ricciardella, Tiziana Polichetti, Maria Lucia Miglietta, Ettore Massera, and Girolamo Di Francia

1 Introduction

Several papers have shown the advantages of direct exfoliation of graphite using organic solvents [1, 2]. So far, the best results for the production of chemically exfoliated graphene are achieved using high-boiling-point, expensive and toxic solvents such as N-methyl-pyrrolidone (NMP) or N,N-dimethylformamide (DMF). On the other hand, using low-boiling-point solvents such as water and/or alcohols for direct graphite exfoliation could be interesting because of their lower cost and intrinsic toxicity together with the simpler handling. However, to date the most widely recognized green solvents, water and ethanol, have both proved to have poor exfoliation and dispersion capability resulting in low graphene concentration and stability [1].

Inspired by the Coleman's work on 2-propanol [3], we used a mixture of that alcohol with 1-butanol for the exfoliation of graphite flakes. The solvent ratio was adjusted in order to approach the Hansen Solubility Parameters of the final mixture to those of graphene, so as to improve the overall quality of the exfoliation process.

F. Fedi
Department of Materials and Production Engineering, University of Naples
"Federico II", Piazzale Tecchio, 80125 Naples, Italy
e-mail: f.fedi@hotmail.com

F. Ricciardella
ENEA-UTTP-MDB Laboratory for Materials and Devices Basic Research,
1, Piazzale E. Fermi, 80055 Portici (Naples), Italy

Department of Physical Sciences, University of Naples "Federico II",
Via Cinthia, 80126 Naples, Italy

T. Polichetti • M.L. Miglietta (✉) • E. Massera • G. Di Francia
ENEA-UTTP-MDB Laboratory for Materials and Devices Basic Research,
1, Piazzale E. Fermi, 80055 Portici (Naples), Italy
e-mail: mara.miglietta@enea.it

C. Di Natale et al. (eds.), *Sensors and Microsystems: Proceedings of the 17th National Conference, Brescia, Italy, 5-7 February 2013*, Lecture Notes in Electrical Engineering 268, DOI 10.1007/978-3-319-00684-0_27, © Springer International Publishing Switzerland 2014

The resulting exfoliated material was characterized by TEM and Raman analysis and further used to fabricate sensing layers for chemiresistor devices. The graphene-based layers were tested towards 350 ppb of NO_2 in nearly environmental conditions (room temperature and 50 % of relative humidity) and compared with those of graphene layers obtained through a similar exfoliation of graphite in NMP.

2 Experimental

Graphite flakes (Sigma-Aldrich, product N.332461) were dispersed in a mixture of 1-butanol/2-propanol (3:2) and in NMP at 2.5 mg/ml each. Exfoliation was achieved by sonication in a low power bath for about 150 h. Unexfoliated graphite flakes were removed by centrifugation for 45 min at 1,000 rpm. Raman analyses were carried out on graphene films prepared by drop casting the suspensions on the top of oxidized silicon wafers with a Renishaw in Via Reflex apparatus in back-scattering configuration at 514.5 nm. TEM images were performed using a FEI TECNAI G12 Spirit-Twin at 120 kV. TEM samples were prepared by casting few microlitres of the suspensions on Cu TEM grid covered with holey carbon film. Alumina transducers with interdigitated Au contacts were used as substrates for the fabrication of chemiresistor devices. A graphene film was directly deposited on top of the electrodes by drop casting. The devices were then mounted in a Gas Sensor Characterization System (GSCS, Kenosistec equipment). Every device was tested under the same conditions: atmospheric pressure with temperature and relative humidity RH set at 22 °C and 50 %, respectively.

3 Results and Discussion

In this work, a mixture of two alcohols was prepared with the aim to improve the solvent capabilities towards graphene. To this purpose, the Hansen Solubility Parameters (HSP) of the system were considered (Table 1). These thermodynamic parameters, derived from the enthalpy of mixing, describe the energetic

Table 1 Hansen Solubility Parameters of alcohols and graphene

Solvent	δ_D [MPa$^{1/2}$]	δ_P [MPa$^{1/2}$]	δ_H [MPa$^{1/2}$]
2-Propanol	15.8	6.1	16.4
1-Butanol	16.0	5.7	15.8
1-Butanol/2-propanol (3:2)[a]	15.9	5.86	16.04
Graphene[b]	18	10	7

[a]HSP values calculated by weighted average
[b]HSP theoretical values [1]

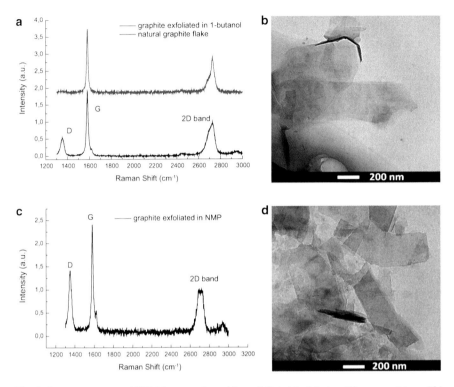

Fig. 1 Raman spectra and TEM images of graphite exfoliated in 1-butanol/2-propanol (**a** and **b**) and in NMP suspension (**c** and **d**), respectively. Raman spectrum of bulk graphite flakes is also shown for comparison (**a**)

contributions of dispersive and dipolar forces and hydrogen bonds between molecules. The nearest these values are for solvent and solute, the more likely solvent will dissolve the solute.

Raman spectra of graphene films exfoliated from alcohol mixture and NMP are shown in Fig. 1. Comparing the 2D peak features of the material exfoliated by the two different solvents with the 2D peak of graphite, a widening of this band can be noticed [4, 5]. This broadening can be ascribed to an effective graphene sheet separation which in turn indicates a good exfoliation degree. Due to the small dimension of the graphene flakes falling within the laser spot area, the D band, which is usually related to structural defects or edge effects, appears in the spectrum. In summary, the Raman analysis suggests that the two solvents lead to the same extent of the exfoliation.

Besides, TEM images of the alcohol-based suspension show the effective exfoliation of graphite flakes, with large amount of few- and bilayer flakes. Also this kind of investigation confirms that the exfoliation degree of the two preparations, namely, the one performed in alcohols and the one in NMP, is comparable.

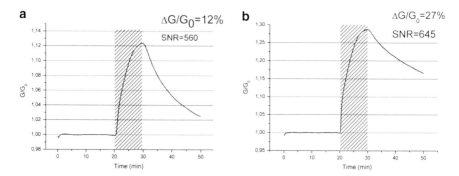

Fig. 2 Devices electrical conductance in 350 ppb of NO₂ for graphite exfoliated in 1-butanol/2-propanol (3:2) (**a**) and in NMP (**b**). The *blue rectangle* is the temporal fraction of exposure to the NO₂ gas

As-prepared materials were tested as sensing layers in chemiresistor devices. The sensitivity of these graphene-based layers to 350 ppb of NO_2 was investigated in nearly environmental conditions (room temperature and 50 % of relative humidity). In Fig. 2 the resulting sensing signals are plotted as G/G_0 that is the ratio of the conductance with respect to the initial value (i.e. the conductance soon before the introduction of the analyte).

The devices show a remarkable sensitivity to the target gas, with appreciable signal-to-noise ratios (SNR). The recovery behaviour after the sensing event appears quite different for the two layers. In particular, the chemiresistor made from graphene exfoliated in 1-butanol/2-propanol recovers more than 66 % of its initial conductance within 20 min. This feature has a considerable interest for room temperature-operating sensors since incomplete or slow recovery may hamper the development of this kind of sensing layers.

4   Conclusions

Our results show that the designed alcohol mixture was effective in exfoliating and dispersing graphene with respect to the pure 2-propanol. In addition, the exfoliation quality obtained with the alcohol mixture and with the more performing high-boiling-point solvent was to a considerable extent equivalent.

The same equivalence can be observed in the sensing performance of the materials. In fact, both the graphene films showed remarkable sensitivity to very low concentrations of NO_2. Better signal recovery behaviour was observed for the graphene film obtained from the alcohol mixture showing the great potential of this material for sensing applications.

References

1. J. N. Coleman. Liquid Exfoliation of Defect-Free Graphene. Acc. Chem. Res. **46** (1), 14-22. (2013)
2. Y. Hernandez, V. Nicolosi, M. Lotya, F.M. Blighe, Z. Sun, S. De, I. T. McGovern, B. Holland, M. Byrne, Y.K. Gun'Ko, J.J. Boland, P. Niraj, G. Duesberg, S. Krishnamurthy, R. Goodhue, J. Hutchison, V. Scardaci, A.C. Ferrari & J.N. Coleman. High-yield production of graphene by liquid-phase exfoliation of graphite. Nature Nanotechnology **3**, 563-568 (2008)
3. A. O'Neill, U. Khan, P. N. Nirmalraj, J. Boland, and J. N. Coleman. Graphene Dispersion and Exfoliation in Low Boiling Point Solvents. J. Phys. Chem. C **115**, (2011), 5422–5428
4. A. C. Ferrari, J. C. Meyer, V. Scardaci, C. Casiraghi, M. Lazzeri, F. Mauri, S. Piscanec, D. Jiang, K. S. Novoselov, S. Roth, and A. K. Geim. Raman Spectrum of Graphene and Graphene Layers. Phys. Rev. Lett. **97**, 187401 (2006)
5. Pimenta, M.A., Dresselhaus, G., Dresselhaus, M.S., Cancado, L.G., Jorio, A. and Saito, R. Studying disorder in graphite-based systems by Raman spectroscopy. Phys. Chem. Chem. Phys. **9**, 1276 (2007)

A High-Resistance Measurement Setup for MOX Sensing Materials Characterization

P.P. Capra, F. Galliana, M. Latino, A. Bonavita, N. Donato, and G. Neri

1 Introduction

The increasing demand for automated devices requires high-performance gas sensors to control living environment conditions, for preserving and processing food, pharmaceutical processing, and healthcare. Metal oxide (MOX) semiconductor-based humidity sensors have received considerable attention in this field owing to their advantages regarding the requirements of low cost, stability, large-scale manufacture, and long-term use.

At room temperature metal oxide semiconductors are generally characterized by high electrical resistance, with evident problems regarding the measurements with conventional instrumentation. Electrical resistance vs. the humidity of the sensors has been evaluated in a system constituted by an enclosure in which the temperature and the relative humidity are externally and remotely controlled. In such scenario the electrical characterization of the sensing films towards long-term stability and electrical reliability is mandatory. The developed system is primarily focused on the electrical evaluation of sensing films which aimed for material development and optimization; in such task, it is important to perform resistance measurement of high value samples in order to reach the right balance between film thickness, baseline resistance, and sensing performance.

The reliability of the screen printing process in order to provide the manufacture of sensors in a single operation on a row of the ceramic substrate containing six

P.P. Capra (✉) • F. Galliana
National Institute of Metrological Research (INRIM),
Str. delle Cacce, 91-10135 Torino, Italy
e-mail: p.capra@inrim.it

M. Latino • A. Bonavita • N. Donato • G. Neri
Department of Electronic Engineering, Chemistry and Industrial Engineering,
University of Messina, Contrada di Dio, 98166 Messina, Italy

C. Di Natale et al. (eds.), *Sensors and Microsystems: Proceedings of the 17th National Conference, Brescia, Italy, 5-7 February 2013*, Lecture Notes in Electrical Engineering 268, DOI 10.1007/978-3-319-00684-0_28, © Springer International Publishing Switzerland 2014

Fig. 1 View of a MOX-based sensor. (**a**) Pt interdigitated conductors. (**b**) The sensor after the Cu-doped SnO$_2$ activation process

interdigited structures has already been tested in a previous paper [1]. In this work, Cu-doped SnO$_2$ was used as sensing element deposited by screen printing on the planar ceramic substrate provided with Pt interdigited electrodes (Fig. 1).

2   Measurement Setup for MOX Sensors Humidity Test

The INRIM group is well experienced in the development of setups able to investigate humidity dependence of high value resistors [2, 3]. These systems allow to maintain the temperature at the level of about ±0.01 °C and to span the relative humidity in the range (10–90)%. Furthermore, the electronic control of the device maintains these settings with long-term stability of the humidity of up to ±1 % suitable for our scope of characterization of standards of electrical quantities vs. relative humidity.

Sampling takes place in a test chamber with two aluminum concentric cylinders with a polystyrene insulation; the external enclosure is a cube of the same insulating material. The temperature sensor in the chamber is a platinum thermo-resistance, mod. PT100: the output is proportional with the temperature at the level of 0.39 Ω/°C, with nominal value of 100 Ω at 0 °C. Tests performed in a laboratory with temperature controlled at (23 ± 1)°C showed that with 10 days of ±0.5 °C ambient temperature, the maximum variation into the enclosure was less than 0.01 °C.

The relative humidity inside the enclosure is obtained by an external system, controlled by a microprocessor. Inside the chamber, a thermo-hygrometer, mod. THGM-880, with digital output is placed. The resolution of the instrument is 1 %, while the accuracy is 3 %. The complete scheme of the control system is reported in Fig. 2.

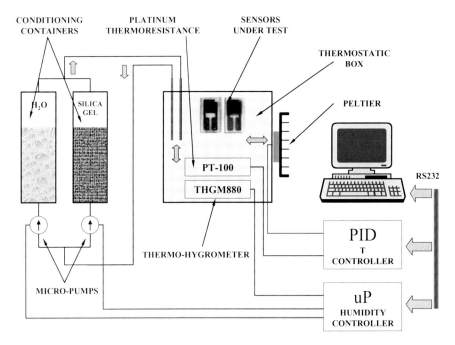

Fig. 2 Scheme of the control system. The units for the thermostatic control (PID) and for the relative humidity control (μP) are external from the chamber

The control circuit, based on the microprocessor, receives the data of the measured relative humidity by the hygrometer in BCD format and compares them with the set values. When the difference by the measured and set relative humidity exceeds the (variable) threshold value, one of the two micro-pumps for the conditioning of the air inside the enclosure is enabled.

Both pumps take the air from the chamber and send it in one of the two conditioning containers, one with silica gel and the other with water (Fig. 2), according to the need to increase or to decrease the value of humidity into the chamber, and then the air is sent backward to reach the set humidity value. For further details, see [2].

3 Experimental Characterization Results

3.1 Humidity Dependence

Measures made in direct current and using the thermostatic cell and the described humidity generator allowed to verify the behavior of sensors vs. humidity variations. The measurements were made with a high-accuracy, high-resistance meter Keithley mod. 6430.

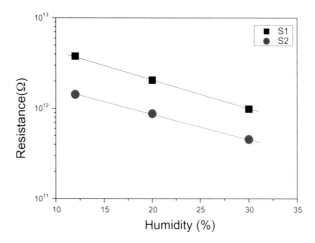

Fig. 3 Behavior of a group of two samples (S1 and S2) under test vs. relative humidity

The possibility to retain water molecules with the consequent long time release might be a limit in the accuracy of the sensors. To investigate this phenomenon, the Cu-SnO$_2$ sensors were measured in repeated cycles at several humidity conditions.

Humidity measurements were carried out in the range between 12 and 30 %, and the measurements are reported in Fig. 3. The investigated samples, starting from different base resistance values, show the same trend towards humidity. Each sensor has a characteristic pattern, which is maintained even after several thermal cycles. The resistance measurements' repeatability, subjecting the samples to humidity and temperature variation cycles, helps to show that the material can be a promising candidate for sensor development.

The uncertainty of the direct readings of resistance was about 0.35 %. For future investigations, higher precision methods could be involved [4, 5].

3.2 Temperature Dependence

A further verification has been made to characterize the behavior of the Cu-SnO$_2$ sensor with respect to temperature. The circuit used is shown in Fig. 4.

A current source supplies power to the heater (placed on side A) of the sensor; the current and the voltage at its terminals are acquired by two DMMs. The sensing element of the sensor (B side) is connected to a high-resistance meter. The temperature of the sensor is acquired in two different ways. Above 200 °C, the temperature is acquired by an infrared camera and evaluated using the temperature coefficient of the heating resistor. Over 200 °C, out of the range of the infrared camera, the temperature is evaluated with the heating resistor. The characterization was performed in the range from 100 to about 350 °C and is reported in Fig. 5 which shows the expected decrease of the resistance with the temperature.

Fig. 4 Measurement setup to
determine the temperature
dependence of the sensor

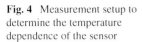
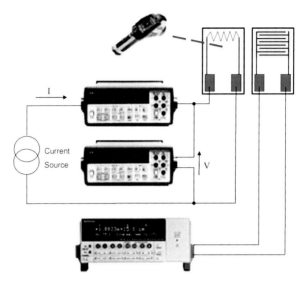

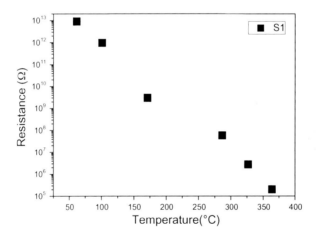

Fig. 5 Behavior of a S1 sample under test vs. temperature

4 Conclusions

This chapter has reported the development of a high-resistance measurement setup
for MOX sensing materials characterization. The developed system is primarily
focused on the electrical evaluation of sensing films aimed for material development
and optimization. In such task, it is important to perform resistance measurement of
high-resistance value samples in order to reach the right balance between film thickness,

baseline resistance, and sensing performance. The system was employed in the characterization of screen-printed Cu-SnO$_2$ sensing films with respect to variations of environmental parameters, namely, temperature and humidity.

References

1. A. Sanson, E. Mercadelli, E. Roncari, G. Cao, R. Licheri, R. Orrù, D. Marzorati, E. Merlone-Borla, A. Bonavita, G. Micali, G. Neri "Influence of processing parameters on the electrical response of screen printed SrFe0.6Ti0.4O3-? thick films" Ceramics International 36 (2010) 521-527
2. F. Galliana, P.P. Capra, and E. Gasparotto, "Behaviour of high value standard resistors versus relative humidity: first experimental results", Proceedings of the 11th IMEKO TC-4 Symp. Trends in Electrical Meas. Instrum., September 13 –14 2001, pp. 48–51.
3. P.P. Capra, F. Galliana, E. Gasparotto, D. Serazio: "Relative humidity calibrator for metrological characterization of electrical quantities standards", IEN technical report no. 619, March 2001.
4. G. Boella, F. Galliana "Analysis of voltage coefficients of high value standard resistors" Measurement 41 (2008), pp 1-9.
5. F. Galliana, P.P. Capra, E. Gasparotto: "Evaluation of two different methods to calibrate ultra-high value resistors at INRIM", IEEE Trans. Meas., 60(6), pp. 965-970. 2011.

A Portable Gas Sensor System for Air Quality Monitoring

Domenico Suriano, Gennaro Cassano, and Michele Penza

1 Introduction

Environmental monitoring is strongly required to protect the public health and save the environment from toxic contaminants and pathogens that can be released into air. Air pollutants include carbon monoxide (CO), nitrogen dioxide (NO_2), and sulfur dioxide (SO_2) that originate from various sources such as vehicle emissions, power plants, refineries, and industrial and laboratory processes. However, current monitoring methods are costly and time-consuming and limitations in sampling and analytical techniques also exist. Clearly, a need exists for accurate, inexpensive long-term monitoring of environmental contaminants using low-cost solid-state gas sensors that are able to operate on-site and real time. Calibrated cost-effective gas sensors are a very interesting solution for networked systems suitable to monitor air pollutants in urban streets and real scenario of smart cities with high spatial and time resolution. In ENEA, at Brindisi Research Center, a handheld gas sensor system called *NASUS IV* based on solid-state gas sensors was designed and implemented [1–3]. This system is the last result of our researches in the area of tiny and portable system building for air quality control based on cost-effective solid-state gas sensors. The main goal of the system designed and built in our laboratory is the development of a handheld device in order to detect some air pollutant gases such as CO, SO_2, NO_2, and H_2S in urban areas at outdoor level, including their specific indoor applications for chemical safety.

D. Suriano (✉) • G. Cassano • M. Penza
Technical Unit for Materials Technologies—Brindisi Research Center, ENEA—Italian
National Agency for New Technologies, Energy and Sustainable Economic Development,
PO Box 51 Br-4, 72100 Brindisi, Italy
e-mail: domenico.suriano@enea.it; michele.penza@enea.it

C. Di Natale et al. (eds.), *Sensors and Microsystems: Proceedings of the 17th National Conference, Brescia, Italy, 5-7 February 2013*, Lecture Notes in Electrical Engineering 268, DOI 10.1007/978-3-319-00684-0_29, © Springer International Publishing Switzerland 2014

2   NASUS IV System

NASUS IV is formed by four modules or, better, by four printed circuit boards (PCBs): the main module, the sensor module, the wireless module, and the power module. The first three modules are packed in the same handheld case, but the power module is arranged in a separate case, as shown in Fig. 1.

The main module is in charge of managing the communications with the wireless module and with the PC via USB port as well as driving the local display, driving the mini-joystick (which is one of the input system, but not the only), and driving the SD card memory for data storage. On the sensor module are arranged six sensors: a temperature sensor, a relative humidity sensor, and four electrochemical solid-state gas sensors (Alphasense Ltd, UK). The wireless module allows to control remotely the portable device by means of commands via short message system (SMS) sent from a mobile phone to NASUS IV. Furthermore, our sensor system can send e-mails containing any sensor data requested by the remote end-user mobile phone. The power module provides the charging of battery inside NASUS IV system. Moreover, it performs a smart management of the available power source: network electricity or solar energy, giving priority to the solar one whenever possible. Currently, the sensors onboard NASUS IV are the electrochemical-type sensors provided by Alphasense Ltd (UK), for the detection of pollutant gases such as CO, SO_2, NO_2, and H_2S. We decided to test those electrochemical-type sensors inside our device because of their interesting features such as very low power consumption, small dimensions, good sensitivity, and improved response to interfering gases.

3   Laboratory Tests and Results

In order to test this machine in our laboratory under environmental conditions similar to the real-world situations, we employed a wide-volume gas chamber (ca. $0.5 \, \text{m}^3$) provided by an input and an output pipe. The test gas is injected in the input pipe, while the output pipe allows the gas mixture to flow out. In this way we simulate an environment containing a steady concentration of various gases to be tested upon

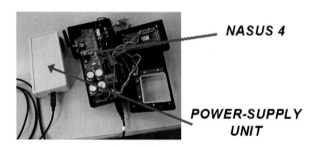

Fig. 1 NASUS IV inside view and power supply unit

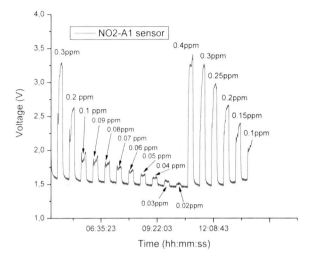

Fig. 2 NO2A1 sensor response to NO₂ gas injected in NASUS IV test cell

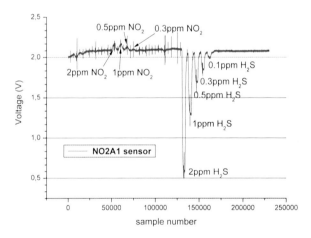

Fig. 3 NO2A1 sensor response to NO₂ gas injected in the chamber

typical environmental conditions of relative humidity (40–50 %) and temperature (12–18 °C). Figure 2 shows the sensor system response to the laboratory tests performed, by directly injecting the gas inside the test cell (volume ca. 50 cm³) of NASUS IV in order to verify the system sensing capabilities.

Figure 3 shows the sensor system response in the simulated environment (volume ca. 0.5 m³). Other similar tests involving CO, H₂S, and SO₂ gases in the concentration range of environmental interest were performed in our laboratory in order to study the NASUS IV system behavior.

4   Conclusions

A handheld sensor system has been designed and implemented for air pollution monitoring. We investigated NASUS IV in our laboratory, and several tests were performed with different kind of air pollutant gases. Some problems experienced during lab tests such as the influence of the interfering gases and the influence of the environmental parameters (humidity and temperature) have to be addressed. Future work concerns about the employment of the NASUS IV in real environment by performing experimental campaigns in collaboration with ARPA-Puglia, Italian public regional environmental agency, which will provide in-field fixed stations in order to compare the performance of our machine with referenced gas analyzers. Moreover, at the same time, we are planning to design a new handheld system by testing a new electrochemical gas sensor with an improved cross-sensitivity response and an enhanced sensitivity towards very low gas concentrations up to a level of tens of ppb.

Acknowledgments This work has been partially funded by the project "PON BAITAH: *Methodology and Instruments of Building Automations and Information Technology for pervasive models of treatment and Aids for domestic Healthcare*" and by project PON Smart Cities RES-NOVAE.

References

1. W. Tsujita, A. Yoshino, H. Ishida, T. Morizumi, Sensors and Actuators B 110 (2005) 304-311.
2. M. Penza, D. Suriano, R. Rossi, M. Alvisi, G. Cassano, V. Pfister, L. Trizio, M. Brattoli, M. Amodio, G. De Gennaro, Lecture Notes in Electrical Engineering (LNEE) 109 (2012) 87-92.
3. M. Penza, D. Suriano, G. Cassano, V. Pfister, M. Alvisi, R. Rossi, Paolo Rosario Dambruoso, Livia Trizio, Gianluigi De Gennaro, Proceedings IMCS 2012, 20-23 May 2012, Nuremberg, Germany.

Part III
Biosensors

Herbicide Analysis Using a New OPIE. A Case Study: Sunflower Oil

E. Martini, M. Tomassetti, and L. Campanella

1 Introduction

Satisfactory determination of several kinds of pesticide in edible oils is an important but not a new problem. In actual fact the quantitative determination of chemical species or real matrices that are scarcely soluble or completely insoluble in aqueous solutions has always posed a serious problem in chemical analysis and has only been partially solved by such techniques as gas chromatography, which are more suitable when employed in a laboratory than in situ. A substantial contribution to solving this problem was the development of organic phase enzyme electrodes (OPEEs), i.e. enzymatic electrodes capable of operating in organic solvents [1–3] and that can also act in situ. One classical example is that of inhibition OPEEs to analyse different types of pesticides that are relatively insoluble in aqueous solution, in the development of which also our team was recently involved [4–6]. The drawback consists of the fact that it is often complained that inhibition biosensors are relatively unselective versus the pesticides belonging to different phytopharmaceutical classes. Immunosensors, on the contrary, are the most selective biosensors, and our team has in recent years fabricated several immunosensors [7–10], including also devices for pesticide determination [11], albeit operating only in aqueous solution. However, we recently developed a new organic phase immuno-electrode (OPIE) [12]. Using this new device, applications were developed to detect triazinic pesticides in extra-virgin olive oil, obtaining good results [12]. Therefore, new applications were recently developed using the novel OPIE to detect several herbicides and pesticides in sunflower oil.

E. Martini • M. Tomassetti (✉) • L. Campanella
Department of Chemistry, University of Rome "Sapienza",
p.le A. Moro, 5, 00183 Rome, Italy
e-mail: mauro.tomassetti@uniroma1.it

C. Di Natale et al. (eds.), *Sensors and Microsystems: Proceedings of the 17th National Conference, Brescia, Italy, 5-7 February 2013*, Lecture Notes in Electrical Engineering 268, DOI 10.1007/978-3-319-00684-0_30, © Springer International Publishing Switzerland 2014

2 Methods

The following working conditions were studied and optimised for this sensor in previous research on the analysis of extra-virgin olive oil [12]: amperometric Clark-type gas-diffusion electrode, as transducer; competitive measurement (ELISA type) in 50 % v/v *n*-hexane/chloroform mixture; BSA peroxidase conjugated labelled antigen (herbicide); and peroxidase OPEE operating in decane, as enzyme probe for the final measurement. In the present research the only condition that changed was the use of 75 % v/v *n*-hexane/chloroform mixture, which replaced the 50 % v/v mixture of the same solvents used in the previous research [12] and employed in the competition step of the immunological method. This was because of the better solubility of sunflower oil in the first mixture than in the latter. Also in this case, anti-atrazine, anti-2,4-D and anti-2,4,5-T antibodies were provided by Dr. S. Eremin (Department of Chemical Enzymology, Faculty of Chemistry, Moscow State University, Russia), while anti-parathion was a commercial antibody and was obtained from Acris Antibodies GmbH, Schillerstraße (Herford, Germany).

3 Results

In view of the good performance of the OPIE developed for triazinic pesticides determined in extra-virgin olive oil [12], we decided to fabricate also an OPIE for the analysis of traces of both triazinic and other types of pesticides in sunflower oil using the same technique. The focus was on organophosphates (for instance parathion) and chlorinated pesticides such as 2,4-dichlorophenoxyacetic acid (2,4-D) and 2,4,5-trichlorophenoxyacetic acid (2,4,5-T). The latter are well-known herbicides used also in modern warfare and known as "agent orange". Figures 1a, b and 2c–e

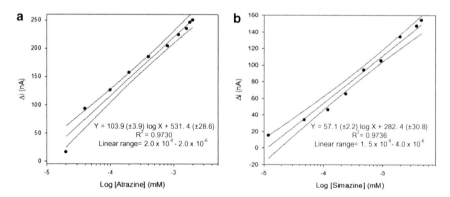

Fig. 1 Atrazine (**a**) and simazine (**b**) determination in sunflower oil: calibration curves and confidence intervals for pesticide determination, obtained using a semilogarithmic scale, employing an Immobilon membrane for antibody immobilisation and an amperometric electrode for O_2 as transducer. Competition in chloroform—*n*-hexane (75 % v/v) and in the presence of sunflower oil. Final enzymatic measurement in decane

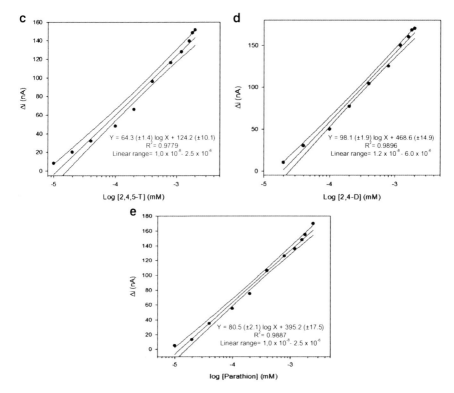

Fig. 2 Pesticides 2,4,5-T (**c**), 2,4-D (**d**) and parathion (**e**) determination in sunflower oil: calibration curves and confidence intervals for pesticide determination obtained using a semilogarithmic scale, employing an Immobilon membrane for antibody immobilisation and an amperometric electrode for O_2 as transducer. Competition in chloroform—n-hexane (75 % v/v) and in the presence of sunflower oil. Final enzymatic measurement in decane

show the calibration curves obtained, together with the corresponding equations and linearity ranges.

In all cases the OPIE resulted suitable, handling and cheaper to test each pesticide that, as reported in the literature, may be contained in sunflower oil [13, 14]. For all the pesticides studied, also the values of the affinity constant were estimated on the basis of the value of the concentration at which half of the maximum response was obtained. k_{aff} were found to be of the order of 10^6 M^{-1} in all cases.

4 Conclusions

The results obtained so far show that an immunosensor which operates the competition step in a solvent mixture such as chloroform–hexane mixture can be developed for the purpose of determining atrazine, simazine, paraoxon, 2,4-D and 2,4,5-T in edible seed oils, such as sunflower oil. It was observed that also in this case the best results were obtained: (1) using a Clark-type transducer instead of an H_2O_2 electrode (see reference [12]), (2) by performing not only the competitive procedure but also the actual final enzymatic reaction in organic solvent, i.e. decane, instead of in aqueous solution (see previous paper [12]) then carrying out the final electro-enzymatic measurement using a classical OPEE and (3), lastly, by using as substrate for the peroxidase-catalysed enzymatic reaction, a solution of tert-butyl hydroperoxide in decane.

References

1. J. Wang, Y. Lin, Q. Chen, "Organic-phase biosensors for monitoring phenol and hydrogen peroxide in pharmaceutical antibacterial products", Analyst, 118 (1993), 277-283.
2. J. Wang, E. Demsey, A. Eremenko, and M. R. Smyth, "Organic-Phase Biosensing of Enzyme Inhibitors" Anal. Chim. Acta. 279 (1993), 203-208.
3. E.I. Iwuoha, D. Saenz de Villaverde, N.P. Garcia, M.R. Smyth, J.M. Pingarron, "Reactivities of organic phase biosensors. 2. The amperometric behaviour of horseradish peroxidase immobilised on a platinum electrode modified with an electrosynthetic polyaniline film", Biosens. Bioelectron. 12 (1997), 749-755.
4. L. Campanella, A. Bonanni, E. Martini, N. Todini, M. Tomassetti, Determination of triazine pesticides using a new enzyme inhibition tyrosinase OPEE operating in chloroform, Sens. Act. B, Chem. 111-112 (2005) 505-511.
5. L. Campanella, R. Dragone, D. Lelo, E. Martini, M. Tomassetti, "Inhibition tyrosinase organic phase biosensor for triazinic and benzotriazinic pesticides analysis", Anal. Bioanal. Chem., 384 (2006) 915-921.
6. L. Campanella, D. Lelo, E. Martini, M. Tomassetti, "Organophosphorus and carbamate pesticide analysis using an inhibition tyrosinase organic phase enzyme sensor; comparison by butyrylcholinesterase + choline oxidase OPEE and application to natural waters", Anal. Chim. Acta, 587 (2007), 22-32.
7. L. Campanella, E. Martini, M. Tomassetti, "Immunochemical potentiometric method for HIgG And Anti-HIgG determination, using a NH_3 probe and the immunoprecipitation as preconcentration procedure", Anal. Lett. 40 (2007), 113-125.
8. L. Campanella, D. Lelo, E. Martini, M. Tomassetti, "Immunoglobulin G determination in human serum and human milk using an immunosensor of new conception fitted with an enzyme probe as transducer", Sensors 8 (2008), 6727-676.
9. L. Campanella, E. Martini, M. Pintore, M. Tomassetti, "Lactoferrin And Immunoglobulin G Determination In Animal Milk By New Immunosensors", Sensors 9 (2009), 2202-2221.
10. L. Campanella, E. Martini, M. Tomassetti, Further development of lactoferrin immunosensor (part III). J. Pharm. Biomed. Anal. 53 (2010), 186-195
11. L. Campanella, S. Eremin, D. Lelo. E. Martini, M. Tomassetti, Reliable new immunosensor for atrazine pesticide analysis, Sens. Act. B, Chem. 156 (2011) 501-515.
12. M. Tomassetti, E. Martini, L. Campanella, New immunosensors operating in organic phase (OPIEs) for analysis of triazinic pesticides in olive oil, Electroanalysis 24 (4) (2012) 842-856.
13. L. Rastrelli, K. Totaro, F. De Simone, "Determination of organophosphorus pesticide residues in Cilento (Campania, Italy) virgin olive oil by capillary gas chromatography", Food Chem. 79(3) (2002), 303–305.
14. N. M. Randolph, H. W. Dorough, G. L. Teetes, "Malathion, methyl parathion, diazinon, and endosulfan residues in sunflower seeds" J. Econ. Entomol. 62(2) (1969), 462-464.

Electrochemical Immunoassay for Mucin 1 Detection as a Diagnostic Tool in Ovarian Cancer

Andrea Ravalli, Giovanna Marrazza, Anca Florea, Cecilia Cristea, and Robert Sandulescu

1 Introduction

Cancer is one of the leading causes of death worldwide. Among this, ovarian cancer represents the fifth cause of death and the most lethal of all gynecological malignancies [1, 2]. Early detection remains the most promising approach to improve long-term survival rate of patients affected by cancer [3]. For this, the analysis of the cancer biomarker represents an ideal tool to determine diagnosis and prognosis and follows the medical treatment of patients [4–6].

MUC1 gene encodes a type I transmembrane glycoprotein that is expressed on the apical surface of various epithelial cells. In malignant neoplasm, Mucin 1 (MUC1) loses its apical distribution, becomes underglycosylated and overexpressed, and is secreted into the blood circulation, serving therefore as potential tumor marker and prognosis factor in different types of cancer [7].

In this paper an electrochemical immunoassay as a screening device for the detection of MUC1 in real samples was presented. The proposed approach uses disposable screen-printed electrodes as transducers and a simple target capturing step by antibody-functionalized magnetic beads.

A. Ravalli • G. Marrazza (✉)
Department of Chemistry "Ugo Schiff", University of Florence, Via della Lastruccia 3, 50019 Sesto Fiorentino, Florence, Italy
e-mail: giovanna.marrazza@unifi.it

A. Florea • C. Cristea • R. Sandulescu
Department of Analytical Chemistry, Faculty of Pharmacy, University of Medicine and Pharmacy, "Iuliu Hatieganu", Pasteur 4, Cluj-Napoca, Romania

C. Di Natale et al. (eds.), *Sensors and Microsystems: Proceedings of the 17th National Conference, Brescia, Italy, 5-7 February 2013*, Lecture Notes in Electrical Engineering 268, DOI 10.1007/978-3-319-00684-0_31, © Springer International Publishing Switzerland 2014

2   Materials and Methods

Dynabeads Protein G-coated magnetic beads were purchased from Invitrogen (Milan, Italy).

MUC1 protein, MUC1 monoclonal mouse antibody (Ab1), MUC1 polyclonal rabbit antibody (Ab2), and polyclonal antibody anti-rabbit IgG labeled with alkaline phosphatase (Ab3-AP) were provided by Novus Biological (Cambridge, UK).

1-Naphthyl phosphate, diethanolamine, potassium chloride, magnesium chloride, and commercial non-pathological human serum were purchased from Sigma-Aldrich (Milan, Italy).

All solutions were prepared using water from Milli-Q Water Purification System (Millipore, UK).

Eight screen-printed cells were used in the experiments. Each cell is based on graphite working electrode (2.0 mm in diameter), each with a graphite counter electrode and a silver pseudo-reference electrode, produced on a DEK 248 (DEK, Weymouth, UK) screen printing machine [8].

Electrochemical measurements were performed with μAutolab type II PGSTAT (Metrohm, The Netherlands) with General Purpose Electrochemical System (GPES) 4.9 software and PalmSens handheld potentiostat (PalmSens BV, The Netherlands).

Differential pulse voltammetry (DPV) was employed as electrochemical technique using the following parameters: potential range of −0.2 to +0.5 V; pulse amplitude, 0.070 V; and scan rate, 0.033 V s^{-1}. The measurements were carried out at room temperature.

3   Experimental Procedure

The sandwich immunoassay for the detection of MUC1 tumor marker was developed following the scheme reported in Fig. 1.

First, anti-MUC1 monoclonal mouse antibody (Ab1) was immobilized on the surface of Protein G-coated magnetic beads followed by blocking step with milk powder 5 % and by an incubation step with MUC1 buffered solution or MUC1 spiked serum sample; after, secondary MUC1 polyclonal rabbit antibody (Ab2) and the third polyclonal anti-rabbit IgG labeled with alkaline phosphatase (Ab3-AP) were added.

Finally, 10 μL bead suspensions were placed onto each working electrode of the array, using a magnetic support to concentrate the magnetic beads on the working electrode. Thus, each well of the array was filled with 60 μL of a solution containing 1 mg mL^{-1} of 1-naphthyl phosphate prepared in 0.1 M diethanolamine buffer pH 9.6 (added with KCl 0.1 M and MgCl$_2$ 1 mM). After 6 min of incubation time, DPV measurements were carried out for each electrochemical cell.

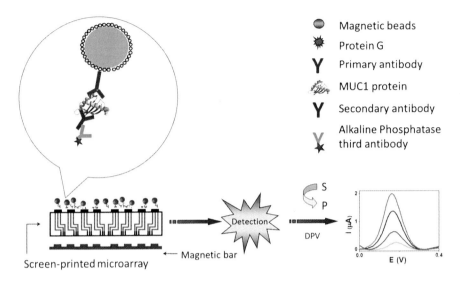

Fig. 1 Schematic representation of MUC1 immunosensor

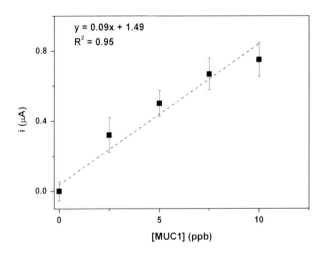

Fig. 2 Calibration curve for the immunosensor in different concentrations of MUC1 in buffered solutions

4 Results and Discussion

Figure 2 shows the calibration curve obtained under the optimized experimental conditions in MUC1 buffered solution. A linear response ($y = 0.09x + 1.49$, $R^2 = 0.95$) was obtained in the range of 0–10 ng mL^{-1} (which is also the important range from the clinical point of view) with a detection limit (calculated as $3S_{Blank}/S_{slope}$) of 2.5 ng mL^{-1}.

Table 1 MUC1 spiked serum sample analysis

Sample	Current (μA)	CV%
Serum	1.89 ± 0.09	10
Serum with MUC1 10 ng mL^{-1}	2.16 ± 0.08	12

Once the suitability of the assay to detect MUC1 in buffered solutions is verified, experiments on serum samples were carried out. Initial experiments concerned commercial serum spiked with MUC1 at a known concentration. The serum sample was filtered (diameter of filter pores is 0.2 μM) and then diluted with phosphate buffer (1:200).

The obtained results show an increase of the current between serum alone and serum spiked with MUC1 10 ng mL^{-1} (Table 1) demonstrating the potential practical applications in biological samples. Good reproducibility in terms of CV% was achieved with both kinds of serum samples (CV% 10 %).

5 Conclusion

In this work, a magnetic beads-based antibody sandwich immunosensor for the detection of MUC1 in buffered solution was presented. Moreover, the proposed immunosensor was used to detect MUC1 in commercial serum samples, offering a promising tool in biomedical applications.

References

1. M. Malvezzi, P. Bertuccio, F. Levi, C. La Vecchia, E. Negri. European cancer mortality predictions for the year 2013. Ann. Oncol. **24**, 792-800 (2013).
2. http://eco.iarc.fr/ (last accessed on 1st May 2013).
3. J. J. Ott, A. Ullrich, A. B. Miller. The importance of early symptom recognition in the context of early detection and cancer survival. European Journal of Cancer **45**, 2743-2748 (2009).
4. A. Ravalli, G. P. dos Santos, M. Ferroni, G. Faglia, H. Yamanaka, G. Marrazza. New label free CA125 detection based on gold nanostructured screen-printed electrode. Sensors and Actuators B **179**, 194-200 (2013).
5. Z. Taleat, A. Ravalli, M. Mazloum-Ardakani, G. Marrazza. CA 125 immunosensor based on poly-anthranilic acid modified screen-printed electrodes. Electroanalysis **25**, 269-277 (2013).
6. I. E. Tothill. Biosensors for cancer markers diagnosis. Seminars in Cell & Developmental Biology **20**, 55-62 (2009).
7. M. Brayman, A. Thathiah, D. D. Carson. MUC1: a multifunctional cell surface component of reproductive tissue epithelia. Reprod. Biol. Endocrinol. **2**, 4 (2004).
8. A. Zani, S. Laschi, M. Mascini, G. Marrazza. A New Electrochemical Multiplexed Assay for PSA Cancer Marker Detection. Electroanalysis, **23**, 91-99 (2011).

Monitoring Photocatalytic Treatment of Olive Mill Wastewater (OMW) in Batch Photoreactor Using a Tyrosinase Biosensor and COD Test

E. Martini, M. Tomassetti, and L. Campanella

1 Introduction

Several studies have been reported [1–4] concerning the abatement or reduction of the pollutant load of olive oil wastewater. The aim of the present research was precisely to test the use of a titanium dioxide/UV light process, as well as of hydrogen peroxide/UV light only, in olive mill wastewater treatment. Experiments were carried out in a bioreactor alternatively containing one of the three different types of heterogeneous membranes with the catalyst. The main parameters (initial and final concentration of the target compounds, amount of oxidising agents and catalysts, etc.) affecting these processes were investigated, while several applications to wastewater treatment are reported in the present paper. In order to monitor photocatalytic treatment, samples periodically taken from the reactor were analysed for their residual total phenolic content (TPh) using a tyrosinase enzyme biosensor developed by the authors [5], and the COD (chemical oxygen demand), i.e. carbon organic concentration, was evaluated colorimetrically according to the standard dichromate method [6].

2 Methods and Samples

2.1 Monitored Process

In the present research, dilute solutions of olive mill wastewater (OMW) were treated in a catalyst containing membrane bioreactor and the heterogeneous photocatalytic treatment of olive mill wastewater investigated. Experiments were conducted in a laboratory scale batch-type photoreactor. UV irradiation was provided

E. Martini • M. Tomassetti (✉) • L. Campanella
Department of Chemistry, University of Rome "Sapienza", p.le A. Moro, 5, 00183 Rome, Italy
e-mail: mauro.tomassetti@uniroma1.it

C. Di Natale et al. (eds.), *Sensors and Microsystems: Proceedings of the 17th National Conference, Brescia, Italy, 5-7 February 2013*, Lecture Notes in Electrical Engineering 268, DOI 10.1007/978-3-319-00684-0_32, © Springer International Publishing Switzerland 2014

Fig. 1 Inhibition tyrosinase biosensor

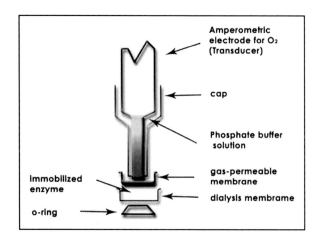

by a 450 W medium pressure Hg lamp, manufactured by Helios Italquartz (Milan) and emitting radiation in the 300–400 nm range, or with a low pressure lamp 36 W supplied by Helios Italquartz (Milan) and emitting radiation in the same range. An experimental model describing the chemical oxygen demand (COD) and total phenols (TPh) from actual OMW was developed to monitor the process. The OMW was provided by a three-phase olive oil mill, located in Lazio, Italy. The effluent was subjected to dilution (1/100) because it had a strong olive oil odour. The solution was slurried with a H_2O_2 0.20 M solution (4 mL in 1.5 L of diluted sample) and the catalytic membrane added. Lastly, its main properties prior to and after the catalytic process were investigated. The total phenolic content was determined using a tyrosinase biosensor (see Fig. 1) previously developed by us [5]. COD was determined colorimetrically, according to the standard dichromate method, using a Hach DR/2010 spectrophotometer.

2.2 Photocatalysis

Photocatalytic processes make use of a semiconductor metal oxide (TiO_2) as catalyst and H_2O_2 as oxidising agent. Several catalysts have been tested so far, although TiO_2 in the anatase form seems to have the most appropriate attributes such as high stability, good performance and low cost.

2.3 Membrane Preparation

Three different catalytic membranes, i.e. ceramic, polymeric and metallic, were fabricated. For the fabrication of the ceramic membrane, several inorganic materials were used, such as quartz, clay, feldspar, sand, sodium carbonate, calcium carbonate and kaolin. These raw materials were mixed and a body slip was prepared. The membrane was then cast in a gypsum mould. Lastly, the dry ceramic membrane was

Fig. 2 Scheme of the photoreactor

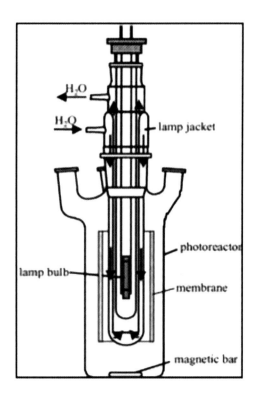

fired at 950 °C. The ceramic membrane was glazed with an enamel containing also 10 % (w/w), or 30 % (w/w) of TiO$_2$, then fired again until 750 °C. To fabricate the polymeric membrane, a homogeneous polymeric solution was prepared containing a polymer, a solvent and an additive such as commercial polysulfone, 1-octanol and N-methyl-pyrrolidone (NMP), respectively. TiO$_2$ was then added to this solution. The alcohol guaranteed a sufficiently fast phase separation in the water containing the coagulation bath. In order to achieve the desired structure of the membrane, polyvinylidene fluoride (PVDF) with TiO$_2$ was used. Lastly, on the metallic membrane, a TiO$_2$ film was immobilised on the stainless steel surface as support.

2.4   Olive Oil Mill Wastewater Degradation

Experiments were conducted in a batch-type laboratory scale photoreactor, as illustrated in Fig. 2. The aerobic olive oil photoreactor (AOP) consisted of a cylindrical vessel with a diameter of 8.0 cm. The working volume was 1.5 L, with provision for UV irradiation and addition of hydrogen peroxide (the concentration of H$_2$O$_2$ was 5 mM in the photoreactor). The catalytic membranes were mounted as required around the UV lamp, so that the TiO$_2$ present in the internal glazed surface was illuminated with UV irradiation to achieve the degradation of pollutants. For the sake of comparison, in addition to this new purpose of ceramic membrane, also two other traditional membranes, either polymeric or metallic (already used in previous research), were used alternatively. Photocatalytic tests were performed using the batch photoreactors shown in Fig. 2. In a typical heterogeneous photocatalytic run, the original OMW was diluted with distilled water as described above, and 1.5 L was loaded into the reaction vessel and H_2O_2 was added as described in Sect. 2.1. 5 mL aliquots of samples periodically taken from the reactor were analysed for their residual organic concentration and total phenolic content (TPh) [5]. COD was determined colorimetrically using the standard dichromate method [6].

3   Results and Discussion

The principal results obtained in all the tests performed are summarised in Table 1.

Several tests were carried out on the photodegradation process performed in the photoreactor described. In particular, the process was investigated above all in the presence of H_2O_2, but without any catalytic membrane, then with H_2O_2 and the catalytic membrane. In the latter case, one of the three different membranes described above was used alternatively. Lastly, the process was investigated using two different UV lamps, at low or at medium pressure. In particular, Table 1, respectively, shows the percentage removal of COD and phenols with a UV lamp at low pressure (36 W) or at medium pressure (450 W) after 24 h, to monitor photocatalytic treatment of olive mill wastewater (OMW) in batch photoreactor using a tyrosinase biosensor and COD test. It was found that the initial effluent organic load and the reaction time have an important effect on COD and phenol removal. After 24 h and in the presence of TiO_2, using the low-pressure UV lamp, almost 37.9 and 80.1 % of COD and phenols were respectively removed in this process for the treatment of olive mill wastewater using a ceramic membrane glazing with an enamel that contained 10 % w/w in TiO_2. The use of a metallic membrane on which titanium had been deposited using an electrolytic anodizing process resulted in COD abatement of 40.9 % and phenol reduction of 89.1 %. Results show that treatment efficiency increased with increasing TiO_2 concentration in the ceramic membrane (about 30 %), achieving 42.4 and 92.3 % in COD and TPh, respectively. This observation indicates that most of the biodegradable compounds initially present in the wastewater were destroyed and/or less biodegradable intermediates were formed. Finally, in the case of the polymeric membrane, the final COD removal was only of the order of 32.5 % and phenols of the order of 64.2 %.

4   Conclusions

Data reported in Table 1 show how the new ceramic membrane containing TiO_2 should provide an excellent combination of thermal, mechanical and chemical stability in addition to good catalytic characteristics. Lastly, it may be concluded that

Table 1 Percentage removal of COD and phenols with UV lamp using: (a) low pressure at 36 W and (b) high pressure at 450 W, both after 24 h

Experimental condition	% COD removal RSD% ≤ 5.0		% phenols removal RSD% ≤ 5.0		pH of the solution after the process RSD% ≤ 5.0	
	(a)	(b)	(a)	(b)	(a)	(b)
Without membrane and without H_2O_2 (blank)	28.10	28.50	50.2	50.2	5.48	5.50
Without membrane and with H_2O_2	29.05	29.15	53.8	53.8	5.85	5.95
With metallic membrane and with H_2O_2	40.85	44.85	89.1	93.2	6.95	6.85
With polymeric membrane and with H_2O_2	32.52	38.82	64.2	71.4	7.70	7.80
With ceramic membrane (10 % TiO_2 on the surface) and with H_2O_2	37.85	41.25	80.1	85.1	7.25	7.25
With ceramic membrane (30 % TiO_2 on the surface) and with H_2O_2	42.42	47.52	92.3	96.3	7.10	7.18

the use of the tyrosinase amperometric biosensor to monitor this process can be considered practical and advantageous. In future investigations, the possibility to develop some kind of sensor device for speedier COD determination is planned.

References

1. C.A. Paraskeva, V.G. Papadakis, D.G. Kanellopoulou, P.G. Koutsoukos, K.C. Angelopoulos, Membrane filtration of olive mill wastewater and exploitation of its fractions, Water Environ. Res. 79 (2007), 421-429.
2. M.G. Di Serio, B. Lanza, M.R. Mucciarella, F. Russi, E. Iannucci, P. Marfisi, A. Madeo, Effects of olive mill wastewater spreading on the physico-chemical and microbiological characteristics of soil, Int. Biodet. Biodeg. 62 (2008), 403-407.
3. R. Molinari, L. Palmisano, E. Drioli, M. Schiavello, Studies on various reactor configurations for coupling photocatalysis and membrane processes in water purification, J. Membrane Sci. 206(1–2) (2002), 399–415.
4. M. Drouiche, V. Le Mignot, H. Lounici, D. Belhocine, H. Grib, A. Pauss, N. Mameri, A compact process for the treatment of olive mill wastewater by combining UF and UV/H_2O_2 techniques, Desalination 169 (2004), 81–88.
5. L. Campanella, E. Martini, N. Todini, M. Tomassetti, Determination of polyphenol 'pool' in olive oil mill waste water using a tyrosinase biosensor operating in aqueous solution or in organic solvent, Int. J. Environ. Anal. Chem. 85(12–13) (2005) 937-957
6. S. Khoufi, F. Aloui, S. Sayadi, Treatment of olive oil mill wastewater by combined process electro-Fenton reaction and anaerobic digestion, Water Research 40(10) (2006) 2007–2016.

Electrochemical Antibody-Aptamer Assay for VEGF Cancer Biomarker Detection

Andrea Ravalli, Giovanna Marrazza, Lourdes Rivas,
Alfredo De La Escosura-Muniz, and Arben Merkoci

1 Introduction

The possibility of performing reliable cancer diagnosis even before any symptom of disease appears is crucial for increasing therapeutic treatment success and patient survival rates [1]. During the last decade, improved understanding of carcinogenesis and tumor progression has revealed a large number of potential tumor markers. Vascular endothelial growth factor (VEGF) is referred to a family of homodimeric glycoproteins which are involved in the development of the blood vascular system (vasculogenesis), of the lymphatic system (lymphangiogenesis), in the formation of new blood vessels from pre-existing one (angiogenesis), and in the vascularization of tumor. VEGF level has been extensively used as biomarker associated with diagnosis and prognosis of different cancer disease [2].

Different detection methods such as enzyme-linked immunosorbent assays (ELISA) and immunohistochemistry have been used for VEGF quantification. Nevertheless these methods do not satisfy the rapidity requirement and the necessity to use simple instrumentation for point of care diagnostics [3]. In this context, rapid non-immunochemical sensors based on electrochemical methods that can use aptamers are emerging. In this work, simple and sensitive approach for VEGF detection using antibody-aptamer assay and screen-printed cells as transducers is presented.

A. Ravalli • G. Marrazza (✉)
Department of Chemistry "Ugo Schiff", University of Florence, Via della Lastruccia 3,
50019 Sesto Fiorentino, Florence, Italy
e-mail: giovanna.marrazza@unifi.it

L. Rivas • A. De La Escosura-Muniz • A. Merkoci
Nanobioelectronics & Biosensors Group, CIN2 (ICN-CSIC), Catalan Institute of
Nanotechnology, UAB Campus, Bellaterra, Barcelona, Spain

C. Di Natale et al. (eds.), *Sensors and Microsystems: Proceedings of the 17th National Conference, Brescia, Italy, 5-7 February 2013*, Lecture Notes in Electrical Engineering 268, DOI 10.1007/978-3-319-00684-0_33, © Springer International Publishing Switzerland 2014

2 Materials and Methods

11-Mercaptoundecanoic acid (MUDA), 6-mercapto-1-hexanol (MCH), diethanol-amine (DEA), 1-naphthyil phosphate, streptavidin-alkaline phosphatase, ethanol-amine (EA), human vascular endothelial growth factor (VEGF), and anti-VEGF antibody were obtained from Sigma-Aldrich (Milan, Italy).

Aptamer used in this work [5] was purchased from MWG Biotech, Germany:
5'-GGGCCCGTCCGTATGGTGGGTGTGCTGGCCTTTTTTTTTTTTTTTT(45) 3'-Biot.

All chemicals were used as received without any further purification. Milli-Q water was used throughout this work.

Electrochemical experiments were performed in a digital potentiostat/galvano-stat Autolab PGSTAT 30(2)/FRA2 controlled with the General Purpose Electrochemical System (GPES) and Frequency Response Analyzer (FRA2) 4.9 software (Eco Chemie, Utrecht, The Netherlands). The immunosensor was assembled using screen-printed cells, comprising of gold working electrode (2.5 mm in diameter), counter graphite electrode, and a pseudo-silver [4].

Differential pulse voltammetry (DPV) was performed using the following parameters: potential range: −0.2 V to +0.5 V; pulse amplitude 0.070 V; and scan rate 0.033 V s^{-1}. The measurements were carried out at room temperature.

3 Experimental Procedure

The scheme of antibody-aptamer assay developed for the detection of VEGF tumor marker was shown in Fig. 1.

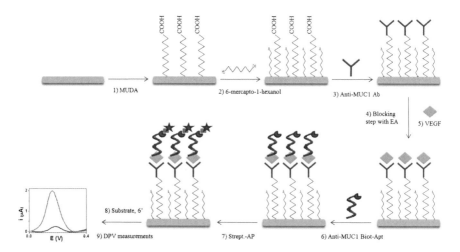

Fig. 1 Schematic representation of antibody-aptamer sandwich assay for VEGF detection

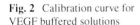

Fig. 2 Calibration curve for VEGF buffered solutions

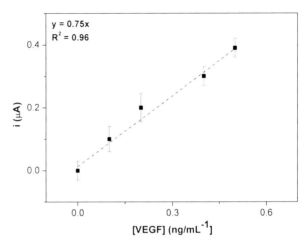

Firstly, 11-mercaptoundecanoic acid (MUDA) was immobilized on the gold working electrode surface of screen-printed cell followed by a blocking step with 6-mercapto-1-exanhol. After the activation of –COOH groups by EDAC/NHS solution, incubation step with anti-VEGF antibody was performed. Then, the unreacted –COOH groups were blocked with ethanolamine (EA), and the sensor was incubated with VEGF protein buffered solution at different concentration. The sandwich assay was later completed by interaction with anti-VEGF biotinylated aptamer. Streptavidin-alkaline phosphatase was incubated at the end with the sensor.

Finally, 50 μL of 1-naphthyl phosphate was placed on the screen-printed cell for 6 min and then DPV measurements were carried out.

3.1 Results and Discussion

Calibration curve for the determination of VEGF protein in buffered solutions by antibody-aptamer assay was reported in Fig. 2.

A linear correlation between the current and the VEGF concentration in the range of 0–0.4 ng mL^{-1} was obtained. The assay was repeated in order to evaluate the reproducibility; at this purpose, three repetitions of each standard solution were carried out. The average coefficient of variation was 10 %, calculated as mean of all the concentrations considered. The limit of detection (LOD), calculated by the ratio between three times the deviation standard of blank and the slope of calibration curve ($3S_{blank}/S_{slope}$), was 0.1 ng mL^{-1}.

4 Conclusion

In this work, simple and sensitive approach for VEGF detection using antibody-aptamer assay and gold screen-printed electrodes as transducers is presented. The assay was performed in a sandwich format. The proposed assay shows a linear calibration curve in the range of 0–0.4 ng mL^{-1} with a LOD of 0.1 ng mL^{-1}.

References

1. J. J. Ott, A. Ullrich, A. B. Miller. The importance of early symptom recognition in the context of early detection and cancer survival. European Journal of Cancer **45**, 2743-2748 (2009).
2. C. Sessa, A. Guibal, G. Del Conte, C. Ruegg. Biomarkers of angiogenesis for the development of antiangiogenic therapies in oncology: tools or decorations?. Nature clinical practice oncology **5**, 378-91 (2008).
3. H. Chen, C. Jiang, C. Yu, S. Zhang, B. Liu, J. Kong. Protein chips and nanomaterials for application in tumor marker immunoassays. Biosensors and Bioelectronics **24** 3399-3411 (2009).
4. F. Bettazzi, F. Lucarelli, I. Palchetti, F. Berti, G. Marrazza, M. Mascini. Disposable electrochemical DNA-array for PCR amplified detection of hazelnut allergens in foodstuffs. Anal. Chim. Acta **614**, 93-102 (2008).
5. H. Hasegawa, K. Sode, K. Ikebukuro. Selection of DNA aptamers against VEGF(165) using a protein competitor and the aptamer blotting method. Biotechnol. Lett. **30**, 829-834 (2008).

Electrochemical Liposome-Based Biosensors for Nucleic Acid Detection

Diego Voccia, Francesca Bettazzi (✉), and Ilaria Palchetti

1 Introduction

MicroRNAs (miRNAs) are a class of small noncoding RNAs with approximately 22 nucleotides in length that play important roles in different biological processes like cell differentiation and prolification and regulation of protein translation. An abnormal miRNA expression (overexpression or down-expression) has been linked to cancer [1, 2] and other diseases. The first evidence of involvement of miRNAs in human cancer came from molecular studies which revealed that two miRNAs, mir-15 and mir-16, were involved in chronic lymphocytic leukemias [3]. Following this initial discover, other researches were focused on the investigation of miRNA expression deregulation in human cancer. For example, mir143 and mir145 are downregulated in colon carcinomas [4] and mir122 is involved in breast cancer [5]. Because of this, the study of miRNAs has become important and necessary for many fields in science, and in particular, they represent good candidates for diagnostic and prognostic biomarkers. Moreover, the ability to selectively regulate protein activity through miRNAs could enable treatment of many forms of cancer and other serious illness. In the previous years, researchers have been challenged to push the sensitivity of analytical bioassays down to subnanomolar values while keeping these procedures as simple, reliable, and cost-effective as possible. In particular, the formation of supramolecular assemblies and the use of enzymes to enhance the sensitivity of genoassays have been the subject of many research efforts [6]; functionalized

D. Voccia • F. Bettazzi • I. Palchetti (✉)

Dipartimento di Chimica, Università degli Studi di Firenze, Via della Lastruccia 3, 50019 Sesto Fiorentino, Firenze, Italy

e-mail: diego.voccia@unifi.it; ilaria.palchetti@unifi.it

C. Di Natale et al. (eds.), *Sensors and Microsystems: Proceedings of the 17th National Conference, Brescia, Italy, 5-7 February 2013*, Lecture Notes in Electrical Engineering 268, DOI 10.1007/978-3-319-00684-0_34, © Springer International Publishing Switzerland 2014

liposomes have been exploited as nanostructures for genosensors [7, 8]. Outstandingly low detection limits were achieved. In this paper, two electrochemical methods based on liposome amplification for miRNA detection are reported. Disposable screen-printed gold electrodes were used as transducers using DPV and EIS. A thiolated DNA capture probe was immobilized onto gold electrode surfaces (the biosensing platform). Biotinylated RNA was then hybridized with the specific capture probes. The biosensing platform was then incubated with the enzymatic label. Biotin-labeled liposomes were used as a functional tether for the enzyme and, owing to their large surface area, are capable of carrying a large number of enzyme molecules. The product of the enzymatic reaction was electrochemically monitored, using α-naphthyl phosphate for DPV and BCIP/NBT Liquid Substrate System for EIS.

2 Experimental

2.1 Chemicals

Streptavidin–alkaline phosphatase, α-naphthyl phosphate, Streptavidin from Streptomyces avidinii, bovine serum albumin (BSA), dimyristoylphosphatidylethanolamine (DMPE), distearoylphosphatidylcholine (DSPC), cholesterol, and BCIP/NBT Liquid Substrate System were obtained from Sigma–Aldrich (Italy). Biotin-XDMPE was obtained from Invitrogen Molecular Probes (Eugene, OR).

2.2 Electrochemical Measurements

Electrochemical measurements were performed using an Autolab PGSTAT10 with the FRA2 module (Eco Chemie, The Netherlands). All potentials were referred to the Ag/AgCl screen-printed pseudo-reference electrode. EIS measurements were performed using an alternating voltage of 10 mV and at a bias potential of 0.13 V in the frequency range from 100 mHz to 50 kHz. The impedance spectra were plotted in the form of complex plane diagrams (Nyquist plots). DPV were performed with a modulation time of 0.15 s, step potential of 5 mV, and a modulation amplitude of 70 mV from −0.1 to 0.6 V.

2.3 Liposome Preparation and Characterization

Liposomes were prepared according to [8]. Light-scattering measurements were made using a Nano ZS90 (Malvern); a distribution of 100 % of monodisperse particles with hydrodynamic diameter of 146 ± 9 nm was obtained.

2.4   Electrode Modification and Analytical Procedure

Electrochemical cells have planar three-electrode strips, namely, a gold working electrode, a carbon counter electrode, and a silver pseudo-reference electrode. The gold surface of the working electrode was modified with 8 µL of 1 µM thiolated DNA probe in 0.5 M phosphate buffer (PB) pH 7.2 overnight. The immobilization step was followed by treatment with 8 µL of 1 mM mercaptohexanol aqueous solution. Biotinylated target solution was diluted to the desired concentration in PB, and 8 µL of these solutions was finally incubated with the probe-modified electrodes for 20 min. After hybridization, a streptavidin 500 µg mL^{-1} solution in PB was incubated for 20 min, followed by 20 min incubation with a 1.5 nM liposome solution. After the introduction of the liposomes, the electrodes were then incubated with 8 µL of a solution containing 0.8 U mL^{-1} streptavidin–alkaline conjugate and 8 mg mL^{-1} BSA in diethanolamine (DEA) buffer for 20 min. Finally, before starting measurements, an opportune substrate was used. For EIS detection the BCIP/NBT Liquid Substrate System was incubated for 20 min and then removed with two washing step in KCl 0.1 M. The electrochemical signal of the enzymatically produced precipitate was measured by EIS through the electron transfer resistance (ETR) measured in the presence of the redox probe. On the other hand, for DPV experiment, a 1 mg mL^{-1} α-naphthyl phosphate solution as substrate was incubated for 20 min. The electrochemical signal of the enzymatically produced α-naphthol was measured by DPV.

3   Results

The calibration curve of miRNA using DPV is reported in Fig. 1a. The results obtained show a linear response in the concentration range 0.1–5.0 nM y = 1.72 x + 3.68 (µA vs. nM) with an R^2 of 0.974 and a LOD of 700 pM (RSD 18 %).

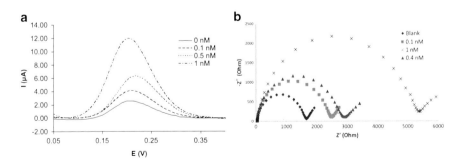

Fig. 1 (a) DPV, Differential pulse voltammograms of different target concentrations; (b) impedance spectra (Nyquist plots) at different target concentrations

The results obtained in EIS experiments (Fig. 1b) show a linear response in the concentration range $0.1–1.0$ nM $y = 3387 x + 114$ (Ohm vs. nM) with an R^2 of 0.994 and a LOD of 65 pM (RSD 10 %).

4 Conclusion

The main aim of the use of biotinylated liposomes is to amplify electrochemical signals. As experimentally observed, liposomes alter the interfacial properties of the electrodes, thus increasing the ETR. This reflects positively on EIS detection but negatively on DPV. Actually, a better LOD is observed in EIS experiments in comparison to DPV (65 pM vs. 700 pM). Moreover, using EIS, a twofold increase of the target/blank signal ratio is observed (data not shown). These preliminary results demonstrated that liposomes coupled to EIS can be used for further investigations in the development of biosensors for miRNA detection. Nevertheless, optimization experiments for sensitive miRNA detection should be performed.

Acknowledgments The author would like to thank Miur (Prin 2009) and Università di Firenze (Fondi di Ateneo) for funding this paper.

References

1. G.A. Calin, C.M. Croce. microRNA signatures in human cancers. Nat Rev Cancer **6**, 857–866C (2006).
2. M.V. Iorio, M. Ferracin, G. C. Liu, A. Veronese, R. Spizzo, S. Sabbioni, E. Magri, et al. microRNA gene expression deregulation in human breast cancer. Cancer Res **65**, 7065–7070(2005).
3. G.A. Calin, C.D. Dumitru, M. Shimizu, et al. Frequent deletions and down-regulation of micro-RNA genes miR15 and miR16 at 13q14 in chronic lymphocytic leukemia. Proc Natl Acad Sci U S A **99**, 15524–9 (2002)
4. M.Z. Michael, S.M. O'Connor, N.G. Van Holst Pellekaan, G.P. Young, R.J. James. Reduced accumulation of specific microRNAs in colorectal neoplasia. Mol Cancer Res **1**, 882–91 (2003).
5. B. Wang, H. Wang, Z. Yang. miR-122 Inhibits Cell Proliferation and Tumorigenesis of Breast Cancer by Targeting IGF1R. PLoS ONE **7**(10), e47053 (2012)
6. Bettazzi F.; Hamid-Asl E.; Esposito C. L.; Quintavalle C.; Formisano N.; Laschi S.; Catuogno S.; Iaboni M.; Marrazza G.; Mascini M.; Cerchia L.; De Franciscis V.; Condorelli G.; Palchetti I. Anal Bioanal Chem **405**, 1025–1034 (2013)
7. Patolsky, F.; Lichtenstein, A.; Willner, I. Angew. Chem. Int. Ed., **39**, 940 (2000)
8. Alfonta, L.; Singh A.K.; Willner, I. Anal. Chem., **73**, 91-102 (2001)

On-Chip Diagnosis of Celiac Disease by an Amorphous Silicon Chemiluminescence Detector

D. Caputo, G. de Cesare, R. Scipinotti, N. Stasio, F. Costantini, C. Manetti, and A. Nascetti

1 Introduction

The early diagnosis and the follow-up of biomarker levels which indicate a pathological state, such as chronic and/or cancer diseases, have gained a huge importance for customized patient therapy. Celiac disease (CD) is a chronic disorder that damages the lining of the small intestine and interferes with the absorption of nutrients that are important for human health. CD is due to an immunoreaction against small sequences of peptides, the epitopes, contained in gluten. CD affects 1 % of the world's population, but only 21 % of cases are diagnosed. It concerns mostly women, with a prevalence of 1.5–2 times than men and with 10–15 % of first-degree relatives with the same disease [1].

Currently CD diagnosis relies on a combination of serologic, clinical, and histological tests. Although a firm diagnosis can only be done using a small intestinal biopsy, screening based on endomysial IgA antibody tests performed on patient serum may reach very high accuracy [2]. However, current diagnostic assays do not allow the drawing, for each patient, of specific immune profile that is of fundamental importance for defining a personalized diet and for improving both therapeutic response and quality of life.

D. Caputo • G. de Cesare • R. Scipinotti (✉) • N. Stasio
Department of Information, Electronic and Telecommunication Engineering,
Sapienza University of Rome, Via Eudossiana 18, 00184 Rome, Italy
e-mail: scipinotti@die.uniroma1.it

F. Costantini • C. Manetti
Department of Chemistry, Sapienza University of Rome, P.le Aldo Moro 5, 00185 Rome, Italy

A. Nascetti
Department of Astronautics, Electrical and Energy Engineering, Sapienza University of Rome,
Via Salaria 851/881, 00138 Rome, Italy

C. Di Natale et al. (eds.), *Sensors and Microsystems: Proceedings of the 17th National Conference, Brescia, Italy, 5-7 February 2013*, Lecture Notes in Electrical Engineering 268, DOI 10.1007/978-3-319-00684-0_35, © Springer International Publishing Switzerland 2014

The goal of our research is to develop a compact and reliable system for the early diagnosis of celiac disease, based on parallel monitoring of patients' specific immune response to the different CD epitopes, thus developing a comprehensive patient-specific profile.

2 System Description

The basic idea is to develop a lab-on-chip device that integrates on a single substrate an array of hydrogenated amorphous silicon (a-Si:H) photosensors [3–5] and a set of selected epitopes that induce an immunoreaction in celiac patients. The epitopes are immobilized on the surface of the device using a layer of poly(2-hydroxyethyl methacrylate) (PHEMA) polymer brushes deposited and patterned on the glass substrate. The diagnosis of CD relies on the detection of immunoreactions by using a chemiluminescent (CL) method that ensures high specificity and sensitivity without the need of external excitation sources [6], thus leading to an extremely compact device. A schematic view of the proposed device is reported in Fig. 1.

The main challenges in the realization of the proposed device are related to the identification of the right combination of materials and technological processes in order to ensure their compatibility along the entire fabrication path.

3 Device Fabrication

In order to avoid unwanted interactions between the analytical part and the on-chip detection, the photosensor array and the functionalized PHEMA pattern are fabricated on opposite sides of the glass substrate. This improves the overall system reliability and simplifies the device processing.

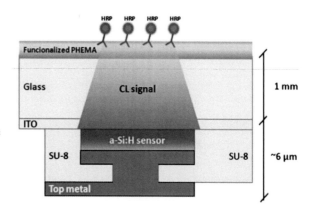

Fig. 1 Schematic view of the proposed lab-on-chip with integrated a-Si:H photosensors for the diagnosis of CD based on a CL method

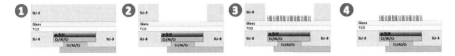

Fig. 2 Deposition and patterning of PHEMA brushes

The first step is the fabrication of the array of a-Si:H photodiodes on the cleaned glass substrate. The main process steps are the following: (1) pattering of the indium tin oxide (ITO) for the definition of the transparent front electrode (mask 1); (2) deposition by PECVD of the p-doped, intrinsic, and n-doped a-Si:H layers that constitute the photodiode; (3) magnetron sputtering of a three-metal-layer stack (Cr/Al/Cr) (back electrode); (4) patterning of the device structure by wet and reactive ion etching for the metal stack and a-Si:H layers respectively (mask 2); (5) deposition of a 5 μm thick SU8 layer acting as insulation layer; (6) opening of via holes on the passivation layer over the diodes (mask 3); (7) sputtering of a Cr/Al/Cr metal stack for the external connections; (8) and patterning of the Cr/Al/Cr external contacts (mask 4).

After the fabrication of the on-chip sensor array, the process continues on the opposite glass side, whose surface is functionalized with the selected epitopes using a PHEMA layer. Sacrificial layer photolithography has been used to pattern the PHEMA polymer brushes following the steps illustrated in Fig. 2. PHEMA polymer brushes were fabricated on the glass surface via atom transfer radical polymerization [7]. The immobilization sites are aligned with the sensor array fabricated on the opposite glass side.

After PHEMA patterning, the polymer film was treated with succinic anhydride (SA) and n-hydroxysuccinimide (NHS) in the presence of 1-ethyl-3-(3-dimethylaminopropyl) carbodiimide (EDC) to obtain NHS ester groups (PHEMA-SA-NHS). Then the epitope buffer solution was poured on the PHEMA-SA-NHS polymer-coated surfaces. The epitopes are immobilized on the polymer layer through formation of amide bonds between the amino groups of the amino acids of the epitopes and the NHS ester moieties.

4 Experimental

The antibodies (Ig1) of a serum of rabbit immunized towards the immobilized epitope and a secondary antibody marked with HRP (Ig-HRP) are brought in contact with the PHEMA-epitope polymer obtaining a chemical stacked structure. Horseradish peroxidase (HRP) is an enzyme which catalyzes CL reaction, that is, the oxidation of luminol in the presence of hydrogen peroxide (H_2O_2).

The lab-on-chip contains 16 detection sites in which different combinations of the epitopes' antibodies have been immobilized to test both specific and unspecific reactions (i.e., binding of the specific or unspecific antibody Ig1 to the epitope). Sites without immobilized epitopes have been also used as controls.

Fig. 3 Experimental results: signal vs. time measured for specific, aspecific, and control sites

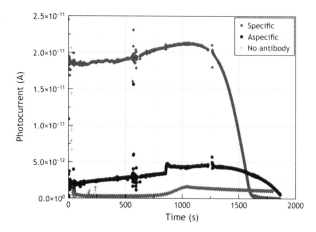

Different experiments have been performed in order to test the response of the device and to verify the specificity of CL reactions. In all the measurements, the specific bonding of the secondary antibody has been successfully detected. Unspecific signal was always at least four times lower than the specific signal but higher than the empty control sites. As shown in Fig. 3, the presented device allows the monitoring of the entire reaction kinetics, thus providing additional information that can be useful, e.g., for verifying that the test was correctly executed, leading to a higher reliability.

5　Conclusions

A novel lab-on-chip device for the diagnosis of CD and for patient-specific profiling has been successfully designed and fabricated. The device is based on the successful integration of an array of a-Si:H photosensors with functionalized PHEMA polymer brushes on a single substrate. The operation of the fully integrated system relies on the use of CL immunoenzymatic reactions to ensure sensitive and specific response without the need of external radiation sources and optical filters, resulting in a compact device suitable for point-of-care testing. Experimental results demonstrate the effective detection of the target antibodies with a good specificity. Future works will integrate all the epitopes of CD for testing real patient serum.

References

1. Peter H.R. Green, C. Cellier, Celiac Disease, New Engl. J. Med. **357**, 1731-1743, (2007).
2. U. Volta, A. Granito, E. Fiorini, C. Parisi, M. Piscaglia, G. Pappas, P. Muratori, F.B. Bianchi, Usefulness of antibodies to deamidated gliadin peptides in celiac disease diagnosis and follow-up, Dig. Dis. Sci. **53**, 1582-1588, (2008).

3. D. Caputo, G. de Cesare, L.S. Dolci, M. Mirasoli, A. Nascetti, A. Roda, R. Scipinotti, Microfluidic chip with integrated a-Si:H photodiodes for chemiluminescence-based bioassays, IEEE Sensors Journal, **13**, 2595-2602 (2013), DOI: 10.1109/JSEN.2013.2256889
4. D. Caputo, G. De Cesare, C. Fanelli, A. Nascetti, A. Ricelli, R. Scipinotti, Amorphous silicon photosensors for detection of ochratoxin a in wine, IEEE Sens. Jour., **12**(8), 2674-2679 (2012)
5. D. Caputo, G. de Cesare, A. Nascetti, R. Negri, Spectral tuned amorphous silicon p-i-n for DNA detection, Journal of Non-Crystalline Solids, **352**, 2004-2006 (2006)
6. D. Caputo, G. De Cesare, C. Manetti, A. Nascetti, R. Scipinotti, Smart thin layer chromatography plate, Lab on a Chip, **7**(8), 978-980 (2007)
7. F. Costantini, E. M. Benetti, D. N. Reinhoudt, J. Huskens, G. J. Vancso and W. Verboom, Lab on a Chip **2010**, *10*, 3407-3412.

Development and Characterization of a Novel Antibacterial Material Based on GOx Immobilized in a PVA Film

M.R. Guascito, D. Chirizzi, L. Giotta, and L. Stabili

1 Introduction

Bacterial infection is one of the most severe concerns for several research fields such as biotechnology, pharmaceutical, textiles, food packaging and storage, shoe industry, water purification, medical devices, and dental surgery equipment [1, 2]. Recently, antimicrobial agents have gained considerable interest from both an industrial and research point of view because of their potential to provide safety benefits to a diverse range of materials. Nowadays scientists need to carry out materials with a surface that has a very broad spectrum of biocide activity, that can be used repeatedly and that kills via a mechanism which will not result in the emergence of resistant strains, and that can be used for development of devices for biomedical applications [3]. This system acts in presence of glucose to generate by-products as oxygen species (H_2O_2, $\cdot O_2^-$, OH) that are well-known endogenous and exogenous toxic products for microbes in vivo [4].

2 Experimental Section

2.1 Reagents and Apparatus

Glucose oxidase (type VII from Aspergillus niger; 174,000 units/g), β-D-glucose, and hydrogen peroxide 30 % were obtained from Sigma. Na_2HPO_4, NaH_2PO_4, and poly(vinyl alcohol) film (PVA, product number Z300381) were purchased from

M.R. Guascito (✉) • D. Chirizzi • L. Giotta
Università del Salento, via Monteroni, 73100 Lecce, Italy
e-mail: maria.rachele.guascito@unisalento.it

L. Stabili
Istituto per l'Ambiente Marino Costiero CNR, via Roma 3, 74100 Taranto, Italy

C. Di Natale et al. (eds.), *Sensors and Microsystems: Proceedings of the 17th National Conference, Brescia, Italy, 5-7 February 2013*, Lecture Notes in Electrical Engineering 268, DOI 10.1007/978-3-319-00684-0_36, © Springer International Publishing Switzerland 2014

Aldrich. Micrococcus lysodeikticus cells were obtained from Sigma. All reagents were analytical grade. Phosphate buffer solution (PB) pH = 7.0, I = 0.2. 1 M stock glucose solutions were allowed to mutarotate at room temperature overnight before use.

XPS analysis was carried out using a Leybold LHS10 spectrometer equipped with an unmonochromatized AlKα source, and a SPECS multichannel detector and UV–vis measurements were carried out with a Cary 50 spectrophotometer (Varian). Mid-infrared spectra were acquired with a PerkinElmer Spectrum One FTIR spectrometer equipped with a deuterated triglycine sulfate (DTGS) detector.

2.2 Preparation of Antibacterial Composite Material

The PVA/GOx composite material was prepared according to the procedure reported in literature [5]. GOx 500 units/ml was dissolved in 1 ml aqueous solution containing 10 % PVA in ultrasonic bath for 5 min. The mixture was kept at room temperature for 6 h and then stored at −18 °C for 48 h.

3 Results End Discussion

3.1 Spectroscopic Characterization

X-ray photoelectron spectroscopy was used to analyze the chemistry of PVA/GOx composite film. Figure 1 shows details of high-resolution spectra of C1s and N1s of composite material that provide binding energy information. For comparison, the same regions for GOx powder and only PVA (data no reported) were acquired. The C1s spectrum of PVA/GOx composite film shows four peak components.

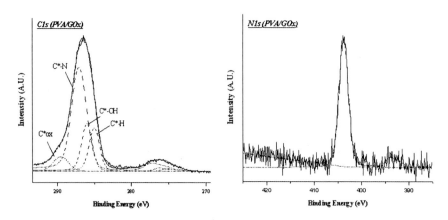

Fig. 1 C1s and N1s detailed spectra for PVA/GOx composite film

Fig. 2 ATR-FTIR spectra of films deposited by casting/ evaporation from the following aqueous solutions: glucose 32 mg/ml (*A*), glucose 32 mg/ml + GOx 80 μg/ml (*B*), glucose 32 mg/ml + GOx 80 μg/ ml + Micrococcus lysodeikticus cells 100 μg/ml (*C*), glucose 32 mg/ml + Micrococcus lysodeikticus cells 100 μg/ml (*D*)

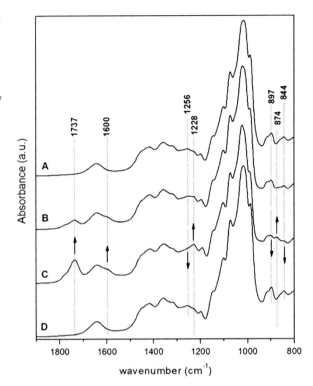

The aliphatic carbon component (C*–C) is at 285.0 ± 0.1 eV and the alcoholic car-bon component (C*–OH) at 286.1 ± 0.1 eV; moreover the third peak (C*–N) at 287.5 ± 0.1 eV is the characteristic of the amidic and aminic carbon in peptide chains. Finally an over-oxidized component (C*ox) is evident at 289.6 ± 0.1 eV. These results confirm that the PVA/GOx material is a composite hybrid film accord-ing to the presence of N1s signal.

The resulting ATR-FTIR spectra, acquired after solvent evaporation, are pre-sented in Fig. 2. It is clear that glucose oxidase addition is responsible for the appearing of a characteristic peak at 1,736 cm^{-1}, since D-glucose does not show a similar feature (trace A) even in the presence of *Micrococcus lysodeikticus* cells (trace D). On the other hand GOx-induced spectral changes, clearly detectable in trace B, are further enhanced in the presence of *Micrococcus lysodeikticus* cells (trace C), indicating a role played by the cellular system in the modulation of the enzyme-catalyzed reaction.

UV–vis characterization was performed to further verify the preservation of GOx integrity when it is embedded in PVA film. The enzyme activity towards glu-cose oxidation was verified by observing on PVA/GOx film (Fig. 3) the expected absorption band modifications after glucose addition due to enzyme reduction. Full reduction of the adenine dinucleotide (FAD) prosthetic group was obtained follow-ing addition of glucose 10 mM. The oxidized form of the enzyme was easily restored when solutions or films were saturated with O_2 [6].

Fig. 3 UV–vis spectra
of glucose oxidase in the
oxidized and fully reduced
states

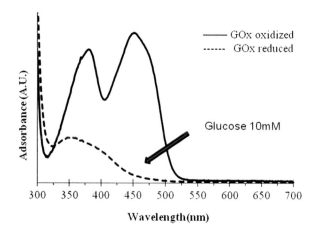

Fig. 4 Standard assay on
Petri dishes with
Micrococcus lysodeikticus
cell walls to detect the
lysozyme-like activity
of PVA/GOx in the presence
of glucose 10 mM (*I*), PVA/
GOx without glucose (*II*),
empty wells as control (*III*)

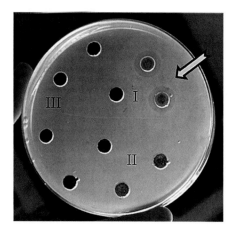

3.2 *Antibacterial Application*

The PVA/GOx composite material showed an antibacterial lysozyme-like activity.
The observed antibacterial activity was strictly affected by pH and ionic strength (I)
of the sample and of the reaction medium as well as by the incubation temperature.
By the standard assay on Petri dishes, a diameter of lysis of 4.2 ± 0.02 mm corresponding to 0.6 mg/ml of hen egg-white lysozyme (cleared zone around each well)
was reported at $I = 0.175$, pH 6.0, and incubation temperature of 37 °C. This activity
was present in all the sampling filled with 30 μl of PVA/GOx in the presence of
glucose 10 mM. On the contrary the wells filled with PVA/GOx without glucose and
the empty ones do not show lysozyme-like activity (Fig. 4).

4   Conclusion

This new antibacterial system based on a simple procedure well dispersed GOx in PVA was prepared and characterized. Thus the findings from this study have implications for future investigations related to employment of PVA/GOx as a compound of pharmaceutical and technological interest. As regards pharmaceuticals, indeed we have to consider that the increasing development of bacterial resistance to traditional antibiotics has reached alarming levels, thus necessitating the strong need to develop new antimicrobial agents. Lastly, the observed antibacterial activity of GOx with glucose could be useful to avoid the settlement of bacteria which is the primary colonizing process in the microbial biofilm development in biomedical implants.

References

1. G. Amitai, J. Andersen, S. Wargo, G. Asche, J. Chir, R. Koepsel, A. J. Russell; Polyurethane-based leukocyte-inspired biocidal materials. Biomaterials **30**, 6522–6529 (2009).
2. K. Reder-Christ and G. Bendas; Biosensor applications in the field of antibiotic research - A review of recent developments. Sensors **11**, 9450-9466 (2011).
3. L. Ferreira, A. Zumbuehl; Non-leaching surfaces capable of killing microorganisms on contact. J. Mat. Chem. **9**, 7796-7806 (2009).
4. R. A. Miller, B. E. Britigan; Role of Oxidants in Microbial Pathophysiology. Clin. Microb. R. **10**, 1-18 (1997).
5. M.R. Guascito, D. Chirizzi, C. Malitesta, E. Mazzotta, Mediator-free amperometric glucose biosensor based on glucose oxidase entrapped in poly(vinyl alcohol) matrix. Analyst **136**, 164-173 (2011).
6. V. Massey. The chemical and biological versatility of riboflavin. Biochem. Soc. Trans. **28**, 283-296 (2000).

Development of a QCM (Quartz Crystal Microbalance) Biosensor to Detection of Mycotoxins

K. Spinella, L. Mosiello, G. Palleschi, and F. Vitali

1 Introduction

In recent years, thanks to advances in electronic technology and biotechnology, biosensors have become an interesting field of research in different areas. In the scientific literature, a biosensor is defined as "an analytical device that incorporates a compact sensing element of a biological nature integrated to a transducer physic-chemical. The main purpose of this device is to provide an electrical signal, analog or digital, proportional to the concentration of a single analyte or a group of analytes" [1]. A particular type of transducer used as biosensor scale is microgravimetric QCM (quartz crystal microbalance) which consists of a thin quartz disk with two gold electrodes, one of which is functionalized so as to be sensitive to the presence of an analyte [2]. Given that the frequency of oscillation of the quartz is proportional to the mass deposited on the functionalized electrode, such devices are able to provide qualitative and quantitative measures concerning the presence/absence of an analyte. One of the more interesting applications of the QCM is to be used as a biosensor in which the gold electrode is functionalized by anchoring of biomolecules on its interface placed in contact with a solution containing the substance or the compound to recognize and measure; in particular in this study, we used antibodies with the aim to determine the presence of mycotoxins in food. Aflatoxin B1,

K. Spinella (✉) • L. Mosiello • F. Vitali
ENEA, Italian National Agency for new Technologies, Energy and the Environment,
Via Anguillarese 301, 00060 Rome, Italy
e-mail: katia.spinella@enea.it

G. Palleschi
Dipartimento di Scienze e Tecnologie Chimiche, Università di Roma Tor Vergata,
Via della Ricerca Scientifica, 00133 Rome, Italy

C. Di Natale et al. (eds.), *Sensors and Microsystems: Proceedings of the 17th National Conference, Brescia, Italy, 5-7 February 2013*, Lecture Notes in Electrical Engineering 268, DOI 10.1007/978-3-319-00684-0_37, © Springer International Publishing Switzerland 2014

ochratoxin A, and fumonisin B1 are mycotoxins, secondary metabolites produced by several species of Aspergillus and Penicillium fungi, and they are frequent contaminants in food and feed commodities. These mycotoxins have well-documented nephrotoxic and immunosuppressive properties, and the International Agency for Research on Cancer has classified mycotoxin as a possible human carcinogen. Regulation of low nanogram per gram range is included in the legislation of many countries, and the European Union regulations have set the acceptable limit of OTA 3 $\mu g/kg^{-1}$ levels, AFLAB1 at 2 ng/g^{-1}, and FB1 at 8 $\mu g/kg^{-1}$. A great interest in developing label-free and less time-consuming online detection methods, such as the use of biosensors, is undiminished. With this purpose, piezoelectric quartz crystals are convenient for affinity-based sensors, particularly for immunosensors. The combination of low cost with increased sensitivity, selectivity, and possible reusability makes piezoelectric immunosensors a valuable alternative to other existing optical and electrochemical immunosensors [3].

2 Experimental

In this study, we have used a direct immunoassay where the simple binding between antigen and an antibody is detected. Immunoassays were performed in a drop system, monitoring the decrease in the frequency of the quartz crystal microbalance device as the mass increases during immunoreaction. The QCM sensor was coated on both sides by gold electrodes, only one side of the crystal (liquid side) was in contact with the solution, the other side (contact side) was always dry. We tested a piezoelectric immunosensor for AFLA-B1, OTA, and FB1 through the immobilization of DSP–anti-mycotoxin antibody (AFLA-B1–Ab anti-AFLAB1, OTA-Ab anti-OTA, FB1-Ab anti-FB1) on gold-coated quartz crystals (AT-cut/5 MHz). For the detection of mycotoxins, we chose to use two different specific antibodies, a monoclonal antibody and a polyclonal antibody, and two different types of immobilization of the antibodies on the surface of the quartz crystals in order to assess different operations and choose the one that has the best performance especially in terms of sensitivity.

We have developed two procedures for the immobilization of antigen-specific Ab on the surface of quartz crystals:

- Direct absorption
- Suspension in the gel and immobilization on a solid support (on the beads)

As regards direct absorption, antigen-specific antibodies were directly added in drops on the surface of the quartz crystals previously coated with DSP. The antibodies suspended in the gel and immobilized on a solid support were extracted from immunoaffinity columns and added in drops on the quartz pretreated with DSP. The DSP was used for the covalent attachment of the proteins. The piezoelectric crystal electrodes were pretreated by DSP for 15 min, rinsed with water, and dried in a gentle flow of nitrogen gas. Then the DSP-coated crystals were installed in a sample holder and exposed to the anti-mycotoxin antibody and to the analyte (mycotoxin).

3 Results

For accurate results of QCM, the characterization of the gold surface of the quartz crystals is highly recommended. Figure 1 shows the AFM (atomic force microscope) image of the surface of the quartz crystal functionalized with DSP.

For the determination of mycotoxins through an immunosensor, QCM was initially developed as a direct immunoassay which used a polyclonal antibody, by which it was possible to identify mycotoxins in solutions at different concentrations (standard solutions: 0.5, 1, 5, 10 ppb (ng/mL)) as shown in Fig. 2.

Frequency and resistance shifts (Δf and ΔR) were measured simultaneously. Δf versus mycotoxins concentrations in the range of 0.5–10 ppb exhibited a perfect linear correlation.

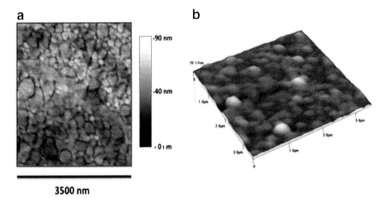

Fig. 1 AFM image of the surface of the quartz crystal functionalized with DSP: (**a**) Phase imaging. (**b**) Topography 3D image

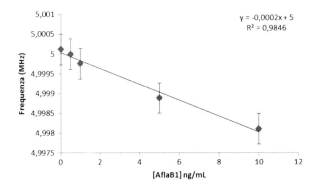

Fig. 2 Standard curve by direct immunoassay using monoclonal antibodies immobilized on beads

Monoclonal antibodies, in fact, generally show a higher and better specificity for antigens against which they are produced, as compared to polyclonal immunoglobulin. QCM-based sensing label-free procedure for mycotoxin detection, developed in our laboratory, can be considered a simple, cost-effective, real-time, and non-time- and labor-consuming technique in comparison with conventional assay procedures such as gas chromatography.

Acknowledgments This work was supported by ENEA, Italian National Agency for New Technologies, Energy and Sustainable Economic Development. The contributions of Dr. Barbara De Santis (Istituto Superiore di Sanità, ISS, Rome, Italy), Ing. Nicola Donato, and Dr. Giovanni Neri (University of Messina, Italy) are also strongly acknowledged.

References

1. A.P.F. Turner, I. Karube, Wilson, G.S. Biosensors Fundamentals and Applications. Oxford University Press, Oxford, UK (1987).
2. J. Curie, P Curie. An oscillating quartz crystal mass detector. Rendu, (1880). vol. 91, pp. 294–297.
3. Vidal JC, Duato P, Bonel L, Castillo JR (2009) Use of polyclonal antibodies to Ochratoxin A with a quartz–crystal microbalance for developing real-time mycotoxin piezoelectric immunosensors. Anal Bioanal Chem 394:575–582.

Label-Free Biosensor Based on Copolymer-Functionalized Optical Fiber Long-Period Grating

F. Chiavaioli, C. Trono, A. Giannetti, M. Brenci, and F. Baldini

1 Introduction

The simple manner for detecting a chemical/biological interaction is undoubtedly the measurement of a change in refractive index (RI) without resorting to luminescence- and absorption-based measurements. Among all the existing optical methodologies, which are used to measure the surrounding RI (SRI), surface plasmon resonance (SPR) is until now the most popular and widespread. However, during the years, different optical technologies have been developed, such as optical resonators, interferometric configurations, and, very recently, optical fiber gratings (OFGs). In particular, OFGs have been effectively proposed as optical tools in the field of biochemical sensing [1], especially long-period gratings (LPGs), thanks to their high RI sensitivity [2]. This kind of devices takes also the typical advantages of optical fibers, such as compactness, lightweight, multiplexing, and remote measurement capabilities.

Broadly speaking, a biochemical interaction along the grating portion induces a change of the RI and thickness of the selective biolayer deposited onto the fiber. Consequently, a change in the transmission spectrum of the fiber (i.e., a change in the spectral position of dips or resonance wavelengths) occurs. Starting from this principle and using the previously published compensated RI sensing system [3], a label-free biosensor based on a copolymer-functionalized optical fiber LPG has been developed and characterized.

F. Chiavaioli (✉) • C. Trono • A. Giannetti • M. Brenci • F. Baldini
Institute of Applied Physics "Nello Carrara", National Research Council of Italy,
Via Madonna del Piano 10, Sesto Fiorentino, FI, Italy
e-mail: f.chiavaioli@ifac.cnr.it

C. Di Natale et al. (eds.), *Sensors and Microsystems: Proceedings of the 17th National Conference, Brescia, Italy, 5-7 February 2013*, Lecture Notes in Electrical Engineering 268, DOI 10.1007/978-3-319-00684-0_38, © Springer International Publishing Switzerland 2014

2 Experimental Setup

When LPGs are used to measure very low changes in RI, as occurs in bioassay, it is crucial to have a reliable and effective thermal and mechanical stabilization of the measuring system. Firstly, a thermostabilized low-volume flow cell was designed and developed [3]. In parallel, a methodology for correcting the LPGs' cross-sensitivities (i.e., temperature and strain changes) was successfully tested [3]. For proving the use of this device in the biochemical field, a homemade bioassay protocol based on the antibody/antigen interaction was implemented.

Instead of the commonly used silanization procedure, the fiber functionalization was carried out by means of a chemistry using Eudragit L100 copolymer (2 mM) in ethanol. After solvent evaporation, polymeric deposition provides the carboxylic functional groups (–COOH) to the surface, useful for antibody hybridization. The first part of the bioassay protocol is related to the preparation of the sensing bio-layer, with the following steps: activation of –COOH groups by means of EDC (2 mM) and NHS (5 mM), covalent immobilization of the antibodies by pumping a solution of 1 mg mL^{-1} mouse IgG in phosphate buffer saline (PBS) solution, washing with PBS to remove the unreacted antibodies, and surface passivation with 3 % bovine serum albumin (BSA) in PBS to reduce nonspecific adsorption. The bioassay was completed by pumping different antigen concentrations of goat anti-mouse IgG, ranging from 100 ng mL^{-1} to 500 µg mL^{-1}.

A schematic representation of the biolayer composition is depicted in Fig. 1 (see the round enlargement). Finally, for comparing the biosensor performance, two different LPGs with distinct periods were manufactured: LPG A with a grating period of 615 µm and coupling with the fourth cladding mode and LPG B with a grating period of 370 µm and coupling with the seventh cladding mode.

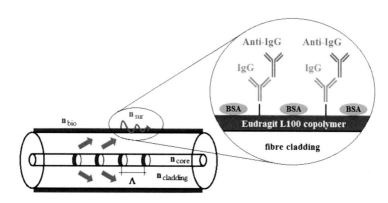

Fig. 1 Sketch of the evanescent-wave biosensor based on a copolymer-functionalized optical fiber long-period grating. The enlargement shows a sketch of the biolayer composition [4]

3   Results and Discussion

All the steps of the bioassay protocol were recorded in real time by monitoring the LPG resonance wavelength using the setup described in [4]. From the response curve of the biosensor (not showed), it is possible to extrapolate the kinetics of the antibody/antigen interaction at each concentration. As shown in Fig. 2 (case of LPG B), a steady state is always reached after 6 min.

It is worth noticing that the important measurand is the RI change induced along the functionalized surface due to the interactions between the capture antibody and the specific antigen, and not the bulk RI change, simply related to the solution RI. This means that a washing step with PBS is necessary after each antigen addition in order to measure the real shift of LPG resonance wavelength, caused by the anti-body/antigen interaction on the fiber surface (see Fig. 3). So the calibration curve of the biosensor is achieved by drawing the concentration of each goat anti-mouse IgG (antigen) as a function of the shift of LPG resonance wavelength. Each experimental point is given by averaging 15 values. Figure 4 details the calibration curves of the bioassay performed with the two LPGs. The fitting with the logistic function, accepted model for describing the sigmoidal curve, is also shown from which can be achieved the characteristic parameters of the biosensor for both the LPGs (see Table 1), such as dynamic signal range (DSR), working range (WR), and limit of detection (LoD).

An improvement of the biosensor performance due to the shorter period of the grating can be seen from both Fig. 4 and Table 1.

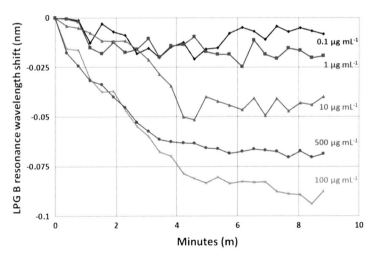

Fig. 2   Goat anti-mouse IgG (antigen) kinetics of LPG B for all antigen concentrations: 0.1 μg mL⁻¹ (*rhombus*), 1 μg mL⁻¹ (*square*), 10 μg mL⁻¹ (*triangle*), 100 μg mL⁻¹ (*star*), 500 μg mL⁻¹ (*circle*)

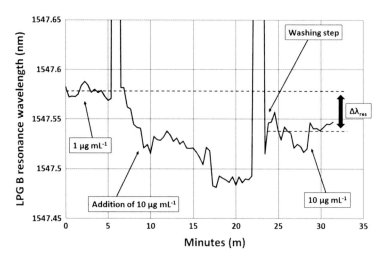

Fig. 3 LPG B resonance wavelength vs. time after the addition of 10 μg ml⁻¹ goat anti mouse IgG, with the clear indication of measured shift of the resonance wavelength

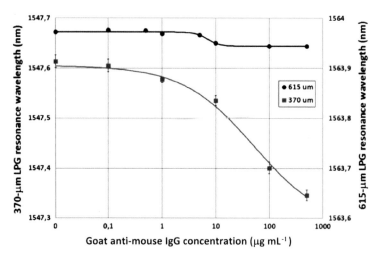

Fig. 4 Calibration curves of the same bioassay carried out with LPG A (circle) and with LPG B (square). The solid lines represent the best sigmoidal curve fitting of the experimental data

Table 1 Comparison of the results achieved with LPG A (615 μm) and LPG B (370 μm)

Parameter	LPG A	LPG B
Correlation	0.94	0.98
DSR (nm)	0.028	0.322
WR (μg mL⁻¹)	4–12.6	1.7–1,450
LoD (μg mL⁻¹)	7.6	0.5

4   Conclusions

Considering an LPG manufactured on a bare optical fiber, in which the coupling occurs with the seventh cladding mode, an LoD of 500 ng mL^{-1} has been achieved, thanks to the use of a platform perfectly stabilized and compensated for small changes of temperature and strain. Obviously, further potential improvements in RI sensitivity and LoD are achievable.

References

1. F. Baldini, M. Brenci, F. Chiavaioli, A. Giannetti, C. Trono. Optical fibre gratings as tools for chemical and biochemical sensing. Anal. Bioanal. Chem. **402** (1), 109-116 (2012)
2. P. Pilla, C. Trono, F. Baldini, F. Chiavaioli, M. Giordano, A. Cusano. Giant sensitivity of long period gratings in transition mode near the dispersion turning point: an integrated design approach. Opt. Lett. **37** (19), 4152-4154 (2012)
3. C. Trono, F. Baldini, M. Brenci, F. Chiavaioli, M. Mugnaini. Flow cell for strain- and temperature-compensated refractive index measurements by means of cascaded optical fibre long period and Bragg gratings. Meas. Sci. Technol. **22** (7), 075204 (9pp) (2011)
4. F. Chiavaioli, C. Trono, A. Giannetti, M. Brenci, F. Baldini. Characterisation of a label-free biosensor based on long period grating. J. Biophotonics (2012). doi:10.1002/jbio.201200135

Part IV
Optical Sensors

Chemiluminescence-Based Micro-Total-Analysis System with Amorphous Silicon Photodiodes

D. Caputo, G. de Cesare, R. Scipinotti, M. Mirasoli, A. Roda, M. Zangheri, and A. Nascetti

1 Introduction

During the last years micro-total-analysis systems (μ-TAS) have shown their relevance as powerful instruments to accomplish complex chemical or biochemical analyses on a single device exploiting various detection principles [1, 2]. However, a detailed analysis of practical lab-on-chip systems, including both commercial and research devices, indicates that most of the proposed systems are just microfluidic chips that allow to perform analytical tasks on tiny volumes of fluids but, nevertheless, need complex external benchtop instrumentation to be operated. On these systems, fragile interconnections between the microfluidic chip and external fluid supplying systems like syringes or pumps are typically needed to actuate the fluidics. These actuation systems are themselves bulky and have non-negligible power consumption. Furthermore, the analysis often relies on optical detection methods (e.g. fluorescence or absorption) that proved to be among the most efficient and sensitive techniques for analytical purposes. However, the necessary radiation sources, lenses, optical filters (e.g. for the rejection of the excitation wavelength and selection of the proper analytical signal) and light detectors (e.g. a CCD coupled with a simple optics or more complex microscopy systems) increase instrumental weight, size and complexity.

D. Caputo • G. de Cesare • R. Scipinotti
Department of Information, Electronic and Telecommunication Engineering,
Sapienza University of Rome, Via Eudossiana 18, 00184 Rome, Italy

M. Mirasoli • A. Roda • M. Zangheri
Department of Chemistry "G. Ciamician", University of Bologna, Via Selmi 2,
40126 Bologna, Italy

A. Nascetti (✉)
Department of Astronautics, Electrical and Energy Engineering, Sapienza University
of Rome, Via Salaria 851/881, 00138 Rome, Italy
e-mail: augusto.nascetti@uniroma1.it

C. Di Natale et al. (eds.), *Sensors and Microsystems: Proceedings of the 17th National Conference, Brescia, Italy, 5-7 February 2013*, Lecture Notes in Electrical Engineering 268, DOI 10.1007/978-3-319-00684-0_39, © Springer International Publishing Switzerland 2014

In order to overcome the above-mentioned limitations, our research group is currently working on the design and fabrication of "true" lab-on-chip systems that integrate in a single device all the analytical steps from the sample preparation to the detection without the need for bulky external components as pumps, syringes, radiation sources or optical detection systems. The resulting device will only need an electrical interface toward the control and readout electronics that also includes the signal processing unit, the data storage and the external communication interface. The three main fields addressed by our research are microfluidic actuation and control, analytical detection mechanism and on-chip transduction to electrical domain.

In the next section, an overview of the proposed approaches in the above-mentioned fields is reported. The following section reports a case study of a system that implements some of the proposed solutions. Details about the system design and fabrication are given first, and then the system calibration is reported comparing the on-chip detection results with those achieved with a scientific cooled CCD camera. Finally conclusions are drawn.

2 Toward Full Integration

Concerning the microfluidic part, in order to avoid both external pneumatic devices and fragile macro- to microfluidic joints, droplet microfluidics actuation based on the electrowetting on dielectrics (EWOD) principle is proposed as the main mechanism to dispense samples from on-chip reservoirs, mixing samples and reagents, moving small quantities along the analytical processing path. This active technique will be combined with passive ones like capillary flow in order to exploit conventional techniques for the capture of target molecules using probes immobilized in microchannels. Thermally actuated valves will be used to selectively connect or isolate different portions of the microfluidic network. The microfluidic system will be based on on-chip components, which require only electrical contacts for their operation.

For what concerns the analytical detection, the main goal is the elimination of external radiation sources. In our approach, this is achieved by means of highly specific chemiluminescent reactions to exploit the analytical task. Indeed, chemiluminescence (CL) is very attractive providing high detectability, even in low volumes; wide linear range of the signal; and specificity and rapidity of the response [3, 4]. CL is well suited for its implementation in miniaturized analytical devices since no radiation sources, wavelength selection filters and background radiation rejection systems are required and the design of the measurement device geometry is only conditioned by the necessity to collect the maximum fraction of emitted photons. In order to complete the lab-on-chip integration process, on-chip transduction of the analytical signal into an electrical one has to be implemented. To this aim, thin-film hydrogenated amorphous silicon (a-Si:H) photosensors can be successfully used as proven by numerous results reported in literature [5, 6]. a-Si:H is a semiconductor material that is grown by using plasma-enhanced chemical vapour deposition (PECVD) with process temperatures below 200 °C that allow the use of substrates like common microscope glass slides or Kapton tape. The main a-Si:H device is the diode based on a p-i-n-stacked structure that can be used both as a photodiode, exploiting the very high absorption coefficient of the material in the visible and near UV range [7], and as thermal sensor by taking advantage from the linear temperature dependency of the voltage measured at constant current bias [8].

3 Experimental Part: A Case Study

The proposed sample device is schematically sketched in Fig. 1. An a-Si:H photosensor is deposited on one side of the glass substrate, while the analytical part of the device is fabricated on the opposite side. The analytical part consists of a black polydimethylsiloxane (PDMS) well (Sylgard 184 + carbon powder) that hosts the CL reaction. The selected CL chemistry is based on alkaline phosphatase (AP), an enzyme able to catalyse a CL reaction (in the presence of a suitable CL cocktail based on a 1,2-dioxetane compound) and often employed as a label in CL bioassays.

In order to evaluate the on-chip detection performances and compare them with the state-of-the-art conventional CL detection, an experimental setup has been implemented on an optical bench. The glass substrates have been put on a holder with the side with bonded PDMS wells facing up, and a CCD has been aligned above the structure focusing on the PDMS wells. In this way the CCD looks into the wells from above without any intermediate interface, while the a-Si:H photodiodes look in the wells from below through the glass substrate of the μ-TAS. The entire system has been enclosed in a dark box. The experiments have been performed by dispensing in the microwells 2 μL of solutions at different concentration (ranging from 0.005 to 0.1 ng/μL) of AP and 8 μL of a suitable CL cocktail. The photodiode signal has been acquired with a custom low-noise electronic readout board using 4 s sample time. The same signal integration time has been used for the CCD in order to ensure the same measurement bandwidth and thus allow a fair comparison between system performances.

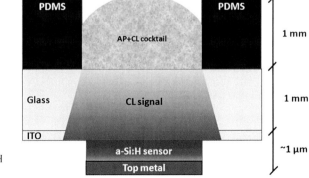

Fig. 1 Schematic cross section of the proposed device. The chemiluminescent reaction occurs in the black PDMS well aligned with the a-Si:H photosensor

Fig. 2 CL signals measured
by the a-Si:H photosensor
and CCD as a function of the
AP concentration in the
PDMS well

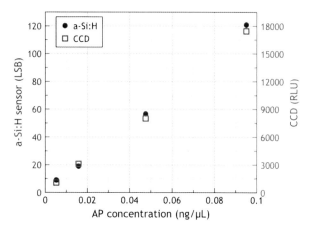

In general a very good agreement between CCD and the on-chip photodiode results has been achieved, confirming the validity of the proposed integrated approach based on a-Si:H technology (Fig. 2). The limit of detection (LOD) for the proposed system has been found to be 4 pg/µL (corresponding to 40 amol of enzyme) that is slightly higher than that of the cooled CCD (1.4 pg/µL).

4   Conclusions

By combining currently available technologies, it is possible to design µ-TAS that integrate on a single substrate all the subsystems needed to accomplish complex analytical tasks in a stand-alone manner. The characterization of a sample device that combines AP-based CL and on-chip a-Si:H photosensors confirms the validity of the proposed approach whose performances are comparable to that of conventional lab equipment.

References

1. L.Y. Yeo, H.C. Chang, P.P.Y. Chan, J.R. Friend. Microfluidic devices for bioapplications, Small, **7**, 12-48 (2011)
2. A. Arora, G. Simone, G.B. Salieb-Beugelaar, J.T. Kim, A. Manz. Latest developments in micro total analysis systems. Anal. Chem., **82**, 4830-4847 (2010)
3. A. Roda, M. Mirasoli, L.S. Dolci, A. Buragina, F. Bonvicini, P. Simoni, M. Guardigli. Portable device based on chemiluminescence lensless imaging for personalized diagnostics through multiplex bioanalysis. Anal. Chem., **83**, 3178-3185 (2011)
4. E. Marzocchi, S. Grilli, L. Della Ciana, L. Prodi, M. Mirasoli, A. Roda. Chemiluminescent detection systems of horseradish peroxidase employing nucleophilic acylation catalysts. Anal. Biochem. **377** (2), 189-194 (2008)

5. T. Kamei, B.M. Paegel, J.R. Scherer, A.M. Skelley, R.A. Street, R.A. Mathies. Integrated hydrogenated amorphous si photodiode detector for microfluidic bioanalytical devices. Analytical chemistry, **75** (20), 5300–5305 (2003)
6. D. Caputo, G. de Cesare, C. Fanelli, A. Nascetti, A. Ricelli, R. Scipinotti. a-Si photosensors for detection of Ochratoxin A in wine. IEEE Sensor Journal, **12** 8, 2674-2679, (2012)
7. D. Caputo, G. de Cesare, A. Nascetti, M. Tucci, Detailed study of amorphous silicon ultraviolet sensor with chromium silicide window layer. IEEE Trans. Electr Dev, **55**(1), 452-456 (2008)
8. D. Caputo, G. de Cesare, A. Nascetti, and R. Scipinotti. a-Si:H temperature sensor integrated in a thin film heater. Phys. Status Solidi A, **207** (3), 708–711, (2010)

Optical Spectroscopy for Hogwash Oil Detection in Soybean Chinese Oils

L. Ciaccheri, A.G. Mignani, A. Cichelli, J. Xing, X. Yang, W. Sun, and L. Yuan

1 Introduction

Stir-frying is the most popular and common cooking technique used in Chinese cuisine, which has become increasingly popular also in the West and around the world. Most Chinese dishes are cooked in oil quickly, and the speed at which the food is cooked preserves its colors, texture, and nutrients. While Chinese recipes are simple and delicious, the implications on health depend to a large extent on the oil quality.

Soybean oil is the most popular cooking oil used in China, because it is cheap and healthful and has a high smoke point. Soybean oil is a healthy food ingredient since it does not contain much saturated fat and cholesterol. Also, soybean oil contains natural antioxidants which remain in the oil even after extraction. These antioxidants help to prevent the oxidative rancidity. Due to the country's booming demand, and shrinking domestic output, China is the biggest soybean importer, with a figure of about 35 million tons in the first 6 months of 2012.

Such a huge need of soybean oil, together with the strong demand for a cheap product, has caused the appearance on the market of lower quality oils, in some cases adulterated by "hogwash" oil, that is, swill oil refined from rotten pork, peroxided oils used repeatedly in frying, and other leftovers from gutters behind restaurants. If used as edible oil, the hogwash oil can cause gastrointestinal sickness, cramps,

L. Ciaccheri • A.G. Mignani (✉)
CNR Istituto di Fisica Applicata "Nello Carrara", Via Madonna del Piano, 10,
50019 Sesto Fiorentino, FI, Italy
e-mail: a.g.mignani@ifac.cnr.it

A. Cichelli
Università degli Studi G. D'Annunzio, DASTA, Viale Pindaro, 42, 65127 Pescara, Italy

J. Xing • X. Yang • W. Sun • L. Yuan
Key Lab. of Fiber Optic Sensors, College of Science, Harbin Engineering University,
145 Nantong Street, Nangang District, Harbin, Heilongjiang Province 15001, P. R. China

C. Di Natale et al. (eds.), *Sensors and Microsystems: Proceedings of the 17th National Conference, Brescia, Italy, 5-7 February 2013*, Lecture Notes in Electrical Engineering 268, DOI 10.1007/978-3-319-00684-0_40, © Springer International Publishing Switzerland 2014

anemia, and toxic hepatic disease, and it is also a potential cancer cause. While there are many licensed collectors of used cooking oils, which are legally recycled into biofuels, the Chinese police has recently discovered an underground production of hogwash oil and arrested several people. Consequently, the Chinese government authorities in charge of food quality, safety, and sanitary controls are looking for new devices capable of spotting out quickly any suspected adulteration of soybean oil [1].

2 Experimental Setup and Processing

Optical spectroscopy proved its effectiveness in distinguishing different types of edible oils and adulteration [2, 3]. This paper shows the first experiment of absorption spectroscopy, carried out in the visible band by means of a white-LED source and a compact low-cost spectrometer, for detecting the adulteration of soybean oil caused by hogwash oil.

Two types of real soybean oils (s1 and s2) were considered and mixed with hogwash oil (h) with adulterant fraction of 5, 10, 25, 50, and 75 % v/v. Another set of samples was prepared for validation purposes, made with 25, 50, and 75 % mixtures, as well as samples of pure substances. Figure 1 synthesizes the experimental setup. The light source is a Multicomp® "warm-white" LED, while the spectrometer is a Cronin® MMS, having a spectral resolution of 7 nm and a step of 3.3 nm between contiguous channels.

The absorption spectra exhibited a small baseline fluctuation, which was removed by applying a Savitsky-Golay first-derivative algorithm. Figure 2 shows the

Fig. 1 Setup for absorption spectroscopy

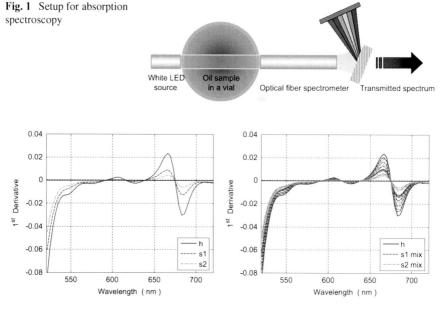

Fig. 2 First-derivative absorption spectra of pure oils (*left*) and mixtures (*right*)

first-derivative spectra of pure soybean and hogwash oils (left) and those of their mixtures (right).

These spectra were then processed using Principal Component Analysis (PCA) [4], for explorative analysis, and with Partial Least Square (PLS) regression [5], for quantitative prediction of the adulterant fraction. PCA processing was made using a MATLAB® code, while for PLS the software package "Unscrambler X" of CAMO Software® was used.

3 Results

Figure 3 shows the score plot of PCA subspace. In the left map two lines were added, showing the trends going from pure soybean samples to pure hogwash. In the right plot, instead, scores were centered on the hogwash sample and expressed in standardized units. Contour levels for the standardized distance were also added. These maps show that the two soybean oils occupy different zones in the PCA space and, at the same concentration of hogwash oil, they lie at different distances from the pure hogwash. Therefore, a dedicated PLS regression model for each oil was necessary, in order to predict the adulterant concentration.

The results of the PLS model are shown in Fig. 4, showing an excellent prediction of the adulterant concentration for both s1 and s2 mixtures. In both cases a single PLS factor was sufficient to achieve a good estimation of adulterant fraction. Regression statistics are shown in Table 1. For each oil the determination coefficient (R-squared), the Root Mean Square of calibration (RMSEC), and the Root Mean Square of Prediction (RMSEP) are listed.

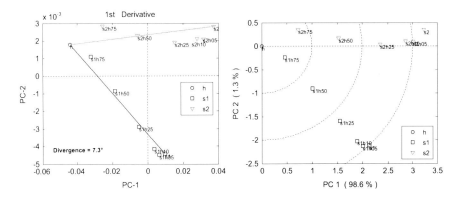

Fig. 3 PCA score plot showing the directions along which adulteration increases (*left*). Scaled score plot showing the contours of standardized distance from the h sample (*right*)

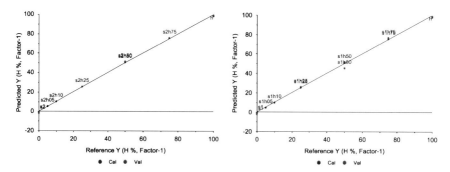

Fig. 4 PLS regression plots of h concentration in s2 (*left*) and s1 (*right*) oil types

Table 1 Regression statistics for quantitative determination of adulterant fraction in soybean oils

Parameter	Oil S1	Oil S2
R-squared	0.999	0.999
RMSEC	1.2 %	1.0 %
RMSEP	2.5 %	1.0 %

4 Conclusions

Visible absorption spectroscopy proved its worthiness in quantitative determination of soybean oil adulteration. Two different kinds of soybean oil were spiked using a low-quality "hogwash" oil with concentration ranging from 5 to 75 %. Processing of visible spectra by means of PLS regression was able to predict the fraction of adulterant in the spiked samples with good accuracy.

Acknowledgments The "High-End Foreign Experts Recruitment Program" of the Chinese Ministry of Foreign Affairs and the "111" Program of the Chinese Ministry of Education are acknowledged for their partial financial support.

References

1. L. Wang, J.H. Hu, X.L. Zhan, "Determination of trace element in the hogwash oil by ICP-MS", Chinese Journal of Health Laboratory Technology", vol. 17, pp. 1993-2014 (2007)
2. A.A. Christy, S. Kasemuran, Y. Du, Y. Ozaki, "The Detection and Quantification of Adulteration in Olive Oil by Near-Infrared Spectroscopy and Chemometrics", Analytical Sciences, vol. 20, pp. 935-940 (2004)
3. A.G. Mignani, L. Ciaccheri, H. Ottevaere, H. Thienpont, L. Conte, M. Marega, A. Cichelli, C. Attilio, A. Cimato, "Visible and near-infrared absorption spectroscopy by an integration sphere and optical fibers for quantifying and discriminating the adulteration of extra virgin olive oil from Tuscany", Analytical Bioanalytical. Chemistry, vol. 399, pp. 1315-1324 (2011)
4. J.E. Jackson, A User's Guide to Principal Components, John Wiley & Sons Inc., Hoboken NJ (2003)
5. S. Wold, M. Sjostrom and L. Erikkson, "PLS-regression: a basic tool for chemometrics", Chemometrics and Intelligent Laboratory Systems, vol. 58, pp. 109-130 (2001)

Ultra-compact Optical Fiber Fabry-Perot Interferometer Based on In-Line Integrated Submicron Silicon Film

A. Micco, G. Quero, A. Crescitelli, A. Ricciardi, and A. Cusano

1 Introduction

Fiber-optic interferometric sensors constitute a long-standing topic of research for both industrial and scientific photonic community. Among them, in-line optical fiber interferometers continue to attract intensive attention in many sensing applications [1, 2]. Basically, in-line fiber interferometers can be divided in two main categories according to [1] Fabry-Perot interferometers (FPIs) and core-cladding-mode interferometers (CCIs). CCIs are based on interference between the fundamental core mode and the higher-order cladding mode that can be achieved with many techniques [3–8]. In general CCIs, although providing high sensitivities, suffer from some disadvantages such as complicated structures, typical sizes of the order of tens of millimeters, high instability, and cross sensitivity [1]. On the contrary, FPIs are much better suitable to be integrated on a single optical fiber as they are much stable and compact (typical sizes of ~tens of microns). In fact they can simply be obtained by creating an internal cavity inside the fiber. In this regard several techniques have been proposed so far. Most of these approaches essentially include the formation of an air gap along the fiber propagation axis. For example, the air cavity can be formed by splicing a section of a hollow-core photonic crystal fiber between two single-mode fibers or by directly drilling the fiber itself by a femtosecond laser [9, 10] or chemical etching [11]. In this work we present a novel, simple, and ultra-compact intrinsic FPI whose cavity is constituted of thin amorphous silicon layer completely embedded between two single-mode fibers by using the electric arc discharge technique. Our choice is motivated by the intent to integrate into the fiber a material whose electrical and optical properties could be successively exploited for the

A. Micco (✉) • G. Quero • A. Crescitelli • A. Ricciardi • A. Cusano
Optoelectronic Division, Department of Engineering, University of Sannio,
Corso Garibaldi 107, 82100 Benevento, Italy
e-mail: alberto.micco@unisannio.it; aricciardi@unisannio.it

C. Di Natale et al. (eds.), *Sensors and Microsystems: Proceedings of the 17th National Conference, Brescia, Italy, 5-7 February 2013*, Lecture Notes in Electrical Engineering 268, DOI 10.1007/978-3-319-00684-0_41, © Springer International Publishing Switzerland 2014

Fig. 1 Fabrication process phases: silicon deposition via PECVD, excimer laser confinement of the silicon, and splicing with another cleaved clean single-mode fiber

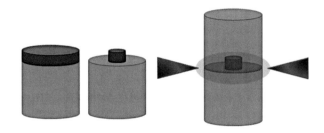

specific application. Moreover the integration of a high refractive index material allows to increase the FP cavity finesse and thus to enhance the interference fringe visibility, essential for most sensing application.

2 Experimental Results

The fabrication process is schematically shown in Fig. 1; it essentially consists of three main steps: (i) α-Si film deposition, (ii) layer confining, and (iii) splicing. The α-Si film is deposited by the following procedure: a standard single-mode (Corning SMF-28) optical fiber tip is rinsed with ethanol and placed on a proper sample holder inside the VHF-PECVD chamber. The deposition takes place at a temperature of 150 °C, pressure of 200 mTorr, power of 5 W, using pure silane, with a growth rate of about 4 A/s. The deposited α-Si layer is successively confined to an area of 20×20 μm around the fiber core through a UV micromachining process. The system used (OPTEC LB 1000) consists of an excimer laser (KrF, $\lambda = 248$ nm) with pulse width of about 5–6 ns. A plastic holder has been used in order to keep the fiber tip perpendicular to the focused laser beam during the ablation process. Finally the α-Si-deposited (and confined) fiber is spliced to a "bare" fiber by using a commercial arc fusion splicer (Fujikura FSM-50S). The splicing process can be considered the most critical phase of our fabrication procedure. In fact the presence of the α-Si on the fiber tip modifies the dielectric rigidity characteristics of the medium interposed between the splicer electrodes causing a significant temperature increase with a consequent failure of the junction. It is thus necessary to reduce both the power and time of the single arc discharge and perform multiple arcs in order to ensure a strong and robust junction.

During the whole fabrication process, both reflection and transmission spectra have been monitored. The spectral measurements have been performed by illuminating the fiber device with a broadband optical source (covering the wavelength range 1,200–1,600 nm) and redirecting the reflected (via a 2×1 directional coupler) and/or transmitted light to an optical spectrum analyzer (Ando AQ6317C). In Fig. 2a the reflectance spectrum before the fusion splicing process is shown (solid line). The fringe visibility is about 30 % due to the high refractive index contrast between the fiber silica (n ~ 1.46) and the α-Si (n ~ 3.3) in the mid-infrared. It is clear that the small cavity size (L ~ 750 nm) results in a slowly varying Fabry-Perot

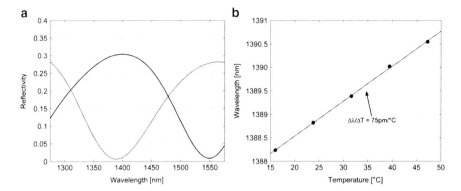

Fig. 2 (**a**) Reflection spectra in comparison before (*solid line*) and after (*dotted line*) the fusion splicing of the interferometer (*on the left*). (**b**) Temperature sensitivity calibration in range 15–50 °C (on the *right*)

envelope in the analyzed wavelength range. In Fig. 2a we also show as dotted lines the reflectance spectrum after the fusion splicing process. The first important thing to notice is that the fringe visibility does not change, meaning that α-Si layer remains fully embedded inside the joined fiber. At the same time, after the splicing, the spectra undergo a slight blue shift of about 60 nm, compatible with a thickness α-Si reduction of ~100 nm. The device insertion loss is estimated to be 1.46 dB.

In order to demonstrate a possible application for sensing, we carried out some preliminary temperature measurements. It is expected that our device exhibits a good temperature sensitivity, thanks to the high α-Si thermo-optic coefficient ($\sim 2 \times 10^{-4}$ °C^{-1}) [12]. The fiber sensor was placed on an aluminum plate heated by a Peltier temperature controlled system in the range 5–60 °C. When the temperature changes from 16 to 47 °C, a wavelength shift of about 2.32 nm is found. In Fig. 2b we plot the reflectance dip wavelengths versus temperature where a linear sensitivity ($\Delta\lambda/\Delta T$) of 75 pm/°C is obtained in the analyzed range.

It is interesting to compare our results with those achieved in the literature. First of all we note that the achieved sensitivity is about seven times higher than that of standard Fiber Bragg Grating sensors that is ~10 pm/°C [13]. On the other hand, in-line fiber FP interferometers typically exhibit lower temperature sensitivities of the order of few pm/°C [1, 2] because the cavity is in general made of air. Much higher sensitivities have instead been demonstrated by using CCI; for example, an MZI showing a sensitivity up to 109 pm/°C (for temperatures in the range 500–1,200 °C) has been recently proposed [8]. Therefore, our device exhibits a temperature sensitivity in line with the best of the state of the art for silica interferometric sensors, also considering that the sensitivity in [8] goes down to 23 pm/°C at lower temperatures (25–200 °C). By taking into account that the α-Si thermo-optic coefficient increases with temperature, we expect to have a considerable sensitivity increase when measuring higher temperatures. Furthermore, what makes our device unique and attractive is undoubtedly its compactness, being its size less than 1 μm and thus more than four orders of magnitude smaller than the typical CCI dimensions that are tens of millimeters. Moreover, differently to the CCI in-line optical fiber sensors, our device is completely insensitive to external refractive index that is desirable for temperature sensing applications. In addition, the proposed interferometer could be in principle effectively used also in high-temperature harsh environments.

3 Conclusions

In this work, a novel ultra-compact FP in-line interferometer, based on the integration of a thin (<1 μm) α-Si layer inside single-mode optical fibers, has been presented. The fabrication process is essentially based on an electric arc discharge splicing technique. A high-temperature sensitivity of 75 pm/°C is achieved in the range 15–50 °C, in line with the highest state-of-the-art values for in-line silica fiber CCI sensors. However, with respect to them the proposed device is about four orders of magnitude more compact. Future investigations deal with multiplexing layouts for multipoint measurement and the integration of different functionalized materials for the specific application.

References

1. Lee, B.H., Kim, Y.H., Park, K.S., Eom, J.B., Kim, M.J., Rho, B.S.and Choi, H.Y., "Interferometric fiber optic sensors," Sensors 12, 2467–2486 (2012).
2. Zhu, T., Wu, D., Liu, M. and Duan, D. W., "In-Line Fiber Optic Interferometric Sensors in Single-Mode Fibers", Sensors 12/8, 10430-10449 (2012).
3. Kim, Y.J.; Paek, U.C.; Lee, B.H., "Measurement of refractive-index variation with temperature by use of long-period fiber gratings." Opt. Lett. 27, 1297-1299 (2002).
4. Swart, P.L., "Long-period grating Michelson refractometric sensor," Meas. Sci. Technol. 15, 1576–1580 (2004).
5. Lu, P., Men, L., Sooley, K. and Chen, Q., "Tapered fiber Mach-Zehnder interferometer for simultaneous measurement of refractive index and temperature," Appl. Phys. Lett. 94, 131110-131112 (2009).
6. Tian, Z.B., Yam, S.S.H. and Loock, H.P., "Refractive index sensor based on an abrupt taper Michelson interferometer in a single-mode fiber," Opt. Lett. 33, 1105–1107 (2008).
7. Tian, Z., Yam, S.S.H. and Loock, H.P., "Single-mode fiber refractive index sensor based on core-offset attenuators," IEEE Photon. Technol. Lett. 16, 1387-1389 (2008).
8. Jiang, L., Yang, J., Wang, S., Li, B. and Wang, M., "Fiber Mach-Zehnder interferometer based on microcavities for high-temperature sensing with high sensitivity," Opt. Lett. 36, 3753-3755 (2011).
9. Rao, Y. J., Zhu, T., Yang, X. C. and Duan, D. W., "In-line fiber-optic etalon formed by hollow-core photonic crystal fiber," Opt. Lett. 32, 2662–2664 (2007).
10. Ran, Z. L., Rao, Y. J., Deng, H. Y. and Liao, X., "Miniature in-line photonic crystal fiber etalon fabricated by 157 nm laser micromachining," Opt. Lett. 32, 3071–3073 (2007).
11. Tafulo, P.A.R., Jorge, P.A.S., Santos, J.L., Araujo, F.M. and Frazao, O., "Intrinsic Fabry-Perot cavity sensor based on etched multimode graded index fiber for strain and temperature measurement," IEEE Sens. J. 12, 8–12 (2012).

12. Della Corte, F. G., Esposito Montefusco, M., Moretti, L., Rendina, I. and Rubino, A., "Study of the thermo-optic effect in hydrogenated amorphous silicon and hydrogenated amorphous silicon carbide between 300 and 500 K at 1.55 μm," Appl. Phys. Lett. 79, 168-170 (2001).
13. Cusano, A., Cutolo, A. and Albert, J., [Fiber Bragg Grating Sensors: Recent Advancements, Industrial Applications and Market Exploitation], Bentham Science Publishers, Oak Park, IL, 2011

Polymer Microflow Cytofluorometer

G. Testa, G. Persichetti, and R. Bernini

1 Introduction

Flow cytometer is a very powerful tool in a number of fields, including cell biology, environmental monitoring, and clinical analysis, as it permits fast and selective analysis of cells or particles suspended in liquids [1]. In the last years, many efforts have been devoted to the development of microflow cytometers by using microfluidic and microfabrication techniques [2]. A major advance in the microfluidic approach is the possibility to integrate various fluidic manipulation components such as pumps, valves, and mixers together with detection system (optical fibers for collecting the emitted light and deliver the light source) on the same platform for complete on-chip analysis. A high degree of integration of the whole microfluidic system can be obtained that permits to reduce the size of the entire fluidic components at a micrometer scale and hence to handle with very small amount of sample and reagent volume. One of the major issues in the development of an integrated flow cytometer is the ability to align the cells or particles along a single line, allowing for single particle analysis [3]. In the following, a simple and innovative technique will be demonstrated to realize a three-dimensional hydrodynamic focusing effect on micrometer scale that constitutes the microfluidic part of the proposed cytometer.

G. Testa (✉) • G. Persichetti • R. Bernini
Institute for Electromagnetic Monitoring of the Environment (IREA),
National Research Council (CNR), Naples, Italy
e-mail: testa.g@irea.cnr.it

C. Di Natale et al. (eds.), *Sensors and Microsystems: Proceedings of the 17th National Conference, Brescia, Italy, 5-7 February 2013*, Lecture Notes in Electrical Engineering 268, DOI 10.1007/978-3-319-00684-0_42, © Springer International Publishing Switzerland 2014

2   Device Design and Fabrication

A schematic view of the proposed device is shown in Fig. 1. A sample fluid injected
from inlet 1 flows in the squared central channel of width w, while sheath fluids are
injected from inlet 2 and 3 and flow through channels of width w and height $h > w$,
orthogonally crossing the main channel. Flowing from higher channels, the sheath
flow causes the sample flow at the junction to drift simultaneously vertically as well
as horizontally to the opposite sides and towards the center of the main channel.
Flow characteristics of the device have been simulated by solving the coupled
Navier–Stokes and diffusion/convection fluid equations using a finite element
method. In the simulation model it has been assumed as a squared cross section for
the main channel with a width of $w = 127$ µm. Numerical simulations predict that
the sample flow is focused in both vertical and horizontal directions when the ratio
between the height of the central channel and of the side channels is $h = d/w = 3.62$.
Moreover, it has been found that the sample fluid presents a circular-shaped cross
section located in the center of the microchannel. The optical characterization of the
three dimensional hydro focusing effect has been reported in a previous work [4].

The cytometer device consists of a two-layer PMMA microfluidic structure
(Fig. 2). Microchannels were fabricated by high-precision micromilling of two
3 mm-thick pieces of PMMA that are then bonded together. A squared groove of
127 µm × 127 µm has been realized orthogonally to the main channel for fiber

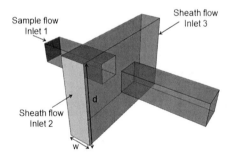

Fig. 1 Schematic of the 3D
flow focusing microfluidic
device

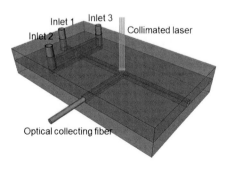

Fig. 2 Schematic of the
microflow cytometer device

positioning. With this arrangement, the optical fiber to collect the signal emitted from the particles flowing in the mainstream can be inserted in a self-aligned configuration for an optimized detection. Fluidic channels and the groove for the optical fiber have been fabricated by using milling tools with a diameter of $127 \pm 12.7 \, \mu m$. The fluidic inlet has been produced with a 500 μm-diameter end mill. For fluid delivering, syringe pumps have been connected to the inlets by Teflon tubing. Due to manufacturing process error, the fabricated core microchannel was 127 mm deep and 135 mm wide. The fabricated device has a final length and width of about 12 mm and 23 mm, respectively.

3   Optical Characterization

The optical performance of the proposed microflow cytometer has been experimentally investigated by detecting the fluorescence light from 10 μm-diameter rhodamine-B-marked particles suspended in the sample stream at a concentration of about $3.9 \times 10^5 \, cm^{-3}$. A beam from an Nd:YAG laser at $\lambda = 532$ nm was collimated on the detection area at the junction between the main channel and the groove for the optical fiber (Fig. 2). The signal from particles is then collected by an optical fiber (50 μm core diameter) orthogonally aligned with the flow and connected to an avalanche photodiode. The signal from photodiode is then displayed on an oscilloscope screen. During measurements, the flow rate ratio between the sample and the sheath flow has been fixed in order to obtain a single line of particles flowing along the center of the channel. The detected optical signal corresponding to the particle transit event in the detection area is shown in Fig. 3.

In Fig. 4 the frequency histogram of the light intensity fluorescence against the number of events is shown. From the data histograms, small variations of the

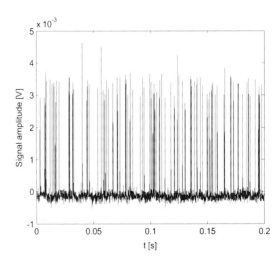

Fig. 3 Detected fluorescence signals from 10 μm-rhodamine-B-marked particles

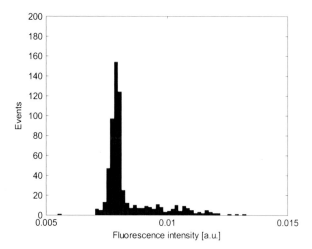

Fig. 4 Frequency histogram of fluorescence light intensity of 10 μm-rhodamine-B-marked particles

fluorescence have been observed, indicating that the particles are well focused with respect to the optical detection system and that they pass one by one through the detection area.

4   Conclusions

An innovative three-dimensional hydrodynamic focusing microfluidic device has been successfully applied for the realization of an integrated cytometer. The results demonstrate that the device permits to align the particles in a single line allowing for single particle analysis with a good signal-to-noise ratio and a high throughput.

References

1. D. A. Ateya, J.S. Erickson, P.B. Howell Jr, L.R. Hilliard, J.P. Golden, F.S. Ligler. The good, the bad, and the tiny: a review of microflow cytometry. Anal Bioanal Chem **391**, 1485-1498 (2008)
2. Z. Wang, J. El-Ali, M. Engelund, T. Gotsæd, I. R. Perch-Nielsen, K. B. Mogensen, D. Snakenborg, J. P. Kutter and A. Wolff. Measurements of scattered light on a microchip flow cytometer with integrated polymer based optical elements. Lab chip **4**, 372-377 (2004)
3. Kennedy MJ, Stelick SJ, Sayam LG, Yen A, Erickson D, Batt CA. Hydrodynamic optical alignment for microflow cytometry. Lab Chip. **11**, 1138-43 (2011)
4. G. Testa and R. Bernini. Integrated tunable liquid optical fiber. Lab Chip **12**, 3670-3672 (2012)

On the Design of a Clad-Etched Fiber Bragg Grating Sensor for Magnetic Field Sensing Applications

A. Saccomanno, D. Pagnano, A. Irace, A. Cusano, M. Giordano, and G. Breglio

1 FBG-Based Magnetic Field Sensor

The design concept of such magnetic field sensor lies in the integration of a clad-etched fiber Bragg grating (FBG), which is sensitive to surrounding refractive index (SRI), with a magnetic-fluid coating as sensing element.

1.1 Magnetic Fluid

Magnetic fluid is a stable colloidal solution of ferromagnetic nanoparticles. The behavior of magnetic fluid depends on the external magnetic field and its refractive index shown to be magnetic field dependent [1–3]. The refractive index of a generic material n_{mat} is given by

$$n_{mat} = \sqrt{\varepsilon_r} = \sqrt{1 + \chi_m} \tag{1}$$

where ε_r is the dielectric constant and χ_m is the electric susceptibility of the material. The selected magnetic fluid consists of an aqueous solution of iron (II, III) oxide (Fe_3O_4—magnetite) magnetic nanoparticles (<50 nm) prepared by the chemical

A. Saccomanno (✉) • D. Pagnano • A. Irace • G. Breglio
Department of Biomedical, Electronic and Telecommunication Engineering,
University of Napoli "Federico II", Via Claudio 21, 80125 Naples, Italy
e-mail: andrea.saccomanno@unina.it

A. Cusano
Department of Engineering, University of Sannio, C.so Garibaldi 107, 82100 Benevento, Italy

M. Giordano
Institute for Composite and Biomedical Materials, National Research Council,
P.le Tecchio 80, 80125 Naples, Italy

C. Di Natale et al. (eds.), *Sensors and Microsystems: Proceedings of the 17th National Conference, Brescia, Italy, 5-7 February 2013*, Lecture Notes in Electrical Engineering 268, DOI 10.1007/978-3-319-00684-0_43, © Springer International Publishing Switzerland 2014

coprecipitation method [4]. Fe_3O_4 has a cubic inverse spinel structure which consists of a cubic close packed array of oxide ions where all of the Fe^{2+} ions occupy half of the octahedral sites and the Fe^{3+} are split evenly across the remaining octahedral sites and the tetrahedral sites. In this material, when the external magnetic field is perpendicular to the propagation direction of light, we have

$$\frac{\partial \chi}{\partial H} < 0 \tag{2}$$

Then, the refractive index of the magnetic-fluid coating will decrease when the magnetic field increases.

1.2   FBG as Surrounding Refractive Index (SRI) Sensor

If the FBG is designed as a SRI sensor, the reflection spectra of the FBG will show a dependence on the refractive index of the media surrounding the fiber. A way to achieve sensitivity to changes in surrounding refractive index (SRI) is to excite evanescent waves in the media surrounding the fiber [5]. It has been proven that by etching the fiber in the grating region to reduce its diameter, SRI significantly affects the effective refractive index of the core, causing the Bragg wavelength to shift [6]. The Bragg wavelength is given by

$$\lambda_B = 2 n_{eff} \Lambda \tag{3}$$

where n_{eff} is the effective refractive index of the core and Λ is the grating period. Studying the dependence of the effective refractive index of the propagating mode, n_{eff}, on the design parameters of the FBG, in particular cladding diameter and SRI, is possible to find the relation between a shift in λ_B and the variation in the SRI that caused the shift. In the case of a magnetic-fluid coating, it is furthermore possible to establish a correspondence between the shift in the Bragg wavelength and the magnetic field the coating is subjected to.

$$\Delta \lambda_B = 2 \left(\Lambda \frac{\partial n_{eff}}{\partial SRI} \right) \Delta_{SRI} \tag{4}$$

2   Device Modeling

In order to study the wavelength shift, a structure with a uniformly thinned cladding along the grating was assumed and a model of doubly clad fiber defined for the cross section [7, 8]. The dependence of the effective refractive index on both the cladding diameter and SRI was evaluated by numerically solving the dispersion equation of the double cladding fiber mathematical model. A two-dimensional mode analysis has been performed on a single-mode optical fiber defined by CAD with a fixed core radius of 4.1 μm and refractive index $n_{core} = 1.460$ RIU, cladding refractive index $n_{clad} = 1.455$ RIU and a variable cladding radius from 4.2 to 62.5 μm, and refractive index of the surrounding media SRI in the range of 1.33–1.45 RIU. The model was simulated in the COMSOL ambient, a commercial multiphysics FEM simulator, connected to MATLAB to sweep in cladding radius and SRI. The outputs of the electromagnetic analysis are the electric field distribution and the effective refractive index of the propagating mode, n_{eff}.

3 Results

The results show the effects of SRI and cladding radius on the electromagnetic properties of the device (Fig. 1). The Bragg wavelength of the thinned FBG shifts to higher wavelengths by increasing the SRI. The Bragg wavelength of the thinned FBG shifts sharply when the SRI is close to the cladding refractive index of the thinned FBG, near 1.460 RIU (Fig. 2). Moreover, the thinned FBG with the smaller radius has a higher sensitivity for the same SRI value (Fig. 3).

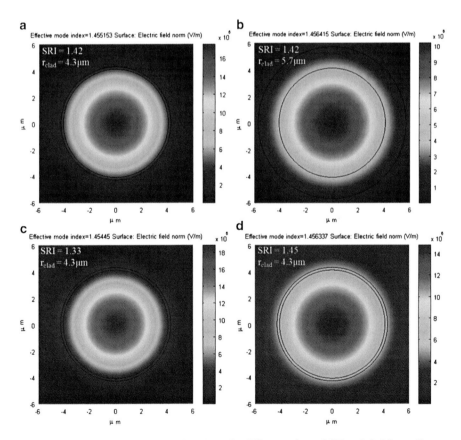

Fig. 1 Electric field distributions and mode n_{eff} for different values of SRI and cladding radius

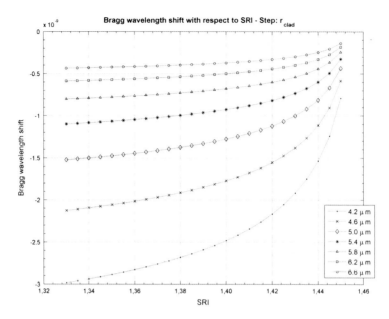

Fig. 2 Simulated Bragg wavelength shift as a function of SRI, for different values of r_{clad}

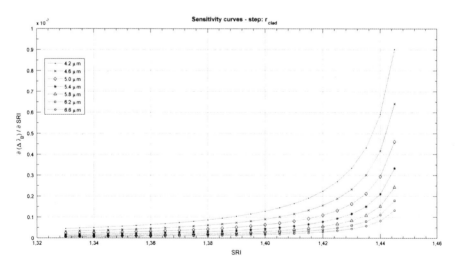

Fig. 3 Simulated sensitivity of a clad-etched fiber Bragg grating as a function of the SRI for different values of the cladding radius

4   Conclusions

The results obtained encourage our design concept and the development of a low-cost class of magnetic FBG sensors suitable for several kinds of applications, for low-intensity and high-intensity magnetic fields. The preliminary design and optimization phases carried out by means of the 2D FEM analysis permit to significantly reduce the cost and the development time of the sensor and to manufacture a customized device for each application.

References

1. S. Y. Yang, Y. F. Chen, H. E. Horng, C. Y. Hong, W. S. Tse, and H. C. Yang, "Magnetically-modulated refractive index of magnetic fluid films," *Applied Physics Letters*, Vol. 81, No. 26 (2002), pp. 4931–4933.
2. H. E. Horng, C. Y. Hong, S. Y. Yang, and H. C. Yang, "Designing the refractive indices by using magnetic fluids," *Applied Physics Letters*, Vol. 15, No. 15 (2003), pp. 2434–2436.
3. C. Y. Hong, S. Y. Yang, H. E. Horng, and H. C. Yang, "Control parameters for the tunable refractive index of magnetic fluid films," *Journal of Applied Physics*, Vol. 94, No. 6 (2003), pp. 3849–3852.
4. Y. Laiqiong, Z. Luji, and Y. Jixiao, "Study of preparation and properties on magnetization and stability for ferromagnetic fluids," *Materials Chemistry and Physics*, Vol. 66, No. 1 (2000), pp. 6–9.
5. A. Cusano, G. Breglio, M. Giordano, M. Russo, J. Nasser, "Optoelectronic refractive index measurements: application for smart polymer processing," *Sensors, 2002. Proceedings of IEEE*, Vol. 2 (2002), pp. 1171-1175.
6. A.N. Chryssis, S.M. Lee, S.B. Lee, S.S. Saini, M. Dagenais, "High Sensitivity Evanescent Field Fiber Bragg Grating Sensor," *IEEE Photonics Technology Letters*, Vol. 17, No. 6 (2005), 1253-1255.
7. M. Monerie, "Propagation in Doubly Clad Single-Mode Fibers," *IEEE Journal of Quantum Electronics*, Vol. QE-18, No. 4 (1982), 535-542.
8. A. Iadicicco, A. Cusano, S. Campopiano, A. Cutolo, M. Giordano, "Thinned Fiber Bragg Gratings as Refractive Index Sensors," *IEEE Sensors Journal*, Vol. 5, No. 6 (2005), 1288-1295.

Moisture Measurement System for Brick Kiln

M. Norgia and A. Pesatori

1 Introduction

Building materials, such as clay bricks, are fundamental elements for the human history. Originally, the bricks were made manually and the worked clay was dried by the sun. Today, this procedure is realized by machines for mixing, cutting, and molding the clay. The moisture level is a very important parameter for an automated system for realizing a good quality end product. Moreover, all raw materials have a different capacity to hold moisture; it depends on the geographical location, the storage procedure, the humidity in the atmosphere, and also the different particle size constituting the clay. Therefore, each batch of bricks requires a different amount of water to be added, in order to ensure the right consistency of the mixing process. The methods for measuring the moisture level are different. For example, the consistency of the clay can be controlled manually by monitoring the power consumption of the mixer utilized to uniform clay constituting materials. This method is unreliable, due to errors related to clay temperature and blades wearing. There are some contact measurement methods, not employable for the on-line monitoring, based on capacitance measurements [1, 2] or on sound-speed measurement [3]. The two methods most used for clay are realized by reflection measurements of radiation, in the near-infrared or in the microwave region. Microwave technique provides a good accuracy and reliability [4, 5], but also optical instruments show excellent performance [6, 7]. The use of this kind of instruments in field production facilities is mainly limited by their high cost; therefore, we propose a simple moisture sensor especially designed for that specific application.

M. Norgia • A. Pesatori (✉)
Dipartimento di Elettronica, Informazione e Bioingegneria, Politecnico di Milano,
Via Ponzio 34/5, Milan, Italy
e-mail: pesatori@elet.polimi.it

C. Di Natale et al. (eds.), *Sensors and Microsystems: Proceedings of the 17th National Conference, Brescia, Italy, 5-7 February 2013*, Lecture Notes in Electrical Engineering 268, DOI 10.1007/978-3-319-00684-0_44, © Springer International Publishing Switzerland 2014

2 Sensor Setup

The purpose of this work is the design and realization of a low-cost system for measuring the clay moisture, by metering the reflection at only two different wavelengths. The classic implementation uses a lamp and a series of optical filters, selected sequentially by a mechanical handling system [7]. In order to simplify the optics and mechanics, the proposed sensors employ two laser sources and a single photodetector, working in time division multiplexing. In this way we avoid moving parts and expensive optics. Moreover, the selected wavelengths, 1,300 and 1,500 nm, are typical values for telecommunication laser sources. Around 1,500 nm there is an absorption peak of the water [8], while around 1,300 nm the water absorption is about 30 times lower.

The optical system is arranged with four laser sources and one receiver, as described in Fig. 1. The two couples of laser diodes at 1,300 and 1,500 nm are placed symmetrically around the receiving lens, in order to minimize the reflection effects of different angles of the clay surface. The receiver is an InGaAs photodiode. The light is collected through a plastic lens with focal length of 5 cm, located about 4 cm from the photodiode, in order to obtain the field of view shown in Fig. 1. The two laser diodes are pulsed with a 10 % duty cycle, thus keeping the system in laser safety class I.

The processing electronics has been entirely built on a circuit board that houses both the analog and digital processing part. All operations are performed by a microcontroller. At the beginning, the system is initialized by measuring the ambient light. Its value is stored as an offset for all the following measurements. The measuring system is pulsed, alternating the lasers at 1,300 and 1,500 nm. To minimize the noise contribution, the digital acquisition is realized after a double integration of the photodiode signal: the signal is positively integrated while the 1,300 nm lasers are on, then the signal is negatively integrated while the lasers are off. The

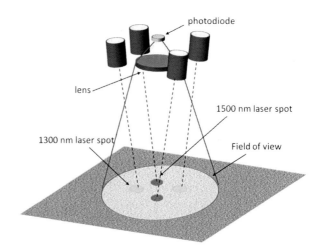

Fig. 1 Optical setup: two couples of lasers, a plastic lens and a photodiode

final value is sampled by the microcontroller. A low-pass analog filter is placed at the entrance of the microcontroller to further improve the signal-to-noise ratio. Double integration provides good noise rejection and excellent instant cancellation of the ambient light contribution. The procedure is then repeated for the 1,500 nm laser. A measurement for the two wavelengths lasts about 3 ms. For the considered application, a measuring frequency of 1 Hz is enough; therefore, an averaging procedure has been implemented. The single measurement is repeated eight times, and the maximum and the minimum values are systematically discarded. This procedure allows discarding erroneous measurements due to electronics disturbances. Finally, the procedure is repeated 30 times and the reflection values at 1,300 and 1,500 nm are averaged.

3 Calibration and Measurements

The two reflection measurements are dependent on moisture, because the reflected power at 1,500 nm with increasing humidity decreases more rapidly than at 1,300 nm. But the function of dependence is not linear and strongly depends on the clay distance and the collimation of laser beams. For our purposes, the moisture sensor should be insensitive to distance changes in a range of approximately 5 cm. The optical setup was arranged in order to maximize the signal variation related to moisture while minimizing the dependence on distance. The best solution is shown in Fig. 1: the lasers are collimated and nonoverlapping, and the optical receiver is placed in order to observe all the spots.

The purpose of this sensor is the measure of moisture change, compared to a set point, determined by the human operators as a good value for working. Therefore, a measurement function was studied that was particularly sensitive to changes in clay moisture, with the ability to be set by the operator to a reference value. Through several experimental attempts, looking at the signals as a function of distance, we have chosen this parametric equation:

$$M = \frac{S_1 - S_2 + a}{S_1 + S_2 + b} \tag{1}$$

where M is the moisture indicator; S_1 and S_2 are the amplitudes of the received signal at 1,300 nm and 1,500 nm, respectively; and a and b are calibration constants. To find the parameters, two samples of clay have been used with known moisture percentage (14 and 18 %). The two equations needed are obtained by equating the value of M at the distances of 15 and 20 cm, for both samples. Implementing the obtained values of the parameters, the measurement becomes stable with a negligible dependence on the distance in a range of about 7 cm. The sensitivity to the clay moisture is very high: the device moves from negative to positive saturation with a relative humidity variation of 5 %. Through a mechanic switch, the operator can set the present measurement M_0 as reference value: the output value is the measured M

divided by M_0. The system has two outputs: a digital RS-485 serial communication and an analog voltage realized by a pulse-width modulation (PWM), with 0–5 V level. When the calibration button is pressed, the value of moisture is brought to the center of the measurement range, 2.5 V of the PWM output.

To evaluate the performances of the sensor, a series of measures was performed, with different moisture levels and different types of clay. The considered moisture level ranges from 10 to 20 %, the distances from 15 to 25 cm. The output standard deviation in stationary conditions is lower than 0.05 V, for measurements done with clay firm in a constant position. Sliding the clay, but keeping the instrument at the same distance, the standard deviation can reach the value of 0.1 V. The maximum output variation shown for a 5 cm change of height was 0.1 V. The influence of ambient light variation is always negligible. The resolution of the sensor was evaluated by changing the moisture level, keeping all the other condition unchanged. In particular, the output voltage has shown a variation of 0.4 V for a variation of 1 % of the moisture level; thus it is possible to estimate a measurement resolution better than 0.5 %, well adequate for the application to a real-time control system. Finally the sensor was tested in a real production line. The moisture level was acquired for 2 days, in comparison with an expensive commercial instrument mounted in close proximity to our sensor. The results were in good agreement, therefore confirming the proper operation of the proposed sensor.

Acknowledgments A special acknowledgment is due to TEA Elettronica S.r.l for collaborating in the measurement activity and the financial support and to Eng. Alessio Roberti and Eng. Matteo Mascia for their support in sensor development.

References

1. S. C. Saxena and G. M. Tayal, "Capacitive Moisture Meter," Ieee transactions on industrial electronics and control instrumentation, vol. Ieci-28, no. 1, pp 37-39, February 1981
2. S. O. Nelson, "Measurement and Applications of Dielectric Properties of Agricultural Products", ISEEE Transactions on Instrumentation and Measurement, vol. 41:1, pp 116 - 122, February, 1992
3. F. Adamo, G. Andria, F. Attivissimo, and N. Giaquinto, "An Acoustic Method for Soil Moisture Measurement," IEEE Transactions on Instrumentation and Measurement, Vol. 53, No. 4, pp. 891-898, August 2004
4. L. Bruckler, H. Witono and P. Stengel, "Near surface soil moisture estimation from microwave measurements," Remote Sens. Environ., 26 (1988), pp. 101–121
5. E. T. Engman, N. Chauhan, "Status of microwave soil moisture measurements with remote sensing", Remote Sensing of Environment Volume 51, Issue 1, Jan. 1995, Pages 189-198
6. R. C. Ehlert, "Infrared reflection and absorption system for measuring the quantity of a substance that is sorbed in a base material", United States Patent 3150264, 1964
7. J. W. Mactaggart, "Infrared moisture measuring apparatus," United States Patent 4171918, 1979
8. J. A. Curcio and C. C. Petty, "The Near Infrared Absorption Spectrum of Liquid Water," J. Opt. Soc. Am. 41, 302-302 (1951)

Breath Figures onto Optical Fibers for Miniaturized Sensing Probes

Marco Pisco, Giuseppe Quero, Agostino Iadicicco, Michele Giordano,
Francesco Galeotti, and Andrea Cusano

1 Introduction

Recently, the realization of miniaturized and advanced optical fiber devices and the consequent development of technological processes, specialized for the optical fibers, led to the definition of the "lab-on-fiber" concept [1], devoted to the realization of novel and highly functionalized technological platforms completely integrated in a single optical fiber for communication and sensing applications. In this scenario, the creation of micro- and nanostructures on the end facet of optical fibers is of great interest because it may yield versatile optical devices well suited to serve as miniaturized probes for remote sensing applications. Several approaches have been recently introduced to fabricate metallic and dielectric structures on the optical fiber end facet. Some approaches rely on the study of appropriate techniques to transfer planar nanoscale structures, fabricated on a planar wafer by means of standard lithographic techniques, onto the optical fiber end facet. These methods exploit well-assessed fabrication processes developed for planar substrates, but they are

M. Pisco (✉) • G. Quero • A. Cusano
Optoelectronic Division - Engineering Department, University of Sannio,
Corso Garibaldi, 107, 82100 Milan, BN, Italy
e-mail: pisco@unisannio.it

A. Iadicicco
Department of Technology, University of Naples "Parthenope", Centro Direzionale Isola C4,
80143 Naples, Italy

M. Giordano
Institute for Composite and Biomedical Materials, Consiglio Nazionale delle Ricerche,
piazzale E Fermi 1, Portici, 80055 Naples, Italy

F. Galeotti
Istituto per lo Studio delle Macromolecole, Consiglio Nazionale delle Ricerche,
via Bassini 15, 20133 Milan, Italy

C. Di Natale et al. (eds.), *Sensors and Microsystems: Proceedings of the 17th National Conference, Brescia, Italy, 5-7 February 2013*, Lecture Notes in Electrical Engineering 268, DOI 10.1007/978-3-319-00684-0_45, © Springer International Publishing Switzerland 2014

limited by the final transferring step that plays a fundamental role in determining both the fabrication yield and the performance of the final device [2]. Alternative approaches are based on direct-write patterning of the fiber tip. These methods, based on conventional lithographic techniques adapted to operate on unconventional substrates such as the optical fiber tip, are able to efficiently provide nano-structured devices on the optical fiber, but they require complex and expensive fabrication procedures with a relatively low throughput [2, 3].

In this work, we propose the creation of periodic metallo-dielectric structures on the optical fiber tip by using self-assembly techniques. Specifically, we selected the breath figure (BF) technique for the preparation of patterned polymeric films directly on the optical fiber tip. After this stage, we employ a simple evaporation technique for the conformal deposition of a thin metal layer of gold. Following this simple approach, we fabricated several prototypes of miniaturized sensing probes.

2 Fabrication Process

To build metallo-dielectric periodic structures directly on the end facet of a single-mode optical fiber in a simple and cost-effective way, we developed a fabrication procedure based on the BF technique. This self-assembly approach, in fact, allows for the preparation of honeycomb-structured films with a high degree of order, which are, at least in principle, suitable for our purpose. BF formation spontaneously takes place when a polymer solution is cast on a substrate, under humid atmosphere. If the solvent has a sufficiently low boiling point, its fast evaporation lowers the temperature of the system, which triggers the condensation of water droplets on the film which is forming. In the following stage, water droplets arrange themselves closely, producing a template for the porous pattern. Finally, once both solvent and water are completely evaporated, a honeycomb imprint is left on the film surface [4].

Our fabrication strategy consisted initially in the preparation of a highly ordered microporous film on the fiber tip and then in the vapor deposition of a thin layer of Au on top of this assembly (see Fig. 1). To reach this goal, we conveniently modified the standard setup which is normally utilized for BFs on a glass substrate. The main drawback of building patterns directly on the fiber is the restricted surface of its facet: 125 μm of diameter for a standard single mode. For that reason, the optical fiber was embedded in a ceramic ferule with diameter of 2.5 mm and then accurately polished. Then this assembly, much easier to handle as compared to the bare fiber, was mounted on an Al holder of $20 \times 20 \times 8$ mm, so that the polymer solution

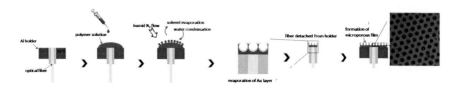

Fig. 1 Schematic view of the fabrication strategy. The detail of the microporous film is a real view taken by confocal microscope, in the central area of the fiber face

can be more easily drop cast on a surface of about 1 cm², much larger than that of the only fiber tip. In this way we overcame also the problem of poor pattern homogeneity, which is often encountered on the edges of BF films.

Films were prepared by drop casting on the holder with the fiber a 4 mg/mL CS₂ solution of a fluorinated fluorescent dye-terminated linear polystyrene, which we had already described as a good candidate for producing highly ordered BFs [4]. A flux of moist nitrogen (60 % R.H. at 25 °C) was directed on the Al surface. The solvent was completely evaporated within 20 s, living on holder and fiber surface an opaque film, which shined in bright iridescent colors when viewed in reflected light. By regulating the rate of the moist nitrogen flux between 1.5 and 2.5 L/min, i.e., by varying the solvent evaporation rate parameter, honeycomb films with cavities ranging from 2.5 to 0.9 μm of external diameter, respectively, were obtained. Once the optical fiber facet was covered with this polymeric pattern, it was placed in a vacuum evaporation chamber and 30–40 nm of Au were deposited on it. By this two-step procedure, prototypes consisting in a single-mode optical fiber end-coated with a double metal pattern, a micrometric Au mesh lying on top of the polymer film and an array of Au cups lying on the bottom of the cavities, were achieved.

3 Experimental Results

In this section the attention is principally focused on the experimental analysis of a representative sample fabricated by means of the fabrication process previously described. With regard to this sample, a complete morphological characterization has been carried out via scanning electron microscope (SEM) and atomic force microscope (AFM) analysis. Figure 2a shows an SEM top view image of the sample where can be appreciated the ceramic ferule with smoothed edge (diameter 2.5 mm). Magnified SEM image (here note reported) and AFM image (Fig. 2b) permit to measure statistic values of the cavity diameters and pitches of the patterned region. From images analyses following average (variance) values were retrieved: diameter of 0.95 μm (8.99×10^{-4} μm²), pitch of 2.67 μm (0.0012 μm²), cavities depth of 1.78 μm (0.0038 μm²), and structure height of 2.5 μm (0.0032 μm²). From these results it is evident that the breath figure technique enables the possibility to realize

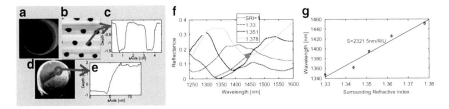

Fig. 2 Morphological characterization via SEM and AFM images. Sample: (**a**) view SEM image; (**b**) AFM image of patterned region; (**c**) AFM profile; (**d**) SEM top view image after excimer laser treatment; (**e**) AFM profile of the pattern edge; (**f**) SRI characterization; (**g**) wavelength of the reflection dip versus SRI

metallo-dielectric structure directly self-assembled on fiber optic tip with highly regular pattern. To show the potentiality of patterned metallo-dielectric lab-on-fiber structures realized via breath figures approach, in the following, we present some preliminary results demonstrating how the fabricated sample is able to sense the refractive index variations in the surrounding environment.

To investigate the surrounding refractive index (SRI) sensitivity, the reflectance spectra were measured while the fabricated samples were immersed in different liquid solutions (n in range 1.333–1.362). The experimental results are shown in Fig. 2f, in which a significant red shift of the curves with increasing values of the SRI is evident. Focusing the attention on the wavelength shift of spectral features, in Fig. 2g, we plot the wavelength of the reflection minimum (at 1,347 nm for SRI = 1.333) as a function of the SRI. The graph demonstrates a sensitivity (S) of ~2,300 nm/RIU for detecting changes in the bulk refractive indices of different chemicals surrounding the fiber-tip device.

The excellent sensitivities versus SRI changes as well as the properties of gold for the binding of suitable bioreceptors make these structures promising candidates for novel miniaturized affinity-based biological nanosensors with the ability of detecting few or even single nanoparticles.

4 Conclusions

In conclusion, the reported experimental results demonstrated the feasibility of the proposed fabrication approach to realize hybrid metallo-dielectric structures directly on the end facet of optical fibers. This enables the fabrication of micro- and nano-structured devices on fiber tip by means of simple and non-expensive fabrication procedures differently from conventional approaches. Also, BF technique could easily permit the realization of patterned structures with a relatively high through-put. The excellent sensitivities versus SRI changes as well as the properties of gold for the binding of suitable bioreceptors make these structures promising candidates for novel miniaturized affinity-based biological nanosensors with ability of detecting few or even single nanoparticles.

References

1. A. Cusano, M. Consales, M. Pisco, A. Crescitelli, A. Ricciardi, E. Esposito, A. Cutolo. Proc. SPIE, **8001**, 800122 (2011).
2. M. Consales, M. Pisco, A. Cusano, "Lab on Fiber Technology: A new avenue for optical nano-sensors", Photonic Sensors, 2, 4, pp 289-314 (2012)
3. A. Ricciardi, M. Pisco, A. Cutolo, A. Cusano, L. O' Faolain, T. F. Krauss, G. Castaldi, V. Galdi. Evidence of guided resonances in photonic quasicrystal slabs, Phys. Rev. B **84**, 085135 (2011)
4. P. Escalé, L. Rubatat, L. Billon, M. Save. Recent advances in honeycomb-structured porous polymer films prepared via breath figures. Euro. Polym. J., **48** (6), 1001-1025 (2012).

Sensitivity of Wood-Rayleigh Anomalies in Metallic Nanogratings

A. Ricciardi, S. Savoia, A. Crescitelli, V. Galdi, A. Cusano, and E. Esposito

1 Introduction

Spectral anomalies in the response of diffraction gratings constitute a subject of long-standing interest in optics [1, 2]. Basically, two types of anomalies can be identified [3]: *sharp* anomalies, due to the passing off of a spectral diffraction order, and *diffuse* anomalies arising from the excitation of surface waves. The former, typically referred to as Rayleigh anomalies (RAs), occur as abrupt intensity variations at Rayleigh wavelengths $\lambda_R^{(m)}$ given by the well-known grating formula [3]:

$$\lambda_R^{(m)} = \frac{\Lambda n_a \left(-\sin\theta_i \pm 1\right)}{m}, \quad m = \pm 1, \pm 2, \dots, \tag{1}$$

where a one-dimensional (1-D) grating of period Λ immersed in a medium with refractive-index (RI) n_a is assumed and θ_i denotes the angle of incidence (measured anticlockwise from the normal to the grating).

In a recent work, it was demonstrated that RA-based nanograting sensors exhibit a *bulk* sensitivity determined solely by the grating period [4]. One may be led to consider RA-based grating nanosensors as a valid alternative to standard surface-plasmon-resonance-based sensors [5, 6]. In label-free chemical and biological

A. Ricciardi (✉) • A. Crescitelli • A. Cusano
Optoelectronics Group, Department of Engineering, University of Sannio,
Corso Garibaldi 107, 82100 Benevento, Italy
e-mail: aricciardi@unisannio.it

S. Savoia • V. Galdi
Waves Group, Department of Engineering, University of Sannio, Corso Garibaldi 107,
82100 Benevento, Italy

E. Esposito
CNR, National Research Council of Italy, Istituto di Cibernetica "E.Caianiello", Pozzuoli, Italy

C. Di Natale et al. (eds.), *Sensors and Microsystems: Proceedings of the 17th National Conference, Brescia, Italy, 5-7 February 2013*, Lecture Notes in Electrical Engineering 268, DOI 10.1007/978-3-319-00684-0_46, © Springer International Publishing Switzerland 2014

sensing applications, however, RI changes are mainly restricted to surface modifications occurring at the sensor interface (e.g., due to the binding of a biological layer). Accordingly, in what follows, we numerically study the *surface* sensitivity of RA-based metallic nanograting sensors, highlighting the connection to a *surface plasmon polariton* (SPP) effect.

2 Results and Discussion

The analyzed device, schematically shown in Fig. 1(left), consists of a gold (modeled according to a Lorentz-Drude model) concentric ring nanograting, approximated as a linear grating characterized by same period $\Lambda = 900\,nm$, $DC = w/\Lambda$, and thickness $t_{Au} = 30\,nm$, laid on a fused-silica substrate ($n_s = 1.45$). A fiber-optic-type illumination (i.e., impinging from the substrate) is considered here. A dielectric overlay ($n_d = 1.33$) is considered.

Figure 1(right-a) shows the numerical reflectivity spectra (computed via a rigorous coupled-wave analysis algorithm) for overlay thickness values $t_d = 0$, 10, 20 nm. The steep rising front at the first-order RA wavelength $\lambda_{R,air}^{(1)} = 900nm$ does not exhibit a sensible modification, while the falling front broadens up, thereby inducing a slight red shift of the reflectivity peak.

For understanding the observed surface sensitivity, it is insightful to look at the numerical reflectivity spectra pertaining to the TM and TE polarizations, shown in Fig. 1b, c (on the right), respectively. Such spectra, which were averaged in Fig. 1a,

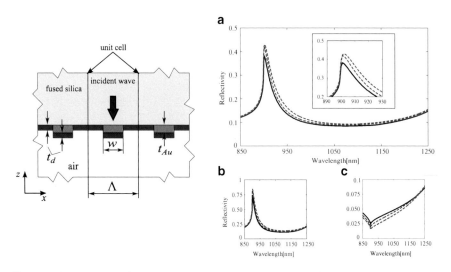

Fig. 1 (*left*) Schematic of the sensor configuration considered in the numerical study. (*right*) Numerical reflectivity spectra (magnified in the inset) for different values of thickness of the dielectric overlay (0,10 and 20 nm): The responses are obtained by averaging those pertaining to the TM and TE polarizations, shown separately in (**b**) and (**c**), respectively

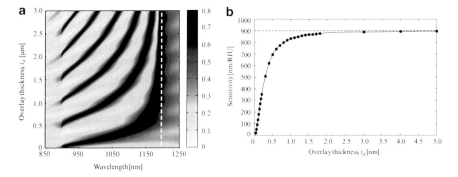

Fig. 2 (**a**) Reflectivity spectra contour plot (TM polarization) for a dielectric (RI = 1.33) with thickness varying within the range 0–3 μm. The *white-dashed reference line* corresponds to $\lambda_{R,d}^{(1)} = n_d\Lambda = 1197nm$. (**b**) Numerical surface sensitivity a function of the thickness of the dielectric overlay. The *dashed horizontal line* corresponds to the theoretical bulk-sensitivity limit (900 nm/RIU)

turn out to be quite different. While both responses exhibit a quite visible spectral feature around the first-order RA wavelength $\lambda_{R,air}^{(1)} = 900nm$, the broadening front that is responsible for the actual surface sensitivity observed in Fig. 1a is visible only for the TM polarization. From this evidence, it is possible to recognize that the small residual surface sensitivity observed is attributable to the interplay with the excitation of a *plasmonic* (SPP) mode at a wavelength close to $\lambda_{R,air}^{(1)}$, which is responsible for the falling front of the spectral reflectivity peak, and cannot occur for the TE polarization.

2.1 Ultimate Sensitivity

In order to gain some insight into the connection between the RA- and SPP-based sensing mechanisms, we carried out an extensive parametric study by varying the dielectric overlay thickness. For the above nanograting, and TM polarization, the contour plot in Fig. 2a shows, in false-color scale, the evolution of the reflectivity spectrum for increasing values of the dielectric overlay thickness t_d (up to $3\,\mu m$), and $DC = 0.55$. We can observe a series of reflectivity peaks (dark ridges in the contour plot) that originate at $\lambda_{R,air}^{(1)} = \Lambda = 900nm$ (corresponding to the first-order RA pertaining to air) and progressively red shift, for increasing values of the overlay thickness, towards a limiting value $\lambda_{R,d}^{(1)} = n_d\Lambda = 1197nm$ (corresponding to the first-order RA pertaining to the dielectric). These peaks are physically associated with bound modes supported by the structure, including the aforementioned SPP mode (which can be observed for arbitrarily thin overlays) and the growing number of guided modes that the overlay can support as its thickness increases. Both SPP and guided (photonic) modes "break down" at $\lambda_{R,d}^{(1)} = 1197nm$, for which the wavevector associated to the diffractive wave needed for their excitation becomes purely imaginary [3].

The above results indicate that the transition between the two limit configurations (i.e., no analyte and bulk analyte) is established, for increasing values of the overlay thickness, via the formation and progressive red shift of spectral peaks associated with the discrete spectrum of bound (SPP and guided) modes that the structure can support, whose merging (for $t_d \to \infty$) eventually gives rise to the sharp spectral peak at the RA wavelength pertaining to the bulk dielectric.

Of particular interest for assessing the surface sensitivity is analysis of the SPP mode behavior. Figure 2b shows the sensitivity associated with the corresponding peak wavelength shift, as a function of the overlay thickness. As it can be observed, the sensitivity initially exhibits a rather steep increase (of nearly two orders of magnitude from the very low levels previously observed in the presence of nanosized overlays) for thickness values up to 300 nm and then approaches (for thickness values ~2 μm) the theoretical RA bulk-sensitivity prediction $S = \Lambda = 900 nm/RIU$. From the physical viewpoint, this ultimate sensitivity is consistent with the complete filling (by the analyte-overlay) of the wavelength-sized sensing volume of SPP modes.

3   Conclusions

In this paper, we have numerically studied the surface sensitivity of RAs in metallic nanogratings. Although extremely sensitive when *bulk* analytes are considered, the spectral features associated to RAs in metal nanogratings are completely insensitive when *local* RI changes are considered. The residual surface sensitivity observed is essentially attributable to interplaying plasmonic (SPP) phenomena spectrally overlapping with RAs. Finally, the ultimate surface sensitivity approaches the theoretical bulk sensitivity (corresponding to the grating period) associated with RAs from wavelength-sized overlays (i.e., entirely filling the typical sensing volume of SPP modes).

Besides shedding light in the physical phenomena underlying RA-based nanodevices, our results above also provide useful quantitative assessments for their applicability to label-free chemical and biological sensing.

References

1. Wood R. W., "On a remarkable case of uneven distribution of light in a diffraction grating spectrum," Philos. Mag. 4, 396 (1902).
2. Lord Rayleigh, "On the dynamical theory of gratings," Proc. R. Soc. Lond., 79, 399, (1907).
3. Maystre D.,[Plasmonics], Springer Series in Optical Science, Berlin-Heidelberg, Germany, (2012).
4. Feng S., Darmawi S., Henning T., Klar P. J., Zhang X., "A miniaturized sensor consisting of concentric metallic nanorings on the end facet of an optical fiber," Small, 8(12), 1937 (2012).
5. Anker J. N., Hall W. P., Lyandres O., Shah N. C., Zhao J., Van Duyne R. P., "Biosensing with plasmonic nanosensors," Nature Materials, 7(6), 442 (2008).
6. Crescitelli A., Ricciardi A., Consales M., Esposito E., Granata C., Galdi V., Cutolo A., Cusano A., "Nanostructured Metallo-Dielectric Quasi-Crystals: Towards Photonic-Plasmonic Resonance Engineering," Adv. Funct. Mater., 22(20), 4185-98, (2012).

Sensored Handheld Fiber-Optic Delivery System for Controlled Laser Ablation

I. Cacciari, A.A. Mencaglia, and S. Siano

1 Fluence at the Target

The laser fluence is monitored by means of a photodiode (PD_1) detecting part of the pulse energy reflected by the imaging lens and a sensor, which reads the position of the fiber tip with respect to the latter, both housed inside the handpiece. The pulse energy, E_p, is calculated by integrating the photocurrent of PD_1 using a suitable electronics triggered by a phototransistor. A preliminary calibration was needed in order to derive the conversion coefficient of the voltage data to laser pulse energies.

The second step for determining the fluence at the target is represented by the association of the laser spot area, A, to the position, p, of the fiber tip with respect to the imaging lens (i.e., the object position). To this goal p is detected using a sensor implemented by a linear potentiometric membrane. In such a way, each position of the fiber tip, p, corresponds to a given position of the focal plane, q, spot diameter, $d(p)$, and area $A(p)$, which are known from preliminary direct measurements and stored into the data processing unit. Finally, the fluence at the target is calculated as $F(p) = E_p/A(p)$.

2 Reflectance of the Irradiated Material Surface

The reflectance (R) is measured by a second photodiode, PD_2, fixed on the handpiece by means of a suitable support, which detects part of the light reflected by the sample surface. As for the fluence case, the photocurrent is integrated by means of

I. Cacciari (✉) • A.A. Mencaglia • S. Siano
CNR-IFAC, Sesto Fiorentino, Florence, Italy
e-mail: i.cacciari@ifac.cnr.it

C. Di Natale et al. (eds.), *Sensors and Microsystems: Proceedings of the 17th National Conference, Brescia, Italy, 5-7 February 2013*, Lecture Notes in Electrical Engineering 268, DOI 10.1007/978-3-319-00684-0_47, © Springer International Publishing Switzerland 2014

a dedicated circuit, and a conversion from voltage to energy is required. In this case, no calibration was required as R is simply given by the ratio between two measurements: on a sample and on a reference surface with known reflectance.

3   Alert

An IR position sensor is placed on the handpiece. It measures the sample-to-lens distance, $q_m(p)$. This value is compared to the estimated image-to-lens distance experimentally related to the fiber-to-lens distance, $q(p)$. When the difference between these two values falls within ±5 mm, no alert is provided. On the contrary, when the difference exceeds such a tolerance, two types of alert are possible depending on the difference sign (yellow LED when the sample is closer and green when it is farther). The handpiece is equipped with this additional position sensor in order to provide the operator (the restorer) with an efficient alert when he is drifting away from the optimal irradiation distance $q(p)$.

4   Technical Details

The device is made of an analogical and a digital core. The first one deals with the area and energy measurements; the latter uses single-chip 16-bit CMOS microcomputer (Mitsubishi M30624FGAFP) to store and analyze the data and to display the results on a 3-digit LCD display (Varitronix VI-321).

The laser light delivery (Fig. 1) is achieved through an optical fiber (Quartz/Quartz, core diameter = 1 mm) and a biconvex imaging lens (BK7, $f = 38$ mm). The fiber is fixed inside a piston, which can slide into a particular lens housing, designed with a tube shape. A screw is placed on the tube to fix the piston position and therefore to set the distance p. Due to its compactness (0.5 mm thick) and excellent repeatability as a positional indicator, an ultra flat potentiometric membrane (Spectra Symbol, TSP-L-0050-103) has been stuck on the piston. The wiper terminal

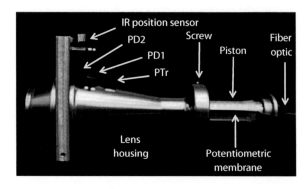

Fig. 1 Laser light delivery system

corresponds to the actuator (the fixing screw on the tube) position: a low pressure on it determines the electrical resistance in the range $100~\Omega$–$10~k\Omega$, allowing a very accurate measurement of the fiber position. The piston position determines the fiber end image plane. The image-to-lens distance is compared with the sample-to-lens distance, which is measured by an infrared sensor mounted on the lens plane (Sharp 3520 2D120x). The system is able to alert if the operator drifts away from the best focusing position.

5 Device Calibration

The irradiated area, A, has been considered as the fiber end face imaged through the focusing lens. We have firstly considered A related to the fiber-to-lens distance according to the lens maker's formula. Measuring the fiber-to-lens distance, p, the image-to-lens distance, $q(p)$, can be determined providing the lens focal length, f. In this picture A is calculated using the magnification factor, i.e., the ratio p/q, on the fiber end face (Fig. 2).

We have measured ten p positions using the potentiometric position sensor and the lens maker's formula in order to determine the corresponding $q(p)$ through the 38 mm focusing lens. An overestimation of more than 25 % has been observed and it has been considered not suitable for the present application. Therefore, we have preferred to find an empirical relation as being more accurate, since the thin lens approximation does not hold and the fiber core cannot be considered as point-like.

The following procedure has been considered to measure the value of $q(p)$ and the corresponding spot diameter to be considered in the fluence calculation. The sample has been placed orthogonally the optical axis. For each laser pulse, the sample has been moved along the lens optical axis to get farther and closer, and the spot diameter of the irradiated area has been measured with a microscope [1, 2].

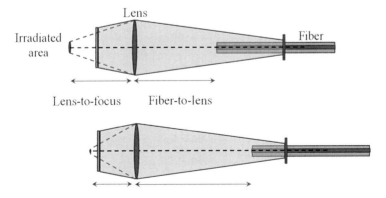

Fig. 2 The fiber-to-lens distance is related to the irradiated area

Fig. 3 Image spot diameter vs. sample-to-lens position relation for a fixed p position. The data has been properly fit with a parabolic curve; its minimum represents the $q(p)$ position. Other parabolic curves have been used to fit the data obtained with other fiber-to-lens positions

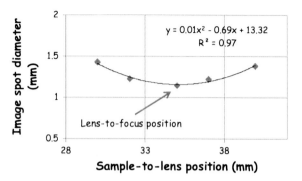

The spot diameter has then been related the sample-to-lens distance. We have observed a parabolic (concave up) trend of these data, and the minimum is located at the focusing position (Fig. 3).

This procedure has been repeated for all the distances p considered in this paper and expressed in terms of resistance of the potentiometric membrane.

A phototransistor (Osram SFH309) has been used for triggering the integrator circuits of the two photodiodes (PD, Centronic BPX65) for the energy-to-voltage measurements. The two PD_s have been placed in the handpiece: as PD_1 is used to measure the fluence, it has been placed directly on the tube before the lens plane to detect part of the internally reflected light; on the contrary, PD_2, which is used to measure the reflectance, has been housed on the infrared sensor mount, to detect part of the light reflected by the surface.

6 Conclusions

A novel handpiece fiber-optic delivery system for controlled laser ablation treatments in conservation of artworks has been developed. It has been equipped with sensors and calibrated in order to provide useful information to the restorers during the laser irradiation. This device can also improve the control and then the quality of the ablation process in other fields of application.

Acknowledgments The device described in this work was developed in the project TEMART (Tuscany Region, POR-CReO/FESR 2007–2013), and it is presently under experimentation within the project Charisma (FP7, Capacities, grant agreement n. 228330).

References

1. I. Cacciari, D. Ciofini, M. Mascalchi, A. Mencaglia, S. Siano. Novel approach to the microscopic inspection during laser cleaning treatments of artworks. Anal. Bioanal. Chem. **402** (4), 1585-1591 (2012)
2. I. Cacciari, A. Mencaglia, S. Siano. Micromorphology of gold jewels: a novel algorithm for 3D reconstruction and its quality assessment. Proc. Spie v. 8790, Optics for Arts, Architecture, and Archaeology IV (2013)

Differential Pulse-Width Pair Brillouin Optical Time-Domain Analysis Employing Raman Amplification and Optical Pulse Coding

M. Taki, M. Soto, F. Di Pasquale, and G. Bolognini

1 Introduction

One of the most adopted distributed sensing techniques exploits Brillouin optical time-domain analysis (BOTDA) due to its capability of providing an accurate simultaneous measurement of strain and temperature. The possibility of this technique, based on stimulated Brillouin scattering (SBS), to achieve long sensing ranges with high spatial resolution using standard single-mode fiber (SMF) enables many practical applications in structural health monitoring, where spatial resolution values ranging in the cm scale are in some cases required over long distances (e.g., crack detection). However, the spatial resolution in standard BOTDA systems is limited to 1 m due to the acoustic-phonon lifetime (~10 ns), inducing a broadening of the Brillouin gain spectrum when pulse widths shorter than 10 ns are used, thus leading to a reduction of the Brillouin peak gain [1] and to inaccuracies in Brillouin frequency shift (BFS) measurements. In order to achieve spatial resolutions smaller than 1 m, one promising technique is given by BOTDA employing differential pulse-width pair (DPP–BOTDA), which has been successfully employed for submeter spatial resolution [1–3], although limited in sensing range due to poor signal to noise (SNR) induced in the DPP subtraction process. Optical pulse coding has been applied to DPP–BOTDA [1, 2] overcoming the SNR limitations, in order to achieve longer distances. In this paper we employ bidirectional Raman amplification together with optical pulse coding to DPP–BOTDA, providing distributed sensing over long ranges together with submeter spatial resolution.

M. Taki • M. Soto • F. Di Pasquale
TeCIP Institute, Scuola Superiore Sant'Anna, Pisa, Italy

G. Bolognini (✉)
IMM Institute, Consiglio Nazionale delle Ricerche, Bologna, Italy
e-mail: bolognini@bo.imm.cnr.it

C. Di Natale et al. (eds.), *Sensors and Microsystems: Proceedings of the 17th National Conference, Brescia, Italy, 5-7 February 2013*, Lecture Notes in Electrical Engineering 268, DOI 10.1007/978-3-319-00684-0_48, © Springer International Publishing Switzerland 2014

2 Experimental Setup and Results

The experimental setup is shown in Fig. 1. The output of a distributed feedback (DFB) laser is split into two branches. In the probe (CW signal) branch, two sideband components are generated (suppressed carrier technique). In the pump (pulsed beam) branch, an erbium-doped fiber amplifier (EDFA) and a Mach–Zehnder modulator (MZM) driven by a waveform generator are used to amplify the DFB laser output and then to generate sequences of Simplex optical codes (127 bit) [4].

Bidirectional distributed Raman amplification [5] has been implemented by coupling two Raman pumps at 1,450 nm (in forward and backward directions) into a ~93 km single-mode fiber (SMF); we employed a depolarized fiber Raman laser (FRL) as forward-propagating pump and a low-RIN depolarized pump based on polarization-multiplexed Fabry–Perot (FP) lasers as backward-propagating pump. We designed the optimal input power levels for Raman pumps and signals employing an in-house developed numerical method [6] based on the solution of a coupled differential equation system. At the receiver, a circulator and a fiber Bragg grating filter have been used, together with a linear-gain EDFA preamplifying the received traces.

The position of another EDFA before the MZM for the Brillouin pump branch avoids unwanted distortions in the coded light stream. Automated feedback controls for both signal and pump MZMs ensure time stability and improved measurement repeatability. A polarization scrambler is used in the signal branch to provide signal light depolarization and to reduce polarization-dependent gain occurring in SBS.

Finally, a 400 MHz PIN photo-receiver and an analogue-to-digital converter (ADC) have been connected to a computer for trace acquisition. A differential pulse-width pair of 60/56 ns has been used, corresponding to a theoretical limit of spatial resolution equal to 40 cm.

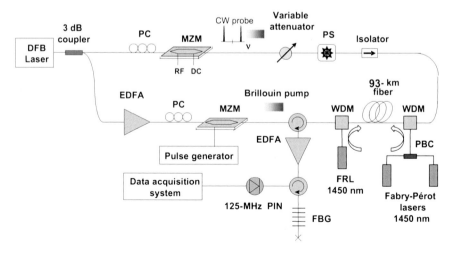

Fig. 1 Experimental setup

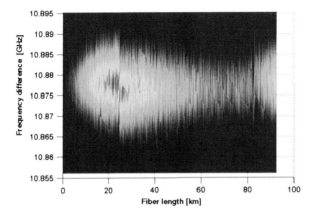

Fig. 2 Contour plot of Brillouin gain spectrum versus fiber length

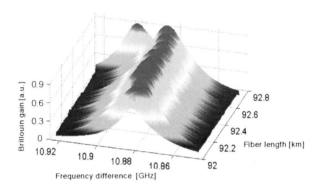

Fig. 3 Frequency spectrum of Brillouin gain versus fiber length

In order to acquire the whole Brillouin gain spectrum (BGS) through the fiber length, the CW probe frequency is tuned through an RF signal generator. For each frequency value, the coded traces for each pulse-width value are then decoded and the trace pairs are then subtracted as required in DPP scheme. The contour plot of BGS (top view) along the fiber length after DPP subtraction is shown in Fig. 2.

Considering the used power values and analyzing the spectral shape of the Brillouin gain throughout the fiber length, pump depletion effects can be considered as negligible. To verify the sensor performance, 10 m of fiber at ~92.7 km distance have been placed inside a temperature-controlled chamber (TCC) heated to 43 °C.

The BGS versus distance is shown in Fig. 3, showing the clear change in BGS peak near fiber end as a consequence to fiber heating. Temperature (or strain) estimations are carried out employing the Brillouin frequency shift (BFS) parameter, i.e., the frequency peak of measured BGS, which exhibits a linear behavior with temperature (or strain) variations [1]. For our experiment, Fig. 4 reports the BFS near the fiber end (92.7 km), also pointing out the attained spatial resolution in

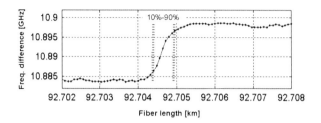

Fig. 4 Brillouin frequency shift versus fiber length (around 93 km)

correspondence to a temperature step, resulting to be ~50 cm (10–90 % response), pretty close to the theoretical limit of 40 cm. The temperature–strain resolution along the fiber length can be calculated from the standard deviation in BFS; in our setup the worst-temperature (strain) resolution has been estimated to be ~1.7 °C (~34 με).

As a conclusion, combining the use of a return-to-zero (RZ)-format Simplex coding together with optimized bidirectional Raman amplification allows for efficient distributed sensing based on Brillouin optical time-domain analysis (BOTDA) employing differential pulse-width pair (DPP) technique; in a long-distance experiment, we showed distributed measurements over ~93 km of SMF, achieving temperature (strain) resolution better than 1.7 °C (34 με) and a spatial resolution better than 50 cm throughout the fiber length.

References

1. Liang, H., Li, W., Linze, N., Chen, L., and Bao, X., "High-resolution DPP-BOTDA over 50 km LEAF using return-to-zero coded pulses," Opt. Lett. **35**(10), 1503–1505, (2010).
2. M. A. Soto, M. Taki, G. Bolognini, F. Di Pasquale, "Optimization of a DPP-BOTDA sensor with 25 cm spatial resolution over 60 km standard single-mode fiber using Simplex codes and optical pre-amplification," Opt. Exp. **20**(7), 6860–6869 (2012).
3. Y. Dong, X. Bao, W. Li, "Differential Brillouin gain for improving the temperature accuracy and spatial resolution in a long-distance distributed fiber sensor," Applied Opt., **48**(22), 4297–4301 (2009).
4. M. D. Jones, "Using Simplex Codes to Improve OTDR Sensitivity", IEEE Photon. Technol. Lett., **15**(7), 822–824 (1993).
5. S. Faralli, G. Bolognini, G. Sacchi, S. Sugliani, F. D. Pasquale, "Bidirectional higher order cascaded Raman amplification benefits for 10-Gb/s WDM unrepeated transmission systems," J. Lightwave Technol. **23**(8), 2427–2433 (2005).
6. G. Bolognini, F. Di Pasquale, "Transient effects in gain clamped discrete Raman amplifiers", IEEE Photon. Technol. Lett., **16**(1), 66–68 (2004).

Water-Jet Waveguide for Fluorescence Spectroscopy

G. Persichetti, G. Testa, and R. Bernini

1 Introduction

This paper reports an optofluidic water-jet waveguide used for laser-induced fluorescence spectroscopy. In this common analytical technique, the sensitivity depends on the ratio between the fluorescent signal and the background signal. The signal intensity is essentially related to the collection efficiency of the detection optics, whereas the background signal (noise) has two main contributions: the excitation source scattering and the fluorescence arising from non-analyte substances that the light source encounters along its path. In order to increase the signal-to-background ratio, recently, a great effort has been addressed towards the development of novel optical excitation/detection configuration [1–4]. In this paper an innovative configuration is proposed, where a high-speed water stream produced by means of a microchannel acts at the same time as the solution to analyse and the collecting optical waveguide.

Water-jet waveguides are sometimes used in laser cutting device [5], and very few examples in fluorescence spectroscopy applications have been proposed, however, without fully exploiting the great advantage of this device [6, 7]. Furthermore, the few configurations proposed in literature require bulky optic and fluidic device with careful alignment procedure. Our configuration allows to minimize the pump contribution in the detected signal and to avoid any problem related to non-analyte fluorescence arising, for instance, from the walls of the flow cell typically used in this spectroscopic technique. In addition the configuration proposed is auto-aligned and integrated as the water-jet waveguide is directly coupled with the receiving optical fibre which collects the fluorescence light to a low-cost minispectrophotometer.

G. Persichetti (✉) • G.Testa • R. Bernini
Institute for Electromagnetic Monitoring of the Environment (IREA),
National Research Council (CNR), Via Diocleziano, 328, Naples 80124, Italy
e-mail: persichetti.g@irea.cnr.it

C. Di Natale et al. (eds.), *Sensors and Microsystems: Proceedings of the 17th National Conference, Brescia, Italy, 5-7 February 2013*, Lecture Notes in Electrical Engineering 268, DOI 10.1007/978-3 319-00684-0_49, © Springer International Publishing Switzerland 2014

2   Liquid-Jet Sensor

Figure 1 shows a schematic of the device. The optofluidic sensor has one inlet, for the liquid analyte, connected to a small microchannel used to create the liquid stream that falls directly onto the collecting fibre optic. The exciting light impinges orthogonally to the water stream in order to minimize the pump component into the detected signal. The fluorescence photons, arising from the analyte, are captured by total internal reflection and transported along the liquid jet to a multimode optical fibre centred with analyte stream.

The fabrication processes have been performed by direct milling a 3 mm thick layer of polymethylmethacrylate (PMMA). A long channel (that acts as a microfluidic channel and as groove for the collecting optical fibre) has been milled using a milling tool with 254 ± 12.7 µm diameter. The fluidic inlet port has been realized by means of an 800 µm diameter drilling tool. Then, the substrate has been bonded to another layer of PMMA by means of a solvent-assisted bonding procedure. The following step has been to cut the chip by a micromilling machine in order to obtain an open region (24 mm wide and 10 mm high) where the jet waveguide flows allowing a length of the jet up to 10 mm. Subsequently, in order to excite the water stream from the backside, a 7 mm hole has been opened at the rear of the device. This exciting direction allows to minimize the scattering of the light refracted by the cylindrical liquid jet. This device layout ensures a perfect self-alignment between the collecting optical fibre and the liquid jet.

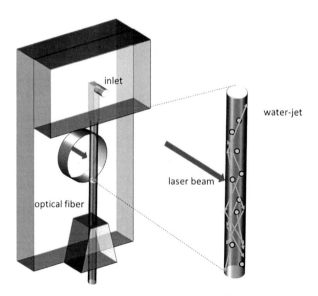

Fig. 1 Schematic of the optofluidic water-jet waveguide. The fluorescence arising from analytes in the water jet is confined in the jet waveguide and collected by a multimodal optical fibre

3   Experimental Results

For the device characterization, a collimated laser diode emitting at 640 nm has been used to excite aqueous solutions containing Cy5 dye concentration ranging over more than three orders of magnitude, from 8.67×10^{-10} M to 1.69×10^{-6} M. The corresponding calibration curve is reported in Fig. 2 where measurements have been obtained considering spectra between 646 nm and 850 nm (the observed fluorescence emission area). The measured limit of detection (LOD) has been 2.56 nM. This LOD value is very competitive also if compared to liquid core waveguides [8, 9] or considering innovative sensors devoted to similar applications [4]. This result is achieved thanks to the very optically smooth side walls due to the surface tension in liquid jet and the high fluorescence collection efficiency of a water stream waveguide. As the current-collecting optical fibre (NA = 0.22) and the minispectrometer (NA = 0.22) do not perfectly match the water cylinder cross section, a further increasing of the collection efficiency by direct coupling the falling jet waveguides to a photodetector by a high-numerical aperture fibre should be possible. In order to explore this solution, in a filter-free approach, the integral of the previously measured spectra has been evaluated on the whole detection range (350–1,000 nm). The obtained values (in Fig. 3) similar to the ones obtained by a silicon photodiode clearly attest that a filter-free detection is possible with a limit of detection of 6.11 nM.

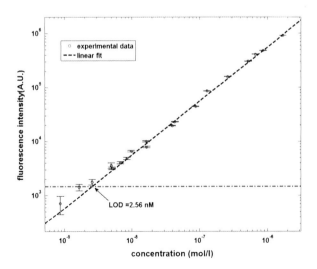

Fig. 2   Calibration curves for Cy5 solutions in the range $[8.67 \times 10^{-10} - 1.69 \times 10^{-6}]$ mol/l

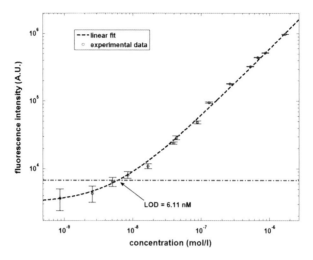

Fig. 3 Calibration curves of the devices in a filter-free detection configuration. The limit of detection is 6.11 nM

4 Conclusions

A water stream waveguide sensor for fluorescence spectroscopy of liquid solution has been developed. The waveguiding nature of a water stream improves the optical performances of the system avoiding usual drawbacks related to the flow cell presence or to the autofluorescence of the chip materials. A direct coupling with the collecting optical fibre allows a perfect integration with pigtail detectors. Fluorescence measurements of Cy5 solution had shown an LOD of 2.56 nM, and proof-of-concept of filter-free measurements has been demonstrated. The reported LOD, the high level of integration, and the compact size of the device are a clear indication that this sensor is potentially useful for environmental monitoring activity, providing online analysis of the concentration of various molecular species.

Acknowledgments The research leading to these results has received funding from the Italian Ministero dello Sviluppo Economico, under Grant Agreement "Industria 2015—New Technologies for the Made in Italy", No. MI0I 00223 (project ACQUASENSE).

References

1. X Fan, IM White, Optofluidic microsystems for chemical and biological analysis. Nature Photon. **5**, 591–597 (2011).
2. R. Bernini, N. Cennamo, A. Minardo, and L. Zeni. Planar Waveguides for Fluorescence-Based Biosensing: Optimization and Analysis. IEEE Sens. J. **6**, 1218-1226 (2006).

3. R. Irawan, C. M. Tay, S. C. Tjin and C. Y. Fu. Compact fluorescence detection using in-fiber microchannels-its potential for lab-on-a-chip applications. Lab Chip **6**, 1095–1098. (2006)
4. R. Gravina, G. Testa, and R. Bernini. Perfluorinated Plastic Optical Fiber Tapers for Evanescent Wave Sensing. Sensors **9**, 10423 -10433 (2009).
5. B Richerzhagen. Chip singulation process with a water-jet guided laser. Solid State Technol., **44** 25-28 (2001).
6. S. J. Mackenzie, J. Hodgkinson, M. Johnson, and J. P. Dakin. A falling analyte stream configuration for spectroscopic analysis of aqueous solutions. Europto EnviroSense, Munich, Germany, 16-20 June 1997 (1997).
7. J. Inczedy, T. Lengyel, and A.M. Ure. Compendium of analytical nomenclature. The orange book, 3rd edn., (Blackwell, Oxford 1998).
8. S. Smolka, M. Barth, and O. Benson. Highly efficient fluorescence sensing with hollow core photonic crystal fibers. Opt. Express 15, 12783 -12791 (2007).
9. W. P. Risk, H.-C. Kim, R. D. Miller, H. Temkin and S. Gangopadhyay. Optical waveguides with an aqueous core and a low-index nanoporous cladding. Opt. Express, 12, 6446 -6455 (2004).

Part V
Microsystems

Moore's II Law and Microsystems Manufacturing

U. Mastromatteo

1 Reducing the Entropy of a System

Moore and Rock came to the conclusions summarized in their statements about the growth of costs in the manufacture of microchips increasingly complex [1], with observations that could be defined in the final balance, probably without an objective analysis of the processes involved in this type of technology, which instead would put highlight the intrinsic reasons behind these observations. In fact, what is observed is that the manufacture of artificial complex systems, such as microchips, is highly energy demanding because it addresses the problems related to the reduction of entropy, which are much more demanding from the point of view of efficient energy use, compared to what one experiences in the transformation of energy from one form to another. The lowering of entropy in a system by creating order starting from material totally disordered is extremely onerous due to the ergodicity of this process, namely, the need to prevent the system you want to order from the occurrence of different configurations coming from the interaction with the process environment, but different from what was requested by the purposive project (which assigns to the complex artefact, the function provided by a specific design).

1.1 Processes for the Creation of Ordered Structures

The working environment and the raw materials used to make microchips are of course subject to specific criteria to avoid introducing, during the processing, geometric configurations and material properties not provided by the project.

U. Mastromatteo (✉)
IMS R&D non conventional technologies, STMicroelectronics,
Via Tolomeo, 1, 20010 Cornaredo, MI, Italy
e-mail: ubaldo.mastromatteo@st.com

C. Di Natale et al. (eds.), *Sensors and Microsystems: Proceedings of the 17th National Conference, Brescia, Italy, 5-7 February 2013*, Lecture Notes in Electrical Engineering 268, DOI 10.1007/978-3-319-00684-0_50, © Springer International Publishing Switzerland 2014

Fig. 1 A sketch of a portion of a microchip architecture and the manufacturing environment

So, let's remind what the main characteristics of microelectronics material are: monocrystalline silicon substrate pure with the surface perfectly planar, ultrapure process gas devoid of any traces of moisture, 18 Mohm resistivity demineralized H_2O in wet chemical benches; and materials and gases for deposition and etching chemical reactions, all with the maximum degree of purity and filtered by the absence of particles and many other precautions to avoid any mismatch with the design.

All this plus a very clean working area determines the ergodicity of the process because the elimination of all the possibilities to create configurations different from the designed one is equivalent to going through all the possible statistical configurations.

Now we ask the question: why, for the microchip system, less complex than a eukaryotic cell, is the need for conditions so special? The response is the following: not only the ingredients but also the environment where it is worked an artefact ordered according to a predefined scheme, in which an entropy decrease is produced (transition from disordered material in ordinate), must have an entropy lower than that of the manufactured artefact; the increase of entropy expected from the second thermodynamics principle must therefore be absorbed from the external environment.

The above statements are confirmed looking in Fig. 1 at the picture of a clean manufacturing area next to the microchip sketch. Hence, to achieve similar complex architectures, by overlapping layers having particular geometric figures defined by photolithographic technique [2], it is necessary to operate in an extremely clean environment (class <1 in the processing area); this is because the mode of operation for the construction of these complex artefacts is primarily designed to prevent all configurations other than those foreseen by the project, and of course the number of these configurations grows more and more as the complexity of the system increases with the reduction of the minimum dimensions of the components integrated in the chip.

Looking again in Fig. 1 at the typical environment for microchip processing, it is intuitive to say that its own entropy must be lower than that you want to get in the objects that are there processed, and as already mentioned, to maintain the entropy of the processing environment low, a higher increase in entropy must be transferred into the external environment of the factory.

Fig. 2 *Left*: Wheat seeds facility. *Right*: Wheat facility at work

1.2 How Living Organisms Create Order

Conversely, if we look at the environment where they grow in living organisms, such as a wheat field before sowing (Fig. 2, left), it is quite clear that the "factory" in this case has much higher entropy than the same wheat field with the living organism that grows up to become a highly complex extremely ordered system (Fig. 2, right). Understanding the reason for this apparent difference can certainly help to make choices that allow an optimization of the processes for the construction of complex artefacts, in order to increase efficiency.

In any case, from the point of view of the mechanisms that make divergent the processes to achieve artificial complexity, compared to those that regulate the growth of living organisms (the wheat in our example), we observe that while in artificial complexity, the processes are addressed to prevent all the possible configurations and select the one planned by the project; in contrast, the instructions present inside living organisms are able to select directly from the environment what they need to achieve the planned configuration, excluding any other without the need to prevent them (non-ergodic system). From the practical point of view, the formulas that describe this situation are the following [3]:

In an ergodic system the entropy decrease for ordering the matter is [3–5]

$$\Delta S = \Delta S_c \ln(w) < 0 \qquad (1)$$

where ΔS_c is the entropy variation in the configured layers; consequently, the work needed to create order can be estimated with

$$L = -T\Delta S_c \ln(w) \qquad (2)$$

Instead, the non-ergodic process of living organisms that grow, despite having a decrease in entropy expressible with the same criterion of artificial complexity (although with a "w" much greater in a comparable size system), needs much less work because the system selects directly only the configuration provided by the internal codes, and therefore,

$$L = -T\Delta S \tag{3}$$

where only the negative entropy change ΔS of the biochemical reactions involved in the growth of the organism has to be taken into account [5].

1.3 Suggested Strategies for Complex Systems

In summary, the strategy suggested by the efficiency of the biological systems in their growth phase is to proceed by parallel processing. In fact, the grain of wheat in the ear generation process proceeds with the generation of more cells and their differentiation to build the parts of the system ranging from the leaves for energy harvesting, chemical synthesis of carbohydrates, proteins and nucleic acids to arrive to the replication of multiple cells identical to the original one.

A similar strategy was adopted in microelectronics for system-in-package devices. In terms of complexity, the traditional strategy of sequential integration of two parts of a process would lead to a number of configurations given by "w = w1 * w2", while proceeding in parallel we will have "w = w1 + w2", whose logarithm value, linked to the reduction of entropy, is less than in the first case.

2 Conclusions

As a conclusion we can mention what was claimed by Moore himself to contain the predictions of his second law: "It would be much cheaper to build large-scale systems from minor functions, interconnected separately. The availability of a variety of applications, together with the design and methods of construction, would allow companies to manage output more rapidly and at lower cost".

References

1. http://it.wikipedia.org/wiki/File:Seconda_legge_di_moore.jpg;
2. U. Mastromatteo, "Complexity management in manufacturing microsystems: remarks on artificial and natural system comparison."; Springer - Sensors and Microsystems, AISEM 2010 proceedings, Vol. 91, pp. 265-269
3. Charles H. Bennet, Diavoletti, Macchine e il secondo principio. Le Scienze – quaderni, n. 85, pp. 8-14, 1995.
4. U. Mastromatteo, P. Pasquinelli, A. Giorgetti. "Thermodynamics, Information and Complexity, in Artificial and Living Systems", International Journal of Ecodynamics. Vol. 2(1), 39-47, 2007.
5. U. Mastromatteo, "II Thermodynamics principle and II Moore's law in a comparison between living and complex artificial systems", Journal of Agricultural Science and Applications, 2013 Vol. 2. No. 1 39-47.

Development of a Low Cost Planar Micro Thermoelectric Generator

S. Pelegrini, A. Adami, C. Collini, P. Conci, L. Lorenzelli, and A.A. Pasa

1 Introduction

Micro thermoelectric generators (µTEGs), in general, can power all low-power electronic devices in situations where battery power is limited [1]. TEGs started to be developed in 1960 to deliver power to space vehicles [2]. Nowadays, there are many reports on devices powered by ambient energy such as battery-free wireless sensor [3] and biomedical and other devices powered by human body energy [4, 5]. This work talks about the microfabrication of a µTEG based on planar technology using constantan and copper deposited by ECD as thermocouples [6]. The constantan is a copper and nickel alloy (CuNi), which was obtained through the low-cost process using electrochemical deposition (ECD) on silicon substrate. The Seebeck coefficient of the constantan and copper are estimated at $-37.9\,\mu V/K$ and $+1.7\,\mu V/K$, respectively, at temperature of 273 K [7].

2 Experimental

2.1 Desing and Microfabrication

The µTEG consists of an array of thermocouples connected in series, measuring the difference between the voltages that appear in the external region (hot) and internal one (cold), as shown Fig. 1f. This difference in temperature between the exterior

S. Pelegrini · A. Adami (✉) · C. Collini · P. Conci · L. Lorenzelli
Fondazione Bruno Kessler, Trento, Italy
e-mail: andadami@fbk.eu

A.A. Pasa
Universidade Federal de Santa Catarina, Florianópolis, SC, Brazil

C. Di Natale et al. (eds.), *Sensors and Microsystems: Proceedings of the 17th National Conference, Brescia, Italy, 5-7 February 2013*, Lecture Notes in Electrical Engineering 268, DOI 10.1007/978-3-319-00684-0_51, © Springer International Publishing Switzerland 2014

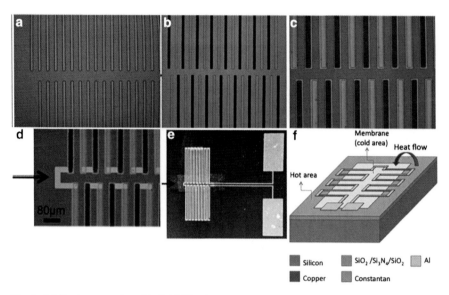

Fig. 1 (**a**) Resist patterning, (**b**) CuNi/Cu electrodeposition, (**c**) protection mask with resist with seed layer removal, (**d**) Al deposition by evaporation, and (**e**) TMAH micromachining

and the interior of the device generates a voltage that can be measured, i.e., the Seebeck voltage. This configuration was selected because it can be manufactured into flexible substrate, large area, and the materials are produced with low cost.

The starting material was a p-type Si wafer (100), where a layered structure of $Si_2O/Si_3N_4/SiO_2$ was grown in both sides of the sample for the future fabrication of the membrane. The first 500 nm thick oxide layer was grown by thermal wet oxidation, which was followed by 150 nm silicon nitride (Si_3N_4) grown by low-pressure chemical vapor deposition (LPCVD), which was capped by an oxide with 500 nm grown again by thermal wet oxidation, as displayed. The seed layer of Cr/Au is sequentially deposited on the polished side of the wafer by e-beam evaporation on top of the $SiO_2/Si_3N_4/SiO_2$ multilayer.

The photoresist AZ4562 (6,000 nm thick) was spin coated on the surface of the seed layer and exposed to light after the alignment of the mask (Fig. 1a). The copper regions with a length of 2 mm and width of 40 μm were open. In the next steps, Cu is electrochemically deposited and AZ4562 layer is removed by solvent and O_2 plasma (Fig. 1b). The same steps are followed for constantan plating and finally the seed layer is removed (Fig. 1c). Al was evaporated on the surface for pads and interconnections and patterned by lift-off as illustrated in Fig 1d. Figure 1e shows the TMAH (tetramethylammonium hydroxide) bulk micromachining of the substrate living the inner part of the insulating multilayer suspended.

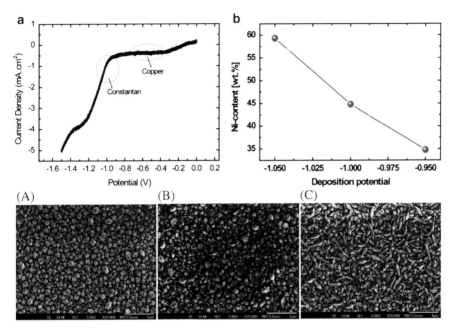

Fig. 2 (**a**) Voltammogram on Au/Cr/SiO2 surface at a scanning rate of 20 mV/s. (**b**) EDX and SEM analysis for Ni concentration of electrodeposited CuNi alloys as function of the deposition potential: (A) −0.95 V, (B) −1.00 V and (C) −1.05 V

2.2 Electrochemical Deposition

The aqueous electrolyte used for the deposition of constantan alloy was 0.171 mol/l $NiSO_4$, 0.019 mol/l $CuSO_4$, and 0.190 mol/l $C_6H_5Na_3O_7$ [8–10], at room temperature and prepared with deionized water with a resistivity of 18 MΩ cm. The Fig. 2 represents the voltammogram, EDX, and SEM analysis for Ni concentration of CuNi alloys ECD as function of the deposition potential: −0.95 V, −1.00 V and −1.05 V. The deposition with larger potential in module presents the higher Ni concentration. A rounded shape of the grains was observed for alloys composed predominantly of copper (Fig. 2A, B), while the grains have elongated shape when nickel prevails (Fig. 2C).

3 Thermoelectric Characterizations

Open-circuit potential and short-circuit current were measured at different flow velocity by using a package to implement a flow channel with cross-sectional area equal to 0.1 cm² and a mass-flow controller to apply a controlled nitrogen flow.

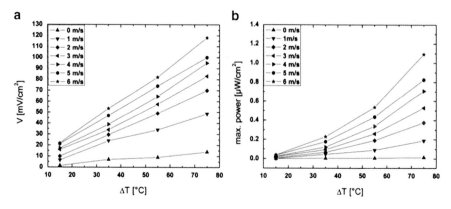

Fig. 3 (**a**) Show the open-circuit voltage (mV/cm²) and (**b**) output power (µW/cm²), both versus temperature difference between the hot plate and room temperature for N_2 velocities from 2 up to 6 m/s and 180 thermocouples

The electrical measurements demonstrated proper ohmic contact and reasonable internal resistance for these devices, which is around 20 Ω per couple and 3 KΩ for 180 couples, which are good values for internal resistance when compared with other TEGs. The maximum generated potential is 118 V/cm² and thermoelectric efficiency (performance) of 1.1 µW/cm² is reached with an airflow velocity of 6 m/s, (Fig. 3), since convection at hot side is very important to increase both output power and potential. With further optimization of line width, number of couples, and chip area, it is possible to improve the specific output power per area; moreover, with the selected technology, it is possible to extend the device to large areas by using substrates like Kapton films and low-cost chemical/electrochemical processing.

4 Conclusions

The internal resistance is not very high for these devices; the value is around 3 KΩ for 180 thermocouples. The maximum generated potential at 6 m/s airflow velocity is 118 mV/cm² and the output power is 1.1 µW/cm². The performance is according to the literature for other TEGs [11].

References

1. W. Glatz, E. Schowyter, L. Durrer, C. Hierold. Bi_2Te_3—Based Flexible Micro Thermoelectric Generator With Optimized Desing, Journal of Micro electromechanical Systems, 18(3), 763–772 (2009)
2. D. M. Rowe (2006) Review Thermoelectric Waste Heat Recovery as a Renewable Energy Source. *International Journal of Innovations in Energy Systems and Power*, 1(1) 13–23.

3. David Salerno, Ultralow Voltage Energy Harvester Uses Thermoelectric Generator for Battery-Free Wireless Sensors, *Journal of Analog Innovation*, Vol. **20**, no. 3 (2010).

4. V. Leonov and Ruud J. M. Vullers, Wearable electronics self-powered by using human body heat: The state of the art and the perspective, *J. Renewable Sustainable Energy*, **1**, 062701 (2009)

5. C.-Y. Sue, N.- C. Tsai, Human powered MEMS-based energy harvest devices, *Applied Energy*, **93** (2012) 390–40.

6. D. G. Delatorre, M. L. Sartorelli, A. Q. Schervenski, S. Güths and A. A. Pasa, J. Appl. Phys, vol. 93, n 10, pp. 6154 – 6158. (2003)

7. M. Genix, P. Vairac, B. Cretin, Local temperature surface measurement with intrinsic thermo-couple, International Journal of Thermal Sciences, vol. 48 1679 – 1682, (2009)

8. T. A. Green, A. E. Russell, and S. Roy, The Development of a Stable Citrate Electrolyte for the Electrodeposition of Copper – Nickel Alloys. *J. Electrochem. Soc.*, Vol. **145**, n 3, March (1998) 875 – 881.

9. M. L. Sartorelli, A. Q. Schervenski, R. G. Delatorre, P. Klauss, A. M. Maliska and A. A. Pasa, Cu-Ni thin films electrodeposited on Si: Composition and current efficiency. Phys. Stat. Sol. A 187, (2001) 91–95.

10. K. Iraj, K. Elham, F. Mansoor, Fabrication and nanostructure study of ultra thin electroplating constantan film on GaAs as a thermopower sensor. Journal of Physics: Conference Series, 100 (2008) 052025.

11. J. Xie, C. Lee, H. Feng. "Design, fabrication, and characterization of CMOS MEMS based thermoelectric power generators". *J MEMS*, (2010) **19** (2) 317–24.

Dielectric Layers for MEMS Deposited at Room Temperature by HMDSO-PECVD

P. Colombi, A. Borgese, M. Ferrari, and V. Ferrari

1 Introduction

Silicon dioxide coatings can be deposited on the surface of different materials by a plasma-enhanced chemical vapour deposition (PECVD) process using metalorganic compounds such as tetraethyl orthosilicate (TEOS) or hexamethylsoxane (HMDSO) as precursors in a safe, cost-effective and green process.

Even though SiO_2 coatings can be also obtained by using silane gas (SiH_4) as silicon precursor, the use of an organosilicon precursor tends to be preferred, especially on an industrial scale because of its better step coverage. Moreover, silane is a toxic and explosive gas. Recently, Pecora et al. [1] compared structural and electrical properties of SiO_2 films obtained through different deposition techniques demonstrating that very high quality SiO_2 films can be obtained by TEOS-PECVD and that SiO_2 layers deposited by TEOS-PECVD exhibit electrical properties similar to those of layers grown by silane ECR-PECVD. On the other hand, plasma polymerized SiO_2 layers deposited by HMDSO PECVD have attracted lots of interest due to their outstanding chemical and electrochemical properties [2]. Moreover, HMDSO/ O_2 plasma offers the possibility to tune the carbon content (and thus film properties) of the film by changing the process conditions [3].

Since typical deposition temperatures are usually lower than 100 °C, the process is not limited to metal and ceramic materials but can be applied to any kind of materials. Moreover, in microelectronics, low processing temperature for the dielectric or passivating layers is particularly suitable since it allows to avoid segregation effects and dopant redistribution.

P. Colombi (✉) • A. Borgese
CSMT Gestione S.c.a.r.l, Via Branze 45, 25123 Brescia, Italy
e-mail: p.colombi@csmt.it

M. Ferrari • V. Ferrari
Department of Information Engineering, University of Brescia,
Via Branze 38, 25123 Brescia, Italy

C. Di Natale et al. (eds.), *Sensors and Microsystems: Proceedings of the 17th National Conference, Brescia, Italy, 5-7 February 2013*, Lecture Notes in Electrical Engineering 268, DOI 10.1007/978-3-319-00684-0_52, © Springer International Publishing Switzerland 2014

2 Experimental

SiO$_2$ films were deposited on both martensitic stainless steel (15-5PH) polished surfaces and metallic interlayers (i.e. chromium, titanium) deposited by DC magnetron sputtering. PECVD processes were performed in a capacitive coupled parallel plate reactor using an HMDSO/oxygen plasmas ignited at mid-low frequency (50 kHz) at 375 W power. Two different configurations were tested, the first having the sample holder grounded and the second having the sample holder connected to RF generator. The deposition parameters for the three most relevant samples are summarized in Table 1.

Film thickness was assessed by stylus profilometry onto step structures obtained by mechanical masks. Chemical and morphological characterization carried out by SEM/EDS adhesion to the substrate was determined by scratch tests using a (400 µm diameter) diamond Rockwell indenter and applying a linearly increasing load from 1 to 15 N while translating at a constant velocity of 10 mm/min. Electrical insulation properties were performed by DC and AC leakage current tests. For this purpose circular gold contacts with an area of about 20 mm^2 were deposited onto SiO$_2$ surface by DC magnetron sputtering and by using mechanical masks for patterning. DC tests were performed by applying the voltage from external source and measuring the current by means of Keithley 6517 electrometer. AC leakage test was performed by generating a 50 Hz sinusoidal signal by means of Agilent 33220A and amplified by a power amplifier Bruel & kjaer 2706 and a high-voltage transformer. A Tektronix DM2510G was employed to monitor the rms voltage delivered to the dielectric film, while an Agilent 34401A was used to measure the voltage on a 1 kΩ shunt resistance in order to calculate the leakage current. Ageing test was performed in humid thermal chamber by applying 50 thermal cycles between −25 and +115 °C. Each cycle duration was 30 min. Adhesion in the presence of substrate bending was assessed by applying pressure cycles between 1 and 500 bar onto specially designed diaphragm structures. The calculated maximum strain during each cycle is about 2,000–2,200 µε.

Table 1 Main deposition parameters

Parameter	Sample S1	Sample S2	Sample S3
Deposition temperature	50 °C	50 °C	50 °C
HMDSO/O2 ratio	1:10	1:5	1:10
Deposition time	90 min	90 min	45 min
Sample holder	Grounded	Grounded	Connected to RF
RF power	375 W	375 W	375 W

3   Results and Discussion

The lower SiO_2 film growth rate was obtained for HMDSO/O2 ratio 1:10 (GR = 11.4 nm/min). It increases by both increasing the amount of metalorganic precursor (17.2 nm/min) and connecting the substrate to the RF source. Film thickness ranged between 3.8 and 6.2 µm; SEM micrograph (polished cross-section view) evidenced dense pinhole-free and uniform microstructure (Fig. 1).

Scratch test shows a good adhesion of SiO_2 film directly grown onto steel substrate. No increase in adhesion was obtained by introducing Ti and Cr metallic interlayers. The main result of film characterization is reported in Table 2.

DC electrical tests show linear I-V curve of up to 200 V when the test was interrupted. Electrical resistance was of the order of 10^{12}–10^{13} Ω, corresponding to a resistivity in the order of 10^{15} Ω cm. Test was performed by applying voltage in the 0–500 V_{rms} range with step of 25 or 50 V_{rms}. Figure 2 is an example of AC test for Sample S1 performed before and after thermal ageing and bending cycles. Noticeably, the leakage current was below 25 µA for the maximum applied voltage.

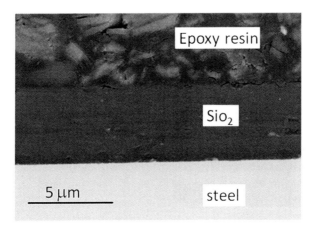

Fig. 1 SEM micrograph (polished cross-section view) of a steel component protected by a thick SiOx layer

Table 2 Main characterization results

Parameter	Sample S1	Sample S2	Sample S3
Thickness	4.1	6.2	3.8
Growth rate	11.4 nm/min	17.2 nm/min	21.1 nm/min
Resistance	$1.93 \times 10^{+12}$ Ω	$1.60 \times 10^{+12}$ Ω	$1.62 \times 10^{+12}$ Ω
Resistivity	$4.71 \times 10^{15} \Omega$ cm	$2.58 \times 10^{15} \Omega$ cm	$4.26 \times 10^{15} \Omega$ cm
AC leakage current at 500 Vrms	20.4 µA	27.6 µA	30.5 µA
Resistance to thermal cycles	Excellent	Excellent	Excellent
Resistance to bending cycles	Good	Good	Poor

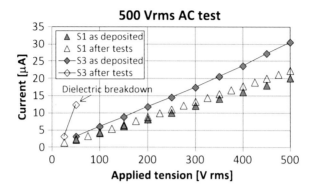

Fig. 2 I–V plot for AC electrical test performed onto SiO₂ layer (samples S1 and S3) before and after thermal and bending cycles

Few breakdown events were registered on some samples (not reported here) and were correlated to the presence of macro/micro defects in the layer.

For all the samples both dielectric and mechanical properties were maintained after thermal ageing. Cyclic bending test caused a reduction of dielectric properties due to the formation of microscopic cracks on the films grown at higher GR (S3), while we assessed that mechanical and electrical properties of samples grown at low GR were not affected by bending.

4   Conclusion

The results allow us to conclude that HMDSO-PECVD SiO_2 films feature outstanding dielectric properties (compared to that silane-CVD SiO_2). High conformality and low-defect densities, together with optimal adhesion to steel and process scalability, make it a good candidate for a wide application to industrial process. Furthermore, our results conclude that electrical properties of SiO_2 remain almost unaltered after thermal ageing and bending of up to 2,000 µε, making them suitable for a variety of MEMS applications.

References

1. E. Vassallo, A. Cremona, L. Laguardia, E. Mesto "Preparation of plasma-polymerized SiOx-like thin films from a mixture of hexamethyldisiloxane and oxygen to improve the corrosion behaviour", Surface & Coatings Technology 200 3035–3040(2006)
2. A. Pecora, L. Maiolo, G. Fortunato, C. Caligiore "A comparative analysis of silicon dioxide films deposited by ECR-PECVD, TEOS-PECVD and Vapox-APCVD", Journal of Non-Crystalline Solids **352** 1430–1433 (2006)
3. A. Patelli et. al "SiOx-Based Multilayer Barrier Coatings Produced by a Single PECVD Process" Plasma Process Polym, **6**, S665–S670 (2009)

Label-Free Detection of Specific RNA Sequences by a DNA-Based CMOS BioMEMS

Lorena Tedeschi, Claudio Domenici, Vincenzo Russino,
Andrea Nannini, and Francesco Pieri

1 Introduction

Resonant mass sensing is a well-established measurement method, widely used in biosensing because of the high accuracy of frequency measurements and the very small mass that can be resolved. As the mass sensitivity is essentially dependent on the ratio between the added mass and the original resonator mass, a common approach to increase the mass sensitivity is downscaling of the device. MEMS (microelectromechanical system) technologies come naturally into play to achieve such downscaling. The microbalance principle, traditionally used in the quartz crystal microbalances, has thus been transferred to MEMS resonators, to exploit the advantages of MEMS technology: small dimensions, batch fabrication, and enhanced sensitivity.

The modification of standard complementary metal-oxide semiconductor (CMOS) technologies allows the co-fabrication on the same silicon chip of MEMS components and the driving and conditioning circuitry, making the design of smart sensors possible without significant increase in cost.

The use of a CMOS-based technology allows the integration of on-chip electronics for sensor driving and signal conditioning, whereas fully electric operation allows direct interfacing to the same electronics.

In our biosensors, the microbalance principle is transferred to MEMS domain by using a CMOS-compatible bulk MEMS technology, leading to the fabrication of a magnetic torsional resonator, with electrical actuation and detection.

L. Tedeschi (✉) • C. Domenici
Istituto di Fisiologia Clinica del CNR (IFC-CNR), Via Moruzzi 1, 56124 Pisa, Italy
e-mail: tedeschi@ifc.cnr.it

V. Russino • A. Nannini • F. Pieri
Dipartimento di Ingegneria dell'Informazione, Università di Pisa,
Via G. Caruso 16, Pisa 56122, Italy

C. Di Natale et al. (eds.), *Sensors and Microsystems: Proceedings of the 17th National Conference, Brescia, Italy, 5-7 February 2013*, Lecture Notes in Electrical Engineering 268, DOI 10.1007/978-3-319-00684-0_53, © Springer International Publishing Switzerland 2014

To obtain a functional biosensor, important issues had to be addressed, spanning from the design and fabrication of microresonators to the surface functionalization and immobilization of bioreceptor molecules compatible with on-chip MEMS and electronics, to the design of optimized and selective biorecognition elements.

2 Resonator Design

The MEMS resonators are based on inductive actuation and detection, and the sensing element is based on a rectangular silicon dioxide plate, suspended over a cavity on the silicon substrate by two beams, acting as torsional springs. Two inductors for electrical driving and sensing are embedded in the plate (Fig. 1).

Since the performance of the sensor is strongly dependent on the size of the plate and springs and from the thickness of the structure, resonators with different layouts and sizes (with different theoretical/simulated resonance frequencies, ranging from 25 to 265 kHz) were designed and tested. The sensitivity is directly proportional to the surface-to-volume ratio of the resonating mass and can be improved by reducing the average thickness of the structural layer. To increase sensitivity, addition of thinner lateral sections (wings) to the plate was also tested to increase the active surface with a negligible increase in mass. The sensors were designed in a standard CMOS technology, and the resonator die includes two general-purpose operational amplifiers to be used to validate the CMOS compatibility.

The MEMS/CMOS approach can be followed for biosensors as long as the specific problems related to the bioactivation of the sensor surface and its compatibility with on-chip MEMS and electronic components are taken into account.

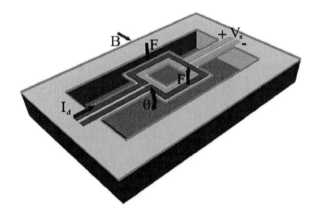

Fig. 1 Resonator structure

3   Experimental and Results

The MEMS resonators are released from the silicon substrate by buffered tetramethylammonium hydroxide (TMAH) etch, following a standard bulk MEMS process described elsewhere [1]. The released dies were subsequently mounted in a standard CLCC68 ceramic package and bonded to the package pins, concluding the MEMS post-processing phase. For the following bioactivation step, they were cleaned in a mild ammonia-based hydroxylation solution and silanized [2] through drop coating with a water-based APTES (aminopropyltriethoxysilane) solution, as preliminary step towards the deposition of a bioactive layer.

The first set of resonators was characterized in terms of mass response by loading them with gold nanoparticles of controlled diameter, to allow independent evaluation of the actual added mass by scanning electron microscope imaging. The frequency shifts were compared to the measured values of the added mass density to obtain an estimate of sensor sensitivity, defined as the ratio between the relative frequency shift and the added mass surface density. The measured values (Table 1) for our resonators, ranging from around 80 to 180 m^2 kg^{-1}, are far higher than the typical values for macroscopic QCM's but also similar or higher with respect to other MEMS resonant sensors [3].

Single droplets of mild reagents (drop coating) were used as an alternative to dip coating (immersion) of the sample, which is not compatible with mechanically sensitive MEMS components and electronic circuits and/or with the package.

By immobilizing a suitable DNA-based biorecognition element, a label-free biosensor for the detection of a specific human messenger RNA was developed.

The sensor surface was activated and an ssDNA probe designed to bind the mRNA sequence of human MGMT (methylguanine-DNA methyltransferase, a suicide enzyme involved in DNA repair). A 25-mer probe (MW 8 kDa), complementary to unique and exposed regions of the target mRNA (835 mer, 275 kDa), designed according to thermodynamics-based algorithms, was immobilized on the resonators surface.

Oligonucleotide probes to be used as bioreceptors were designed by using a previously developed computational method [4] whose selection criteria are based on the accessibility of target region to probe, on the stability of probe-target duplex and on the uniqueness of selected targets over all known expressed sequences from a genome data base.

Table 1 Resonance frequency before (f_0) and after (f_D) exposure to the analyte, quality factor (Q), and sensitivity (S), for four different resonator geometries

Device	f_0 (Hz)	f_D (Hz)	Q	Sensitivity (m^2 kg^{-1})
400-M3	31,749	31,612	157	111.8
200-M3	94,581	94,291	286	83.7
200-M3-W	74,701	74,167	116	156.3
200-M3-W	75,854	75,354	124	176.7

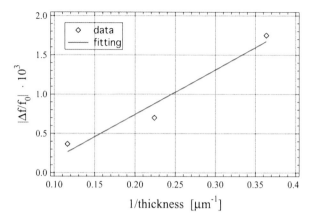

Fig. 2 Relative shift of the resonance frequency after exposure to the analyte from three different resonator geometries

After the computational selection, the probe sequences (25-mer oligonucleotides) were synthesized and chemical modifications (a thiolated tail) were introduced to allow the immobilization of probes on the resonant surface.

The functionalized resonators with immobilized probes were then characterized in terms of resonance frequency and other parameters and subsequently exposed to a solution containing the target mRNA. After the hybridization the resonators were characterized again and the frequency shift caused by the binding was evaluated by resonators different in thickness and size, showing a good concentration/shift relationship.

Plotting the measured relative frequency shift due to the probe-target interaction versus the inverse thickness of three different resonators (Fig. 2) shows the thickness/sensitivity relationship was experimentally verified with a good fit.

These results open interesting perspectives for the label-free detection of several molecular species, by using oligonucleotide-based probes (aptamers, decoys, or ssDNA/RNA/LNA).

Acknowledgments This work was partly financed by Italian Ministry of Education, University and Research (MIUR) through PRIN grant no. 2008SXRHWL.

References

1. D. Paci, F. Pieri, P. Toscano, A. Nannini. A CMOS-compatible, magnetically actuated resonator for mass sensing applications. Sens. Act. A **129** (1), 10-17 (2008).
2. S. Lenci, L. Tedeschi, F. Pieri, C. Domenici. UV lithography-based protein patterning on silicon: towards the integration of bioactive surfaces and CMOS electronics, Appl. Surf. Sci. **257** (20), 8413-8419, 2011.
3. W. Pang, H. Zhao, E.S. Kim, H. Zhang, H. Yu, X. Hu, Piezoelectric microelectromechanical resonant sensors for chemical and biological detection. Lab Chip **12** (1), 29-44 (2012).
4. L. Tedeschi, A. Mercatanti, C. Domenici, L. Citti. Design, preparation and testing of suitable probe-receptors for RNA biosensing. Bioelectrochemistry **67** (2), 171-179 (2005).

Thin Film Device for Background Photocurrent Rejection in Biomolecular Analysis Systems

D. Caputo, G. De Cesare, R. Scipinotti, and A. Nascetti

1   Introduction

Lab-on-chip (LoC) device [1] is an example of system where several laboratory functions are integrated onto a single substrate, yielding a sensor-like system requiring minimal quantities of biological samples with a fast response time and high stability. Miniaturization of the system is obtained by micro fluidic structures, while the detection, in most cases, is done off-chip.

On-chip optical detection is still a challenge for improving sensitivity and compactness. One of the most promising materials to this aim is amorphous silicon (a-Si:H) and its alloy. The low deposition temperature (below 250 °C) and its physical characteristics prompt the use of this material in different device such as solar cells [2], electronic switching [4], strain sensors [3], and photosensors [5]. The use of thin film a-Si:H photosensors for the detection of biomolecules has been already developed by different research groups [6]. In particular, detection of biomolecules by optical absorbance measurements in the UV range [7] or by measuring the analyte fluorescence [8] has already been demonstrated. In these experiments, a very low current variation (in the order of picoamps) had to be measured with a background current of several orders of magnitude higher. In this case a trade-off between effective dynamic range and resolution has to be considered. Differential measurement is extensively used to reject large common signals and to amplify only their difference. Here, we present an amorphous silicon balanced photosensor structure, integrated with a microfluidic network to perform on-chip detection with high dynamic range in biomedical applications.

D. Caputo • G. De Cesare (✉) • R. Scipinotti • A. Nascetti
"Sapienza" University of Rome, Via Eudossiana 18, Rome 00184, Italy
e-mail: decesare@die.uniroma1.it

C. Di Natale et al. (eds.), *Sensors and Microsystems: Proceedings of the 17th National*   281
Conference, Brescia, Italy, 5-7 February 2013, Lecture Notes in Electrical Engineering 268,
DOI 10.1007/978-3-319-00684-0_54, © Springer International Publishing Switzerland 2014

2   Amorphous Silicon Balanced Photodiode

In Fig. 1a the schematic structure of the amorphous silicon balanced photosensor is
reported. The device is constituted by two a-Si:H/a-SiC:H n-i-p stacked junctions.
A network of metal lines allows the series configuration of the two diodes and the
in/out connections. Each sensor is biased at the same reverse voltage, while the
output signal is the difference between the two diode current, which is amplified by
a transimpedance circuit (Fig. 1b).

Thanks to the differential current measurement, the structure is able to reveal
very small variations of photocurrent in a large background current signal. In addition, the differential approach reduces the effect of the common mode signal due to
temperature variations and/or instability of light source.

We have fabricated several balanced structures for detection with different geometries utilizing a four mask-step process:

1. Vacuum evaporation of a three-layer metal stack Cr/Al/Cr and its patterning by
   wet etching for the bottom electrode of the diodes.
2. PECVD deposition of the a-Si:H layers and evaporation of the top metal electrodes, mesa patterning by wet, and reactive ion etchings.
3. Deposition by spin coating of a 5 µm thick SU-8 passivation layer and opening
   of the window over the diodes and of the via holes for the electrical
   connections.
4. Sputtering of titanium/tungsten layers and pattering of the series connection of
   the two diodes, the connection of the diode electrodes to the contact pad area,
   and the metal top grid of the two devices.

Five balanced structures are fabricated on a glass substrate (Fig. 2a). Each of
them has three electrical contacts toward the edge of the substrate for the connection
to an electronic read-out circuit.

Experimental results, performed both in dark conditions and under 365 nm
monochromatic ultraviolet radiation, have demonstrated the ability of our structure

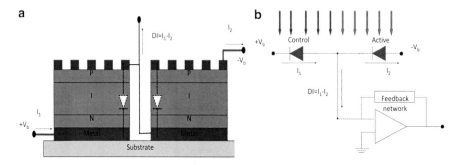

Fig. 1 (**a**) Structure of the amorphous silicon balanced photosensor. (**b**) Operation mode of the
balanced photodetector

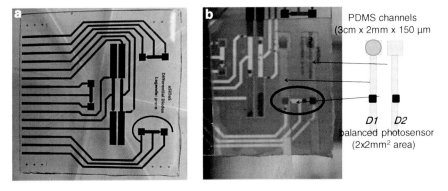

Fig. 2 (a) Picture of the fabricated devices; (b) two PDMS channels aligned on a three terminal balanced photodiode

to detect differential currents three orders of magnitude lower than the current of each sensor. The achieved common mode rejection ratio keeps constant with variations both the reverse bias voltage and the radiation intensity and increases with increasing wavelengths, varying from 30 dB nm at 254 nm to 42 dB at 365 nm [9].

3 Microfluidic Network

Polydimethylsiloxane (PDMS) is the most widely used silicon-based organic polymer for the fabrication of microfluidic network and is particularly known for its optical properties and its low manufacturing costs.

Two (PDMS, SYLGARD 184 from Dow Chemicals) channels, with dimensions of 3 cm $\times$ 2 mm $\times$ 150 μm (L$\times$W$\times$H) for each channel, have been fabricated by pouring of the mixed PDMS on Kapton tape mold. After curing at 80 °C for 90 min, the internal surface of the channels has been treated with polyethylene glycol (PEG) to turn the PDMS hydrophobic properties into hydrophilic properties and thus to allow capillarity flows in the channels.

4 Experiment and Results

The microfluidic network has been positioned directly on the glass surface over the balanced photodiode aligning each diode with a channel, as shown in the picture of Fig. 2b. The squares at one end of the channels are the inlets, while bibula papers have been used as outlet. The black squares (D1, D2) beneath the microfluidic channels are the balanced photodiodes whose current has been monitored during the experiments. This configuration guarantees optimal optical coupling between luminescence events occurring in the channels and the photosensor.

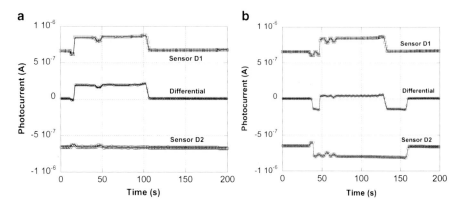

Fig. 3 Sensor current and the differential current of the balanced structure measured under a background light, with (**a**) water flowing only in one channel and (**b**) in both channels

Figure 3 reports the single sensor current and the differential current of the balanced structure measured under a large background light intensity, with water flowing only in one channel (Fig. 3a) and in both channels (Fig. 3b).

We have found that under identical channel conditions, the differential current is at least a factor 20 lower than the current flowing in each diode and that the balanced photodiode is able to detect the presence or absence of water in the channel.

5   Conclusion

In this paper we have integrated a two-channel microfluidic network with an a-Si:H balanced photodiode suitable for detection of small variations of photocurrent. We have found that the proposed device is able to detect the presence or absence of water flow in the channel. These preliminary results demonstrate the successful integration of microfluidic network with a-Si:H photosensors for on-chip detection in biomedical applications.

References

1. A.G. Crevillén, M. Hervás, MA. López, MC. González, A. Escarpa. Real sample analysis on microfluidic devices. Talanta, **74**, 342 (2007)
2. D. Caputo, G. de Cesare, F. Irrera, F. Palma, M. Tucci. Characterization of intrinsic a-si: H in p-i-n devices by capacitance measurements. J. of applied physics **76**(6),3534-3541 (1994).
3. G. De Cesare, M. Gavesi, F. Palma, B. Riccò. A novel a-si: H mechanical stress sensor. Thin solid films **427**(1), 191-195 (2003)
4. N. Ibaraki. a-si tft technologies for am-lcds, Materials Research Society Symposium Proceedings, **336**,749-749, Cambridge Univ Press (1994)

5. D. Caputo, G. de Cesare, A. Nascetti, M. Tucci, Detailed study of amorphous silicon ultraviolet sensor with chromium silicide window layer. IEEE Trans. Electr Dev, **55**(1), 452-456 (2008).
6. F. Fixe, V. Chu, D. Prazeres, J.P. Conde. An on-chip photodetector for the quantification of DNA probes and targets in microarrays. Nucleic Acids Res., **329**, 70-75 (2004).
7. G. de Cesare, D. Caputo, A. Nascetti, C. Guiducci, B. Riccò, "a-Si:H ultraviolet sensor for deoxyribonucleic acid analysis", Appl. Phys. Lett., **88**, 083904, (2006).
8. D. Caputo, G. de Cesare, C. Fanelli, A. Nascetti, A. Ricelli, R. Scipinotti, a-Si photosensors for detection of Ochratoxin A in wine, IEEE Sensor Journal, **12** 8, 2674-2679, (2012)
9. D. Caputo, G. de Cesare, A. Nascetti, Amorphous Silicon Differential Photodiode Structure. Sensor & Actuators: A. Physical, **153**, 1–4 (2009).

FEM Modeling of Nanostructures for Sensor Application

Adriano Colombelli, Maria Grazia Manera, Roberto Rella,
and Lorenzo Vasanelli

1 Introduction

In the last years, many theoretical and experimental studies have been made on metal nanoparticles because of their possible applications as chemical and biological sensor [1]. Significant attention has been paid in particular to Localized Surface Plasmon Resonance modes (LSPR) that gold and silver nanostructures are able to support when excited by incident light under specific conditions [2–5]. When the frequency of the light photons matches the frequency of surface plasmon polaritons (SPP), strong enhancements of the absorption, scattering and local electric field around the metal particles arise and the feature strongly depends on the material, the particle size, shape and distribution onto the substrate. These non-propagating plasma oscillations are strictly related to the optical properties of the local environment, allowing real-time monitoring of molecular adsorption occurring at metal/dielectric interface with potential applications in gas and biosensing.

Extensive research has been reported in the area of LSPR-based optical sensor, with the aim of sensitivity enhancement and detection limit reduction owing to the great enhancement of electromagnetic field near the surface of these nanostructures. New design and fabrication strategy of plasmonic transductors have been proposed in the last years in order to enhance the sensing performances and reproducibility of this optical sensors [6, 7].

A. Colombelli (✉)
CNR, Institute for Microelectronics and Microsystems, Unit of Lecce, Lecce, Italy

Department of Innovation Engineering, University of Salento, Lecce, Italy
e-mail: adriano.colombelli@le.imm.cnr.it

M.G. Manera • R. Rella
CNR, Institute for Microelectronics and Microsystems, Unit of Lecce, Lecce, Italy

L. Vasanelli
Department of Mathematics and Physics, University of Salento, Lecce, Italy

C. Di Natale et al. (eds.), *Sensors and Microsystems: Proceedings of the 17th National
Conference, Brescia, Italy, 5-7 February 2013*, Lecture Notes in Electrical Engineering 268,
DOI 10.1007/978-3-319-00684-0_55, © Springer International Publishing Switzerland 2014
287

Since advances in manufacturing techniques lead to a very precise control of structural dimensions, periodic arrays of metal nanoparticles have recently received considerable attention. When metal nanoparticles are brought into close proximity to each other, the resonance modes they support may interact modifying both the resonance shape and frequency of the LSPRs [8]. As recently demonstrated [9], also single and periodic array of nano-holes generated on the surface of thin metallic films may support LSPRs showing, under appropriate conditions, extraordinary optical transmission properties. These periodic arrays of nanostructures provide a very interesting platform for the development of detection devices with high sensitivity. However, the enhancement of LSPR sensors' performances needs a careful optimization of the nanostructured transductors fabricated with precise geometries and sharp resonances. Following this aim, a numerical comparison of the electric field distribution and the optical properties of gold nanostructures characterized by complementary geometry is proposed in this work.

2 COMSOL Simulation

A variety of theoretical techniques has been developed in order to understand how the optical response of metal nanostructures depends on their geometrical properties. In this work, the electric and optical properties of periodic array of gold nano-disks and gold nano-holes on glass substrate are investigated through 3D finite element simulations with COMSOL Multiphysics platform. We use the harmonic propagation mode of the RF module to calculate the optical transmission and reflection coefficients of the structures, when illuminated by a linearly polarized plane wave propagating along the z direction.

Square arrays of cylindrical disks and cylindrical holes with radius of 45 nm and period of 200 nm were compared. The height of these structures was set at 20 nm, and their optical response was compared with that of a planar thin film of the same height. Even though the geometry of the structures has cylindrical symmetry, it is still necessary to do 3D simulations when we use a linearly polarized light as incoming wave (Fig. 1a).

Different layered domains characterize the simulation structure (Fig. 1a). The higher domain simulates the homogeneous environment in which metal nanostructures are embedded, while the lower one represents the glass substrate. The middle layer was designed in order to use a single model as starting point for the investigation of three different transductors (nano-disk, nano-hole, thin film) changing only few parameters. The analysis of single and periodic arrays of nanoparticles was performed by setting different boundary conditions for the sidewalls. Perfect electric conductor (PEC) and perfect magnetic conductor (PMC) were set for the analysis of a single nanoparticle, while periodic boundary conditions (PBC) were set for the investigation of periodic arrays.

Port boundary conditions were set for the top and bottom surfaces of the model in order to define the linearly polarized incident wave and also to calculate the transmission and reflection coefficients.

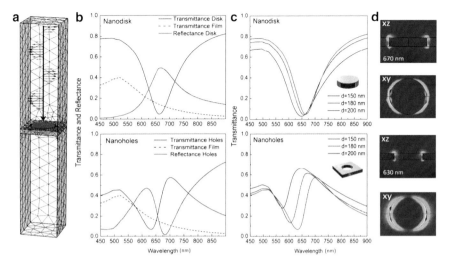

Fig. 1 (**a**) Simulation geometry; (**b**) transmittance and reflectance spectrum of the structures; (**c**) transmittance of array with different inter-particle distance; (**d**) electric field norm distributions

The optical properties of the system were investigated in a particular frequency range, corresponding to a free space wavelength from 300 to 800 nm. At these frequencies, gold can be modelled as having a complex-valued refractive index, with real and imaginary components. In the model, these components are described by an interpolated version of the often-used experimental data from Johnson and Christy [10].

3 Results and Discussions

Theoretical data from both structures show a pronounced minimum in their wavelength-dependent transmittance (Fig. 1b) due to the excitation of localized surface plasmon resonances associated with the nanoparticles/nano-holes. As expected, great enhancements of electromagnetic field arise in proximity of metallic nano-structures when resonance conditions are satisfied (Fig. 1d). LSPR conditions were examined with parametric study, varying inter-disks distance as well as nano structures shape. These periodic arrays exhibit shift of the LSPR condition as the holes or disk density increases (Fig. 1c), as confirmed by literature [11]. The same shift toward lower wavelengths was also noticed increasing the height of the nanostructures.

A change in the shape of the considered geometries, from nano-cylinder to nano-truncated cones, produced opposite shift of the LSPR conditions.

Results from numerical modelling show that nano-hole arrays are more sensitive to possible variations of the simulation geometry. This result can be attributed to a coupling effect between the local resonance of nano-holes and the propagating plasmons generated on the thin metallic film. For the complementary nanoparticle arrays, only a little shift can be noticed, since SPPs are not supported by this kind of structure.

Despite their geometric similarities, these complementary structures show the marked differences in their optical response.

4   Conclusions

The performed numerical study could represent a good starting point for the design of new LSPR-based optical sensor. Localized resonance and electric field enhancement can be obtained in the desired spectral region with an appropriate optimization of the transductors geometry. Further simulations with COMSOL Multiphysics module will help to better understand the optical response of these complementary nanostructures, allowing the fabrication of new LSPR transductors characterized by innovative design and better performances in the gas and biosensing field of research.

Acknowledgments This work was partially funded by FIRB-Futuro in ricerca (Nanoplasmag) MIUR RBFR10OAI0.

References

1. A. Lesuffleur, Hyungsoon Im, N. C. Lindquist, Sang-Hyun Oh. Periodic nanohole arrays with shape-enhanced plasmon resonance as real-time biosensors. Appl. Phys. Let. **90**,243110 (2007)
2. M. Futamata, Y. Maruyama, M. Ishikawa. Local Electric Field and Scattering Cross Section of Ag Nanoparticles under Surface Plasmon Resonance by Finite Difference Time Domain Method. *J. Phys. Chem. B* 107, 7607-7617 (2003)
3. Z. Liu, A. Boltasseva, R.H. Pedersen, R. Bakker, A.V. Kildishev, V.P. Drachev, V.M. Shalaev. Plasmonic nanoantenna arrays for the visible. Metamaterials **2**, 45–51 (2008)
4. K.S. Lee, M.A. El-Sayed, Dependence of the Enhanced Optical Scattering Efficiency Relative to That of Absorption for Gold Metal Nanorods on Aspect Ratio, Size, End-Cap Shape, and Medium Refractive Index. *J. Phys. Chem. B* 109, 20331-20338 (2005)
5. Olivier J. F. Martin, in Optical Nanotechnologies: The Manipulation of Surface and Local Plasmons, ed. by J. Tominaga and D. P. Tsai (Springer, Verlag Berlin Heidelberg, 2003) 183
6. K.S. Lee, M.A. El-Sayed. Gold and Silver Nanoparticles in Sensing and Imaging: Sensitivity of Plasmon Response to Size, Shape, and Metal Composition. J. Phys. Chem. B, **110**, 19220-19225 (2006)
7. J.C. Banthí, D. Meneses-Rodríguez, F. García, M. U. González, A. Garcia-Martin, A. Cebollada, G. Armelles. High Magneto-Optical Activity and Low Optical Losses in Metal-Dielectric Au/Co/Au-SiO$_2$ Magnetoplasmonic Nanodisk. Adv. Mater. **24**, OP36–OP41 (2012)
8. J. Parsons, E. Hendry, C. P. Burrows, B. Auguié, J. R. Sambles, and W. L. Barnes, "Localized surface-plasmon resonances in periodic non-diffracting metallic nanoparticle and nanohole arrays", Physical Review B **79** (2009), 073412

9. W. Ebbesen, H.J. Lezec, H.F. Ghaemi, T. Thio, P. A. Wolff. Extraordinary optical transmission through sub-wavelength hole arrays. Nature **391**, 667-669 (1998)

10. P. B. Johnson, R. W. Christy. Optical Constants of Noble Metals. Phys. Rev. B **6**, 4370 (1972)

11. J. Prikulis, P. Hanarp, L. Olofsson, D. Sutherland, M. Käll. Nano Lett. **4**, 1003 (2004)

Part VI
Sensor Electronics and Sensor Systems

Supervised Machine Learning Scheme for Wearable Accelerometer-Based Fall Detector

Gabriele Rescio, Alessandro Leone, and Pietro Siciliano

1 Introduction

The problem of falls in the elderly has become a healthcare priority due to the related high social and economic costs [1]. Many solutions have been proposed in the detection and prevention of falls, and some excellent reviews have been presented [1, 2]. In this paper, a fall detector through a wearable triaxial MEMS accelerometer is presented. The proposed solution overcomes the limitation of well-known threshold-based approaches [3] in which several parameters need to be manually estimated according to the specific features of the end user. In particular, a machine learning scheme has been used, and high generalization capabilities in the fall detection discrimination process have been recovered. The expert system uses robust features extracted taking into account important constraints and/or requirements of mobile solutions (workload). The extracted features are (quasi-)invariant both to specific characteristics of the mounting set-up (device on chest, on waist, on abdomen) and specific characteristics of the end users in terms of age, weight, height and gender.

2 Materials

The algorithmic framework has been developed by using the wearable device [4] composed by commercial discrete circuits. The system integrates an ST LIS3LV02DL triaxial MEMS accelerometer with digital output, an FPGA for computing functionalities and a ZigBee module for wireless communication up to 30 m,

G. Rescio (✉) • A. Leone • P. Siciliano
CNR, Institute for Microelectronics and Microsystems, Via Monteroni, Lecce 73100, Italy
e-mail: gabriele.rescio@le.imm.cnr.it; alessandro.leone@le.imm.cnr.it;
pietro.siciliano@le.imm.cnr.it

C. Di Natale et al. (eds.), *Sensors and Microsystems: Proceedings of the 17th National
Conference, Brescia, Italy, 5-7 February 2013*, Lecture Notes in Electrical Engineering 268,
DOI 10.1007/978-3-319-00684-0_56, © Springer International Publishing Switzerland 2014

suitable for indoor contexts. The wearable device can operate in streaming mode (raw data is sent to an external computing platform for data analysis with a 10 Hz frequency). Raw data is in hexadecimal format and represents the acceleration values with full scale in the range ±2 g for higher sensitivity. The accelerometer is DC coupled and it measures both static and dynamic acceleration along the 3 axes. The accelerometer measures the projection of the gravity vector on each sensing axis. If the accelerometer relative orientation is known, the acceleration data can be used to determinate the angle of the user posture in respect to the vertical direction.

3 Feature Extraction and Supervised Classifier

The acceleration data on three axes (A_x, A_y, A_z), coming from the device worn by a user during the data collection, is filtered out by a low pass 8-order, 8 Hz cut-off finite impulse response (FIR) filter, to reduce the noise due to electronic components, environment and human tremor. A calibration procedure has been accomplished by recovering the initial conditions after the user wears the device for the first time. The acceleration data (A_{x0}, A_{y0}, A_{z0}) is registered and used as reference in the feature extraction phase. Robust features are extracted in the time domain by considering both quick and relevant acceleration changing along each axis (due to the fall) in a 5 s sliding window and the change in posture registered after the shock. The aim is to produce robust features taking all the information into account and enables the distinction of falls from other events. It is also important that such features have a low dependence on both the position of the sensor (whether it is placed on the waist, on the chest or on the abdomen) and the human body characteristics of the user. In the former, the shock is measured due to the impact towards the floor plane and a dynamic acceleration changing is registered. In the latter (the body is already lying) the static acceleration value records a great change due to the new position of the individual with respect to the calibration phase (A_{x0}, A_{y0}, A_{z0}). Hence the difference between the value of the 3D-static acceleration after the fall and the one stored in the calibration phase will result in an offset, called Changing Position Offset (CPO), which is proportional to the user displacement. In this way, a study of posture was not made; only the relative varying posture analysis was considered, causing a computational cost reduction and improving the robustness of the set-up. The feature vector is made up by three parameters, one for each acceleration component. It makes sense to consider the acceleration signal on each axis singularly, because a fall event leads to a change in the value of the static acceleration in at least two of the three acceleration axes (due to the orientation change of sensing axes). The difference among the features of the falling, sitting and lying events is apparent in Figs. 1, 2, and 3. Finally it was verified that the features obtained discriminate the falls from typical activities of daily living (ADLs) also when the device is placed in other areas of the torso.

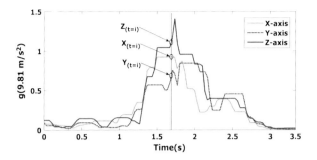

Fig. 1 Features extracted of the forward fall

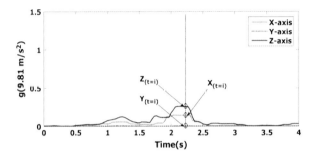

Fig. 2 Features extracted of the sitting event

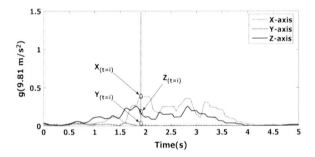

Fig. 3 Features extracted of the lying event

Once features are extracted, the fall events are detected by a One Class Support Vector Machine (OC-SVM) which is less computationally intensive than other algorithms like neural networks [5]. SVM is a robust classification tool (in the presence of outliers too) with a good generalization ability. For the extracted features, the optimal kernels are Gaussian radial basis function (GRBF) [6] and polynomial. Finally a simple post-processing step has been implemented in order to reduce false-negative events, according to a high-scoring approach.

4   Experimental Results

The OC-SVM classifier has been trained by using about 40 falls and 50 ADLs belonging to a large dataset in which more than 250 falls and 200 ADLs were performed compliant to the specifications proposed in [7]. The remaining 210 falls and 150 ADLs have been used for testing. The performances of the system have been evaluated considering two normally used metrics, the sensitivity and specificity [7], respectively. The algorithm is tested when the device is placed on the waist, abdomen or chest. GRBF and polynomial kernel functions give the best results in terms of performance (more than 95 %); even the polynomial kernel shows a lower computational cost (relative number of support vectors and relative execution time are considered). The implemented SVM improves in specificity and sensitivity with respect to the threshold-based approach detailed in [8]; see Table 1. As already discussed, the two fall detection systems have common hardware, benchmark dataset and training/test sets. Of course, the computational cost of the threshold-based is lower than the implemented expert systems, but it suffers in detection rate due to high false positive and true negative. Since the number of features is compact and the computational cost of the extracted features is low, the overall system workload is compatible with an integration in embedded low-power solutions.

5   Conclusions

The proposed supervised scheme overcomes the limitation of well-known threshold-based approaches in which a heuristic choice of the parameters is accomplished. A specific study on postures was not made in order to make a low computational power system. The calibration step guarantees the generalization of the approach in terms of invariance to the specific characteristics of the end users, during the fall detection process. High performances in controlled conditions (simulated events) in terms of sensitivity and specificity were obtained using only the 20 % of dataset for training. Performance metrics of different kernels in OC-SVM have been compared and the best results are obtained with the polynomial function and GRBF. Future work will be devoted to validate the solution in real conditions, test the methodology with a large set of different MEMS accelerometers and port the framework on embedded mobile solutions.

Table 1 Comparison of the proposed OC-SVM method and threshold-based algorithm [8]

	Sensitivity (%)	Specificity (%)	Relative execution time
OC-SVM	97.7	94.8	1×
Threshold-based	88.4	79.3	0.4×

References

1. N. Noury, P. Rumeau, A.K. Bourke, G. ÓLaighin, J.E. Lundy, "A proposal for the classification and evaluation of fall detectors". IRBM; **29**(6):p. 340–9 (2008).
2. X. Yu, "Approaches and principles of fall detection for elderly and patient". In: Proceedings of the 10th IEEE HealthCom: p. 42–7 (2008).
3. F. Bagalà, C. Becker, A. Cappello, L. Chiari, K. Aminian, J.M. Hausdorff, W. Zijlstra, J. Klenk, "Evaluation of Accelerometer-Based Fall Detection Algorithms on Real-World Falls", PLoS One: p. e37062 (2012).
4. A. Leone, G. Diraco, C. Distante, P. Siciliano, M. Malfatti, L. Gonzo, M. Grassi, A. Lombardi, G. Rescio, P. Malcovati, V. Libal, J. Huang, G. Potamianos, "A Multi-Sensor Approach for People Fall Detection in Home Environment", Proc. Of M2SFA2-ECCV: p. 1-12 (2008).
5. L.M. Manevitz, M. Yousef, "One-class SVMs for document classification.", J Mach Learn Res **2**(1): p.139–154 (2001).
6. T. Zhang, J. Wang, L. Xu, P. Liu, "Fall Detection by Wearable Sensor and One-Class SVM Algorithm.", Lecture Notes in Control and Information Science **345**: p. 858–863 (2006).
7. N. Noury, A. Fleury, P. Rumeau, A.K. Bourke, G.O. Laighin, V. Rialle et al. "Fall detection—principles and methods" In: Proceedings of the 29th IEEE EMBS: p. 1663–6 (2007).
8. M. Grassi, A. Lombardi, G. Rescio, P. Malcovati, M. Malfatti, L. Gonzo, A Leone, G. Diraco, C. Distante, P. Siciliano, V. Libal, J. Huang, G. Potamianos, "A Hardware-Software Framework for High-Reliability People Fall Detection", Proc. Of IEEE Sensors 2008: p. 1328-1331 (2008).

A Compact Architecture for Heartbeat Monitoring

V. Stornelli, P. Mantenuto, G. Ferri, and P. Di Marco

1 Introduction

As well known, ECG (electrocardiogram) is an electrical test that measures and records the electrical activity of the heart. The signals that make the heart muscle fibres contract come from the sino-atrial node, which is the natural pacemaker of the heart. In an ECG test, the electrical impulses generated by a beating rhythm can be promptly known.

The electrical signal produced by the human heart can be modelled by the "Tokyo standard" curve, depicted in Fig. 1 [1, 2]. This signal is characterised by a series of local maximum (P and T: atrial depolarisation and ventricular repolarisation, respectively) and local minimum (Q and S) nearby the pulse peak (R) (this last sequence represents the ventricular depolarisation, thus the systole).

In the literature, there are several works reporting a heartbeat detector using various kinds of solutions (Doppler radar sensor [3], electrocardiography (ECG) measurement [4] and capacitive sensing of ECG [5] for non-invasive cardiac or activity measurement). The most of the circuitries implementing ECG devices currently require a huge number of blocks, such as operational amplifiers employed for differential gain amplification and common mode rejection both in voltage or current mode [6, 7], filtering stages [8–10], peak detector stages, failure electrode connection circuits and oscillators, to be implemented together with quite expensive devices such as radio frequency components, antennas [3] and conductive fibres [4]. Furthermore, they have to be applied using up to 12 self-adhesive electrodes which must be attached into selected locations of the skin such as the arms, legs and chest which make the system quite uncomfortable and often not friendly to move with.

V. Stornelli (✉) • P. Mantenuto • G. Ferri • P. Di Marco
Department of Industrial and Information Engineering and Economics,
University of L'Aquila, L'Aquila, AQ, Italy
e-mail: vincenzo.stornelli@univaq.it; paolo.mantenuto@univaq.it;
giuseppe.ferri@univaq.it; patrizio.dimarco@univaq.it

C. Di Natale et al. (eds.), *Sensors and Microsystems: Proceedings of the 17th National Conference, Brescia, Italy, 5-7 February 2013*, Lecture Notes in Electrical Engineering 268, DOI 10.1007/978-3-319-00684-0_57, © Springer International Publishing Switzerland 2014

Fig. 1 "Tokyo standard"
curve

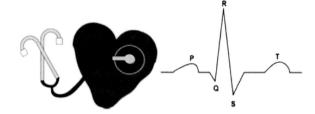

Fig. 2 Complete schematic
block circuit for the read-out
and display of the heartbeat
rate

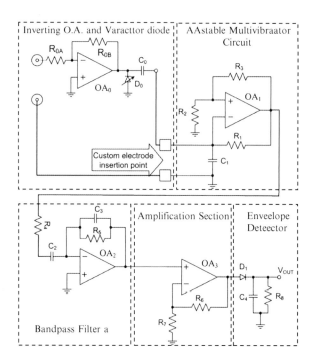

2 Implemented Architecture and Simulation Results

This work arises with the intent of creating a circuit solution consisting of a non-invasive detection of cardiac (and respiratory) activity without removing clothing [11, 12], which is desirable both in emergency circumstances and in long-term home wireless monitoring [13–16]. A new approach in determining heartbeat rate, through carrier signal frequency modulation technique, is obtained by employing a capacitive sensor used for detecting heartbeat rate without a direct contact with the skin. More in detail, the system is based on an astable multivibrator circuit, and the cardiac activity modulates the oscillator frequency through the capacitive change [14] of a varactor diode biased by the heart pulse peak (see Fig. 2).

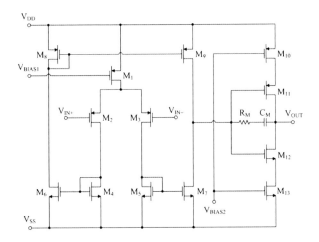

Fig. 3 OA at transistor level, standard CMOS technology

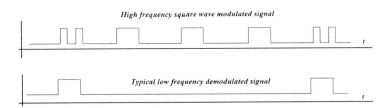

Fig. 4 High frequency signal, modulated by the heartbeat peak (*upper* graph) and typical low frequency demodulated signal (*lower* graph)

The active blocks shown in Fig. 2 have been implemented as operational transconductance amplifiers (OTA), designed in a standard AMS 0.35 μm CMOS technology, whose internal topology is reported in Fig. 3.

The free oscillation frequency of the astable oscillator is determined by the capacitor C_1 (see Fig. 2) while a parallel time-varying heartbeat-dependent capacitance modulates it, as shown in Fig. 4, upper picture. The heartbeat signals can be therefore obtained by demodulating the oscillating signal (see Fig. 4, lower picture). The sensor output is filtered and amplified using a non-inverting amplifier and a bandpass filter. The result is a highly accurate, low-cost front-end device showing good energy efficiency (low voltage, low power), has small and light portable size and easiness of use for everyone (either professional or not).

Concerning the measurement results, the patch electrode of the capacitive heartbeat sensor has been placed on a cloth near the chest for cardiac measurement. The sampling frequency is 1 kHz with 12-bit resolution. The considered frequency of the heartbeat signal is 1.41 Hz. Some harmonics of the signal due to the nonlinearity of the human body have been encountered.

3 Conclusions

First tests conducted through a discrete PCB design have shown the expected behaviour results. In particular the following advantages of the proposed system can be highlighted:

- Capability of placing the electrode on the whole body
- Low-voltage and low-power design
- Reduced dimension of the system

The project can be improved according to the following future activities:

- IC layout at CMOS transistor level
- Digital acquisition and elaboration data
- Energy autonomous system design (vibration harvesting)
- Comparative tests with commercial devices

References

1. J.G. Webster. Design of cardiac pacemaker, IEEE Press (1995), ISBN: 978-0780311343.
2. JognG. Webster. Medical Instrumentation, Application And Design, JohnWiley& Sons inc., 2009.
3. Lin, James C., et al.: 'Microwave apex cardiography', IEEE Trans. Microw. Theory Tech., 1979, MTT-27.
4. Ishijima., M.: 'Monitoring of electrocardiograms in bed without utilizing body surface electrodes', IEEE Trans. Biomed. Eng., 1993, **40**, (6), pp. 593-594
5. Ueno, A., Akabane, Y., Kato, T., Hoshino, H., Kataoka, S., and Ishiyama., Y,: 'Capacitive sensing of electrocardiographic potential through cloth from the dorsal surface of the body in a supine position: a preliminary study', IEEE Trans. Biomed. Eng., 2007, **54**, (4), pp. 759-766
6. V. Stornelli. Low Voltage Low Power Fully Differential Buffer, Journal of Circuits, Systems, and Computers (JCSC),**vol.18**, no.3, pp. 497-502, 2009
7. G. Ferri, V. Stornelli and M. Fragnoli. An integrated improved CCII topology for resistive sensor application, AnalogIntegr. Circuits Signal Process., **vol. 48**, 247-250, 2006.
8. V. Stornelli, L. Pantoli, G. Leuzzi, G. Ferri. Fully differential DDA-based fifth and seventh order Bessel low pass filters and buffers for DCR radio systems, Analog Integrated Circuits and Signal Processing, May 2013, **vol. 75**, Issue 2, pp 305-310.
9. V. Stornelli, G. Ferri. A 0.18μm CMOS DDCCII for portable lv-lp filters, Radioengineering, **Vol. 22**, n°2, June 213.
10. P. Colucci, G. Leuzzi, L. Pantoli,"Third order integrable UHF bandpass filter using active inductors", Microwave And Optical Technology Letters, **Vol. 54**, n° 6 pp. 1426-1429, June 2012.
11. Y.T. Li. Pulse Oximetry, Stanford Undergraduate Research Journal **2**, 10 (2007), 11-15.
12. G. Ferri, V. Stornelli, A. di Simone. A ccii-based high impedance input stage for biomedical applications, Journal of Circuits, Systems and Computers **Vol. 20**, 08, pp 1441-1447, 2011.
13. J.H. Oum, S.E. Lee, D.W. Kim, S. Hong. Non-contact heartbeat and respiration detector using capacitive sensor with Colpitts oscillator, Electronics Letters, **44** (2008), 87-88.

14. Gabriel, S., Lau, R.W., and Gabriel, C.: 'The dielectric properties of biological tissues: 3 models for the frequency dependence', Phys. Med. Biol., 1995, **41**, (11), pp. 2271-2293.
15. G. Bucci, E. Fiorucci, C. Landi, G. Ocera. The Use of Wireless Network for Distributed Measurement Applications, Proc. IEEE Instrumentation and Measurement Technology Conference, Anchorage, USA, May 21-23, 2002.
16. G. Bucci, E. Fiorucci, C. Landi, G. Ocera. Architecture of a digital wireless data communication network for distributed sensor applications, Measurement, Journal of International Measurement Confederation, **vol. 35**, issue 1, (2004), pages 33-45.

MEMS Flow Sensor Based on a Sigma–Delta Modulator Embedded in the Thermal Domain

M. Piotto, A.N. Longhitano, F. Del Cesta, and P. Bruschi

1 Introduction

MEMS flow sensors are gaining increasing interest due to the development of more and more complicated micro-fluidic devices, to be employed in a large variety of applications, such as drug delivery control, reagent monitoring in micro-reactors, and propellant control in miniaturized aerospace thrusters. Recently, micro-flow sensors have been used to build an innovative 2-dimensional anemometer with ultralow power consumption and very compact size [1]. The excellent compatibility of thermal flow sensors fabrication technology with CMOS processes has stimulated the design of compact versatile interfaces, which have been successfully integrated on the same die as multiple sensing structures [2]. The intrinsic low-pass characteristics of thermal flow sensors suggested the implementation of $\Sigma-\Delta$-based readout interfaces that incorporate part of the $\Sigma-\Delta$ modulator into the thermal domain of the sensor [3]. The limits of the device proposed in [3] is that the whole chip is heated, resulting in relatively high power consumption and long response and warm-up times.

In this work, we propose the application of the thermal $\Sigma-\Delta$ approach to MEMS flow sensing structures that, due to their effective thermal insulation from the substrate, represent the ideal solution to reduce power dissipation and obtain faster response times.

M. Piotto
CNR IEIIT—Pisa, via G. Caruso, 16, Pisa 56122, Italy

A.N. Longhitano · F. Del Cesta · P. Bruschi (✉)
Dipartimento di Ingegneria dell'Informazione, via G. Caruso, 16, Pisa 56122, Italy
e-mail: bruschi4.35@gmail.com

C. Di Natale et al. (eds.), *Sensors and Microsystems: Proceedings of the 17th National Conference, Brescia, Italy, 5-7 February 2013*, Lecture Notes in Electrical Engineering 268, DOI 10.1007/978-3-319-00684-0_58, © Springer International Publishing Switzerland 2014

2   Description of the Sensing Structures

An optical micrograph of the sensing structure used in this work is shown in Fig. 1a. The device consists of two n-polysilicon heaters (H) placed between two n-polysilicon/p-polysilicon thermopiles (TP1 and TP2). The sensing structure is fabricated starting from a 4×4 silicon chip, produced with the BCD6s process of STMicroelectronics. Selective anisotropic etching of the bulk silicon is subsequently applied to suspend the sensor elements over a cavity, obtaining effective thermal insulation. A proper plastic adapter, provided of stainless steel inlet and outlet pipes, is applied to the chips in order to convey the gas flow to the on-chip sensing structures [4]. A photograph of the complete flow sensor is shown in Fig. 1b.

It has been shown [4] that the relationship between the thermopile differential output voltage (V_d) and the heater powers is linear. Driving the heaters with a constant common mode voltage, the power difference is proportional to the heater differential voltage ΔV_H, so that the voltage V_d can be written as:

$$V_d = f(Q) + k_H \Delta V_H \tag{1}$$

where $f(Q)$ is a function of flow rate Q, and k_H is a constant. Double-heater structures have been previously exploited to perform drift-free offset compensation [5] and closed-loop readout procedures [6].

3   Sensor Interface

The block diagram of the interface is shown in Fig. 2. The analog switch S_H (74HC4066N) alternatively connects heaters H_1 and H_2 to the two voltage sources V_H and V_L, with $\Delta V_{H0} = V_H - V_L > 0$. The thermopile output voltage, V_d, is amplified by the instrumentation amplifier (In-Amp, AD627) and then compared with zero by

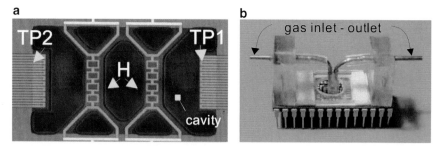

Fig. 1 (**a**) Optical micrograph of a sensing structure after post-processing; (**b**) photograph of the assembled flow sensors with the gas inlet-outlet connections indicated

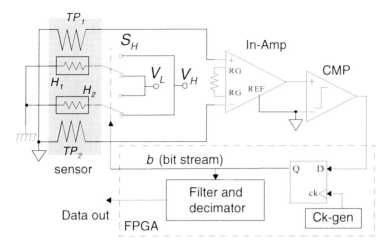

Fig. 2 Block diagram of the Σ–Δ interface

comparator CMP (MAX942). Considering (1) and using a linear approximation of the flow dependent component $f(Q)$, we can write:

$$V_d = -mk_1 \Delta V_{H0} + k_2 Q \qquad (2)$$

where k_1 and k_2 are constants and m is 1 for $b=1$ and -1 for $b=0$.

Voltage V_d is filtered by the thermal frequency response of the sensor, which is dominated by a pole at nearly 70 Hz for the devices used in this work. Since the clock frequency was set to a much higher frequency (20 kHz), the thermal response can be approximated to that of an integrator. As a result, the scheme shown in Fig. 2 is equivalent to a first-order Σ–Δ modulator. The bit stream is filtered by a sinc³ digital filter implemented on an FPGA (Altera Cyclone II). The oversampling ratio of the converter was set to 1,024, obtaining, after decimation, an update rate of nearly 20 Hz. The output data, represented with 16 bits codes, are sent to a personal computer through an RS-232 serial interface synthesized in the FPGA.

4 Experimental Results

The experimental characterization of the flow sensor was performed using a reference gas line, equipped with precision flow controllers. The output data codes were read by a personal computer through a program developed with the LabVIEW (National Instrument) environment. The results of the tests performed with nitrogen flows between 0 and 10 sccm are shown in Fig. 3a. Negative flow values are obtained by swapping the inlet and outlet connections. An excellent linearity across the

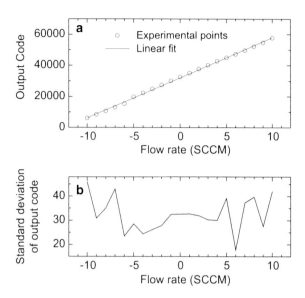

Fig. 3 Output code (**a**) and its standard deviation (**b**) as a function of nitrogen flow rate

whole flow interval can be observed. The standard deviation of the measurements (code units), calculated from repeated readings performed at constant flow rate, is shown in Fig. 3b. From these data, an effective resolution of nearly 9 bits can be estimated.

5 Conclusions

The application of a Σ–Δ interface to MEMS double-heater thermal flow sensors has been described. The experiments performed in nitrogen flow demonstrated that the proposed approach is suitable for the implementation of simple interfaces for thermal flow sensors. A further study is required to optimize the system in order to improve the effective resolution.

Acknowledgments The authors thank STMicroelectronics R&D group of Cornaredo (MI) for fabricating the sensor chip.

References

1. P. Bruschi, M. Dei, M. Piotto, A Low-Power 2-D Wind Sensor Based on Integrated Flow Meters, IEEE Sensors Journal. **9**(12), 1688-1696 (2009)
2. M. Piotto, M. Dei, F. Butti, G. Pennelli, P. Bruschi. Smart Flow Sensor With On-Chip CMOS Interface Performing Offset and Pressure Effect Compensation. IEEE Sensors Journal, **12**(12), 3309-3317 (2012)

3. K.K.A. Makinwa, J.H. Huijsing A smart wind sensor using thermal sigma-delta modulation techniques. Sensors and Actuators A, **97-98**, 15-20 (2002).
4. P. Bruschi, M. Piotto, Design Issues for Low Power Integrated Thermal Flow Sensors with Ultra-Wide Dynamic Range and Low Insertion Loss. Micromachines, **3**, 295-314 (2012)
5. P. Bruschi, M. Dei, M. Piotto. An Offset Compensation Method With Low Residual Drift for Integrated Thermal Flow Sensors. IEEE Sensor Journal, **11**(5), 1162-1168 (2011)
6. M. Dijkstra, T. S. J. Lammerink, M. J. de Boer, J. W. Berenschot, R.J. Wiegerink, and M. Elwenspoek. Low-drift flow sensor with zero offset thermopile-based power feedback. Proc. of DTIP 2008, Apr. 9–11 2008, Nice (France), 247–250 (2008)

Optimal Parameters Selection for Novel Low-Voltage Vibration Energy Harvesters

A. Giuffrida, F. Giusa, C. Trigona, B. Andò, and S. Baglio

1 Introduction and Motivation

One of the most popular techniques to scavenge energy from ambient vibrations uses the piezoelectric transduction [1]. An electrical interface is needed in order to rectify the alternate piezoelectric voltage and thus to store the energy. This is the main limitation of existing interfaces, usually based on diode-bridge rectifiers, due to the random and low amplitude voltage (typically under diode threshold). The most important approaches in this subject are based on the SSHI interface (*Synchronized Switch Harvesting on Inductor*) [2] and its main variations [3, 4]. All these solutions represent synchronized approaches, so they cannot work correctly in many real situations, where the vibrational input is random, low amplitude and broadband [5].

In this paper, we discuss about an innovative solution, called Random Mechanical Switching Harvester on Inductor (RMSHI) [6]. This system succeeds in overcoming the highlighted problems.

2 RMSHI Working Principle: Brief Description

The RMSHI system and its main subsystems are shown in Fig. 1. The vibrational input acts both as excitation for the piezoelectric transducer (to generate electrical energy) and driving signal for the mechanical switch. When the switch is closed (state on), the electrical charge produced by the piezoelectric transducer is transferred only on the inductor L because the voltage is under the diode threshold so the

A. Giuffrida • F. Giusa • C. Trigona (✉) • B. Andò • S. Baglio
Dipartimento di Ingegneria Elettrica Elettronica e Informatica, University of Catania,
V.le A. Doria 6, 95125 Catania, Italy
e-mail: carlo.trigona@diees.unict.it

C. Di Natale et al. (eds.), *Sensors and Microsystems: Proceedings of the 17th National Conference, Brescia, Italy, 5-7 February 2013*, Lecture Notes in Electrical Engineering 268, DOI 10.1007/978-3-319-00684-0_59, © Springer International Publishing Switzerland 2014

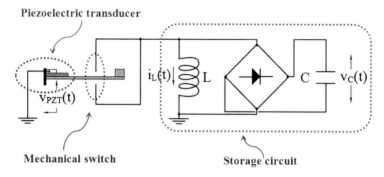

Fig. 1 The random mechanical switch harvester on inductor (RMSHI) system

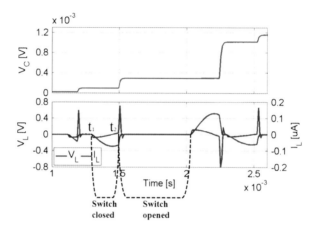

Fig. 2 RMSHI system: electrical signals produced by the circuit during a working cycle. When a voltage peak occurs across the inductor, the output voltage on C rises

current cannot flow through the Graetz bridge. When the switch turns off and the current on, the inductor cannot be instantly interrupted (according to the Lenz's law); so, via an auto-adaptive process, it is forced to flow through the diodes on the storage capacitor. In Fig. 2, the typical electrical signal for RMSHI system is shown. As it can be seen when the mechanical switch turns off, a characteristic voltage peak (having over-threshold amplitude) appears on the inductor. At this instant the current (i_L, current flowing through the inductor L) is given by the following equation:

$$i_L(t_2) = \frac{1}{L}\int_{t_2}^{t_1} v_L(t)\,dt + i_L(t_1)$$ (1)

where $v_L(t)$ represents the voltage across the inductor, while t_2-t_1 is the integration range (see Fig. 2).

3 Experiments and Results

A prototype has been realized to experimentally validate the working principle. As it can be seen in Fig. 3, the mechanical switch is composed of a cantilever with a proof mass and two stoppers to close the electrical contacts. The piezoelectric transducer has been placed at the point where the beam mostly deforms. The set-up used in order to perform measurements consists of an electrodynamical shaker with a signal generator to stress the system with desired vibrational accelerations, an accelerometer as feedback element and an oscilloscope for data acquisition.

In order to evaluate the performances, a vibrational input acceleration (Gaussian white noise) having a band of 700 Hz and a root mean square (RMS) amplitude of 1.6 g has been used. Several studies have been carried out varying the configuration of the switch (in terms of distance d of the stoppers from the beam and position X of the stoppers along the beam direction). In Fig. 4 some experimental results obtained

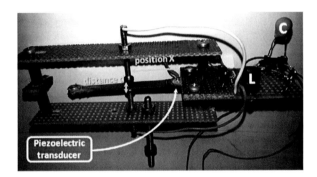

Fig. 3 The laboratory prototype of the RMSHI system realized to perform the characterization with respect to parameters d and X

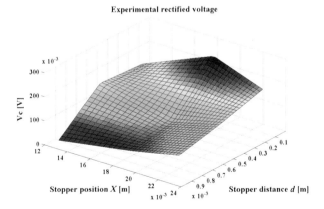

Fig. 4 Experimental results obtained for nine different configurations of the mechanical switch. Results confirm that the system has an optimal set of parameters that maximizes the performances

for nine different configurations of the switch are shown. As it can be seen, the optimal performances (in terms of rectified voltage across output capacitor) provided by the experimental set-up have been obtained for $d = 0.1$ mm and $X = 18$ mm (41 % of the beam's length).

4 Conclusions

In this work, a novel strategy to scavenge energy from environmental vibrations (assumed to be random, broadband and under-threshold amplitude) has been presented. In particular, in order to maximize the output power, a study of geometrical configuration of the switch has been accomplished. The analysis has been performed considering the position X and the distance d of the stoppers from the beam: an optimal set of parameters has been found for $d = 0.1$ mm and $X = 41$ % of the beam's length.

The development of a PZT-MEMS prototype and the investigation of nonlinear dynamics is in progress.

References

1. A. R. Anton and H. A. Sodano, "A review of power harvesting using piezoelectric materials," *Smart materials and Structures*, vol. 16, pp. R1-R21, 2007.
2. D. Guyomar, A. Badel, E. Lefeuvre and C. Richard, "Toward Energy Harvesting Using Active Materials and Conversion Improvement by Nonlinear Processing," IEEE Transactions on Ultrasonics, Ferroelectrics, and Frequency Control, vol. 52, no. 4, pp. 584-595, 2006.
3. J. Liang and W. H. Liao, "Improved Design and Analysis of Self-Powered Synchronized Switch Interface Circuit for Piezoelectric Energy Harvesting Systems," *IEEE Transactions on Industrial Electronics*, vol. 59, no. 4, pp. 1950-1960, 2012.
4. L. Garbuio, M. Lallart, D. Guyomar, C. Richard and D. Audigier, "Mechanical Energy Harvester With Ultralow Threshold Rectification Based on SSHI Nonlinear Technique," *IEEE Transactions on Industrial Electronics*, vol. 56, no. 4, pp. 1048-1056, 2009.
5. F. Neri, F. Travasso, R. Mincigrucci, H. Vocca, F. Orfei and L. Gammaitoni, "A real vibration database for kinetic energy harvesting application," *Journal of Intelligent Material Systems and Structures*, vol. 23, no. 18, pp. 2095-2101, 2012.
6. F. Giusa, A. Giuffrida, C. Trigona, B. Andò, A. R. Bulsara and S. Baglio, "Random Mechanical Switching Harvesting on Inductor": A Novel Approach to Collect and Store Energy From Weak Random Vibrations with Zero Voltage Threshold, Sensors and Actuators A: Physical, vol. 198, pp. 35-45, 2013.

Wireless Sensor Networks for Distributed Measurements in Process Automation

A. Flammini and E. Sisinni

1 Wired and Wireless Communications in Process Automation

Wireless Sensor Networks advantages are well known (e.g. Smart Dust is more than 10 years old [1]); the advent of standards as IEEE802.15.4 [2] makes them a reality. Referring to industrial applications [3], and particularly to process control, main requirement is reliability, especially in large plants. Wired solutions, e.g. field-buses, are intrinsically unsafe, because they can be the transmission medium of electrical spikes. Specially designed physical level, as IEC 61158-2, has been adopted by fieldbuses operating in process control in order to increase safety.

For this reason, wireless technologies have been introduced at the field level as soon as possible [4]. It should be highlighted that the most widespread wireless technologies like WiFi (IEEE802.11) and Bluetooth (IEEE802.15.1) are unsuitable for this application field because of the high power consumption. Wireless nodes mean not only wireless communications but also no wires for power supply. Instruments for process automation allow powerful batteries, because miniaturization is not the first requirement. For this reason, IEEE802.15.4 is the technology used to develop solutions for this application field. IEEE802.15.4 is the physical level of ZigBee that has been designed for event-based communications, being unsuitable for the real-time cyclic communications required at the field level. It should be noticed that new profiles of ZigBee purposely designed for metering could have some points in common with some industrial monitoring applications, but the required real-time behaviour is still different.

A. Flammini (✉) • E. Sisinni
Department of Information Engineering, University of Brescia, Via Branze 38,
25123 Brescia, Italy
e-mail: alessandra.flammini@ing.unibs.it

C. Di Natale et al. (eds.), *Sensors and Microsystems: Proceedings of the 17th National Conference, Brescia, Italy, 5-7 February 2013*, Lecture Notes in Electrical Engineering 268, DOI 10.1007/978-3-319-00684-0_60, © Springer International Publishing Switzerland 2014

A European commission, IEC-65C-WG16, Wireless for Industrial Communications, is studying proposals. At the moment, three proposals have been submitted:

- WirelessHART, proposed by HCF Consortium (HART Communication and Foundation), released as the standard IEC62591
- WIA-PA (Wireless for Industrial Automation-Process Automation), proposed by the Chinese National Committee, released as standard IEC62601
- ISA100.11a, proposed by ISA (International Society of Automation), under review as IEC62734cdv (Committee Draft for Voting)

Some other proposals have been considered, like OCARI [5], all based on the physical level of IEEE802.15.4, but only WirelessHART and ISA100.11a [6] seem to be supported by recent real products. According to some approach, WirelessHART can be considered as a special case of ISA100.11a, but no coexistence is guaranteed at this moment.

2   WirelessHART

WirelessHART is the standard that, at the moment, has been more experienced. It is based on IEEE802.15.4-2006 allowing to keep low the cost of the nodes, thanks to the many vendors of transceivers. OQPSK (offset quadrature phase shift keying) and spread spectrum modulation make it a robust solution, while AES128 support guarantees security. The MAC (media access control) level has been modified both in WirelessHART and in ISA100.11a, in order to achieve a slow frequency hopping, ensuring a better coexistence with other wireless networks (e.g. WLAN) and, therefore, reliability. Reliability is one of the most important feature in industrial applications, especially if measurements are used to close control loops [7, 8], and, for this reason, the support of mesh topologies and redundant routing schemes are mandatory. While ISA100.11a provides a slow Internet connectivity, thanks to 6LowPAN support, WirelessHART avoids IP traffic, filtered by the gateway. The mesh topology is completely managed by the network manager, providing routes and organizing the communications in slots arranged in a superframe.

Figure 1a shows the information transfer between node A and node B, while Fig. 1b highlights the slow frequency hopping. This mechanism allows blackchannel listing, avoiding interference with other wireless communications (e.g. WiFi).

Every information transfer takes a "slot" (10 ms) and provides ACK. Nodes are normally in low-power mode, and synchronization allows both nodes to wake up correctly to communicate within the slot. Shared slot is provided for asynchronous communication.

One of the key points is real-time behaviour and maintenance. The single node can awake just before its slots, being in low-power mode for the remaining time. A node with no active communication keeps alive the link communicating just one time every 30 s. As IEEE802.15.4 in active state (transmission or reception) allows

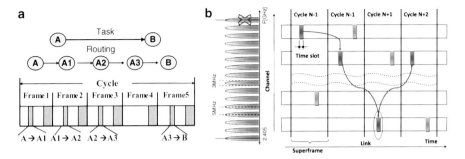

Fig. 1 (**a**) Cycle hosting communication slots. (**b**) Slow frequency hopping

a current consumption below 30 mA, the mean current consumption can be on the order of few tens of μA, avoiding battery replacement or battery recharge for several years. Obviously power consumption strongly depends on the communication cycle, normally of some seconds, and on the power consumption of the instrument itself (which sensing element, which actuator and so on).

A special attention has been paid to the self-organizing mechanism, that is, the procedure allowing a node to join the WirelessHART network. A security manager is responsible for the distribution of encryption keys for Join, Session and Network. The network manager is the heart of a WirelessHART network. It is responsible for the overall management, scheduling, routing and optimization of the wireless network; provides mechanisms for devices to join and leave the network; assigns the 16-bit nickname within the network; and is responsible for collecting and maintaining diagnostics about the overall health of the network. The network manager, the security manager and the gateway could be in the same device.

WirelessHART devices are on the market since a few years, and only now real plants can be experimentally analysed, although very few tools and instruments are available for performance characterization [9, 10].

3   Conclusions

With respect to the idea of Smart Dust, WirelessHART is one of the first real multi-vendor standard based on a low-power wireless technology, but many limitations are still present:

- Miniaturization is not optimized, but it is not a strict requirement of process automation.
- Maintenance is quite good, thanks to batteries, although some harvesting approaches are under study for some applications.
- The real-time behaviour is ensured for process automation needs, and control loops on the order of seconds (e.g. temperature) can be closed, but WirelessHART is impractical for factory automation.

- WirelessHART supports mesh topology but nodes are not self-organized, and everything, including routing, is managed by a network manager, which is one of the bottleneck of this standard.
- Self-localization is not supported, because nodes are normally in fixed positions.

WirelessHART is a very important test bench for wireless sensor networks, but there is still a long way to go to, both on the research on the hardware and at the system level [11, 12]. The real and pervasive execution of the Smart Dust idea probably will wait still for some years, until inexpensive, 0-power MEMS, with ranging and communication capabilities, will be available.

References

1. http://robotics.eecs.berkeley.edu/~pister/SmartDust/SmartDustBAA97-43-Abstract.pdf
2. IEEE802.15.4-2006 Spec.
3. Wired and wireless sensor networks for industrial applications, P. Ferrari, A. Flammini, D. Marioli, E. Sisinni, A. Taroni, Microelectronics Journal, September, 2009, Vol. 40, No. 9, pp. 1322-1336, DOI: 10.1016/j.mejo.2008.08.012.
4. Radmand, P.; Talevski, A.; Petersen, S.; Carlsen, S., Comparison of industrial WSN standards. 2010 4th IEEE International Conference on Digital Ecosystems and Technologies (DEST), IEEE, Dubai, United Arab Emirates, 2010, pp. 632–637, DOI: 10.1109/DEST.2010.5610582.
5. T. Dang, E. Sisinni, The Industrial Electronics Handbook, Industrial Communication Systems, Chapter 53: WirelessHART, ISA100.11a, and OCARI, 2nd Edition, February, 2011, pp. 53.1-53.17, CRC Press, Boca Raton, FL, DOI: 10.1201/b10603-57
6. Petersen, S.; Carlsen, S., WirelessHART Versus ISA100.11a: The Format War Hits the Factory Floor, Industrial Electronics Magazine, IEEE, Vol. 5, Issue: 4, 2011, pp. 23-34, DOI: 10.1109/MIE.2011.943023
7. Muller, I.; Netto, J.C.; Pereira, C.E., WirelessHART field devices, Instrumentation & Measurement Magazine, IEEE, Vol. 14, Issue: 6, DOI: 10.1109/MIM.2011.6086896, 2011, pp. 20 - 25
8. Xiuming Zhu; Lin, T.; Song Han; Mok, A.; Deji Chen; Nixon, M.; Rotvold, E., Measuring WirelessHART against wired fieldbus for control. 2012 10th IEEE International Conference on Industrial Informatics (INDIN), 2012, pp. 270-275, DOI: 10.1109/INDIN.2012.6300852
9. On the Implementation and Performance Assessment of a WirelessHART Distributed Packet Analyzer, Ferrari, P.; Flammini, A.; Marioli, D.; Rinaldi, S.; Sisinni, E., Instrumentation and Measurement, IEEE Transactions on, Vol. 59, Issue: 5, 2010, pp. 1342-1352, DOI: 10.1109/TIM.2010.2040907
10. P. Ferrari, A. Flammini, S. Rinaldi, E. Sisinni, Performance assessment of a WirelessHART network in a real-world testbed. 2012 IEEE International Instrumentation and Measurement Technology Conference (I2MTC), Graz, Austria, May 13-16, 2012, pp. 953-957, DOI: 10.1109/I2MTC.2012.6229177
11. Xiuming Zhu; Song Han; Mok, A.; Deji Chen; Nixon, M., Hardware challenges and their resolution in advancing WirelessHART. 2011 9th IEEE International Conference on Industrial Informatics (INDIN), 2011, pp. 416-421, DOI: 10.1109/INDIN.2011.6034913
12. C. M. De Dominicis, P. Ferrari, A. Flammini, S. Rinaldi, E. Sisinni, On the improvement of WirelessHART Access Points by means of Software Defined Radio. 2010 8th IEEE International Workshop on Factory Communication System (WFCS), Nancy, France, May 18-21, 2010, pp. 71-74, DOI: 10.1109/WFCS.2010.5548620

An Ultrasonic Human–Computer Interface

L. Capineri, M. Calzolai, and A. Bulletti

1 Introduction

Portable devices have reached a remarkable level of miniaturization and other less portable products are following the trend. Due in part to human physiology, human–machine interfaces have evolved much more slowly. Moreover, with the increasing complexity of computers, interaction has become often cumbersome and limited, even in conventional laptops and desktops. The enabling factor of such interfaces will be a family of devices for capture of and interaction with three-dimensional realities and human features. This work describes the development of ultrasonic airborne transducers and related analog and digital electronics to demonstrate a human–machine interaction paradigm of the next generation. Through the devices supporting it, the new paradigm will remove any physical contact between the user and machine or between the user and a surface of any kind, thus reducing the interaction hardware to zero. The research started with the architecture design for a programmable analog front-end electronics capable of managing an array of 16 airborne ultrasonic transducers. The developed platform is shown in Fig. 1.

The implementation of the system required careful characterization of the transducers made with a flexible piezopolymer film (PVDF) with hemicylindrical shape. The resonating characteristics and equivalent transducer electrical impedance have been previously studied [1, 2]. The compact dimension of the array and ultrasonic field sampling required an innovative design of the mounting of the airborne transducers. The developed method is based on printed circuit board laminate material. The choice of central frequency determines the diameter of the cylinder that can be realized with appropriate thickness choice of multilayer FR4. The final assembly is obtained by blocking the whole stack with micro rivets with two external PCBs.

L. Capineri (✉) • M. Calzolai • A. Bulletti
Department of Information Engineering, University of Florence,
Via S. Marta 3, 50139 Florence, Italy
e-mail: lorenzo.capineri@unifi.it

C. Di Natale et al. (eds.), *Sensors and Microsystems: Proceedings of the 17th National Conference, Brescia, Italy, 5-7 February 2013*, Lecture Notes in Electrical Engineering 268, DOI 10.1007/978-3-319-00684-0_61, © Springer International Publishing Switzerland 2014

Fig. 1 Assembly of the 16 elements PVDF film array with 150 kHz ($\lambda_{air} = 2.2$ mm) airborne piezopolymer film transducers and the electronic platform

Fig. 2 Block scheme of the ultrasonic airborne real-time target tracking system

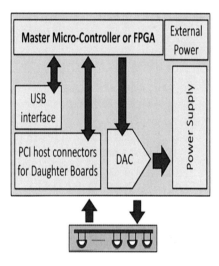

These two external PCBs can be thin to minimize the distance between transducers; the transducer pitch must be at maximum half wavelength for full image reconstruction without loss of information.

- Operating central frequency: 150 kHz
- Impedance modulus at central frequency: 5 kΩ
- Horizontal beam divergence: ±70° at −6 dB
- Vertical beam divergence: ± 7° at −6 dB
- Bandwidth: 19.5 kHz at −3 dB
- Transducer capacitance: 262 pF +/− 1 %

The designed architecture (see Fig. 2) exploits the processing power of advanced microcontrollers. In our case, the ATMEL 8 Bit ATXMEGA128A1 on board microcontroller is programmed by firmware to generate the excitation sequences to drive a MOSFET bridge power circuit efficiently at frequency up to 450 kHz. The same microcontroller does the acquisition at 12 bits resolution synchronously with the transmitted burst. Then the real-time target detection is implemented digitally by cross-correlation algorithm or by envelope calculation based on the phase and quadrature signals. The latter method requires more computational resources but it provides more accurate range estimation.

The analog front-end boards are then controlled by a master board that transfers setting parameters to the analog front-end board (operating frequency, burst duration, sampling window, VGA setting, etc.) and also does the data collection by polling the boards.

This platform allowed experimentation with different applications for target detection by synthetic aperture imaging or real-time target tracking for new human–computer interface.

2 Mother Board

With reference to Fig. 2, the main characteristics of the mother board electronics are reported below:

- I/O: USB connector for getting 5 V power supply and to communicate with the PC; in alternative banana connectors to get the power from external power supplies (both 5 and 100 V).
- Power: onboard non-isolated flyback to provide 100 V (software controllable) from the 5 V input.
- Master CPU: connector for Xilinx Spartan3 to drive all the channels with real-time capabilities; connector for Atmel USB keyboard, which hosts the AT90USB1287, an 8-bit microcontroller with integrated ADC, FLASH, SRAM and USB interface.

3 Daughter Boards

The analog front-end electronics of each daughter board is shown in Fig. 3. A daughter board controls two complete channels TX + RX made of 100 V H-bridge for bipolar driving; RX analog front-end; first configurable stage + second inverting stage; third

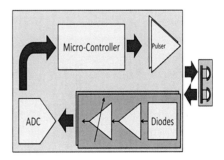

Fig. 3 Block scheme of the programmable ultrasonic front-end electronics

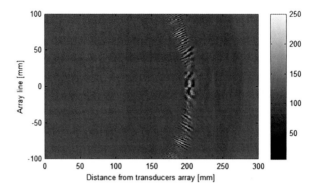

Fig. 4 Resulting imaging

	−3 dB	−6 dB
Table 1 Estimated spatial resolution after beamforming with radiofrequency signals		
Axial resolution	1 mm	2 mm
Lateral resolution	1 mm	2 mm

digitally controlled variable gain stage; analog to digital converter selectable, 12 bit ADC inside the microcontroller; CPU Atmel 8 bit/32 MHz ATXMEGA256A3B; and I/O connections.

4 Results

The imaging (see Fig. 4) is obtained by beamforming with integration of radiofrequency signals reflected from a 50 mm diameter target at distance 200 mm from the array.

The estimated spatial resolution is reported in Table 1. These values support the theory [3] and depend on transducer array characteristics.

Acknowledgments The authors wish to acknowledge the research grant from Texas Instruments for the development of the project.

References

1. L. Capineri, A.S. Fiorillo, L. Masotti, S. Rocchi, "Piezo-polymer transducers for ultrasonic imaging in air", IEEE Transactions on UFFC, vol. 44, No. 1, January 1997, pp. 36-43
2. L. Capineri, L. Masotti, S. Rocchi, "A 3D airborne ultrasound scanner", Measurement Science and Technology, 9, (1998), pp. 967-975
3. Gordon S. Kino, Acoustic waves: devices, imaging, and analog signal processing, Prentice-Hall, Lebanon, Indiana, 1987

Power Management Circuit Analysis for an Inductive Energy Harvester

A. Cadei, E. Sardini, and M. Serpelloni

1 Introduction

Energy harvesting represents an alternative power source technique to realize battery-less implantable medical device. In this paper, a specific kinetic energy harvester has been considered [1]. The electrical signal produced by a kinetic energy harvester is typically in AC form; therefore, the scavenged energy cannot directly power a device or a circuit and it cannot even be stored in a capacitor. For this reason, the energy harvester must be connected to a power management circuit that performs different functions: AC/DC conversion, DC/DC conversion, energy storage, and/or battery charging. In the literature, power management circuits can be divided in two categories: classic rectifiers and switched rectifiers [2, 3]. The first ones are typically composed of diodes and capacitors, while the second ones employ, in addition, switches to transfer energy from the harvester to the load. With respect to the classic rectifiers, the switched rectifiers have higher efficiency. However, this type of circuits is usually rejected for energy harvesting applications because they need smart switch drivers, which require additional energy consumption. Since the harvested energy is very low, the power management circuits should have high efficiency in order to maximize the output energy. The efficiency is influenced by several factors: impedance mismatch, loss related to the parasitic resistances in the capacitors, and diode losses.

In this paper, a comparison between power management circuits connected to a specific generator is presented. In particular, simulation results of three different power management circuits are described. Furthermore, experimental measurements with the most efficiency circuit have been performed.

A. Cadei (✉) • E. Sardini • M. Serpelloni
Department of Information Engineering, University of Brescia,
via Branze, 38, Brescia, Italy
e-mail: andrea.cadei@ing.unibs.it

C. Di Natale et al. (eds.), *Sensors and Microsystems: Proceedings of the 17th National Conference, Brescia, Italy, 5-7 February 2013*, Lecture Notes in Electrical Engineering 268, DOI 10.1007/978-3-319-00684-0_62, © Springer International Publishing Switzerland 2014

2 The Electromagnetic Energy Harvester

The abovementioned harvester is a non-resonant electromagnetic generator implantable in human total-knee prosthesis. As shown in Fig. 1, it is composed of six prismatic magnets placed in each condyle of the knee prosthesis and two copper coils disposed between the condyles, in a prominence of the tibial plate. Figure 3a shows the equivalent model of the harvester; the values of R_H and L_H, obtained by experimental measurements, are $280\,\Omega$ and 200 mH, respectively. The generator produces a voltage that depends on the cycle step. In particular, the harvester is able to generate a quasi-sinusoidal signal with an open-circuit voltage of about 2 peak-peak V with a gait frequency of 1 Hz.

3 Simulation Results

Three different power management circuits have been simulated through the PSpice software. The simulated circuits are the three-step Villard circuit, three-step Cockcroft–Walton [4] circuit, and two-step Cascode circuit [5] (Fig. 2b–d). All the circuits have capacitors with a value of 500 μF and they were simulated with an

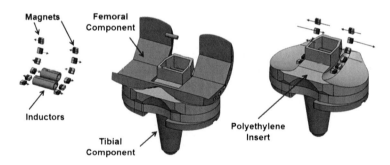

Fig. 1 Structure of the electromagnetic harvester described in [1]

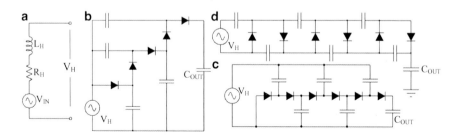

Fig. 2 (a) Model of the electromagnetic generator described in [1]. (b) Two-step Cascode circuit. (c) Three-step Cockcroft–Walton circuit. (d) Three-step Villard circuit

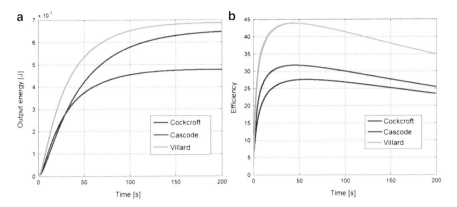

Fig. 3 Simulation results: (**a**) output energy and (**b**) efficiency

infinite load resistance. For each capacitor, a model with a series and parallel resistors was considered; the values of the series and parallel resistor, determined by means of an impedance analyzer (HP4194A), are of 25 mΩ and 3 MΩ, respectively. Each employed diode is a Schottky diode with a threshold voltage of about 200 mV. According to the above-said electromagnetic harvester, the V_{IN} generator has the following characteristics: sinusoidal signal, frequency of 10 Hz, and peak voltage of 1 V. The input energy is calculated by means of a time integration of the power generated by the harvester while the output energy is the energy stored in the output capacitor.

Figure 3a, b summarize the obtained simulation results. All the circuits convert the AC voltage, related to the electromagnetic harvester, into DC voltage; in particular, they achieve a steady-state voltage more than 4 V. In each circuit, the output voltage increases up to a certain value after which it remains constant. The output energy has the same trend (Fig. 3a) because of the relation between the voltage and stored energy in the capacitor: $E = \frac{1}{2}CV^2$. The three-step Villard circuit has the higher output energy; in steady-state condition, this circuit stores an energy of about 6.8 mJ in the output capacitor. In Fig. 3b, the efficiency of each circuit is reported. The three-step Villard circuit, the two-step Cascode circuit, and the three-step Cockcroft–Walton circuit have, respectively, a maximum energy efficiency of about 43 %, 32 %, and 27.5 %. Considering the obtained simulation results, the three-step Villard circuit is the most efficient circuit for this particular application.

4 Experimental Results

Experimental measurements with the Villard circuit have been performed. In particular, this circuit has been connected to a charge pump (Seiko S-882Z2) that stores the harvested energy in a capacitor, said start-up capacitor, and releases this

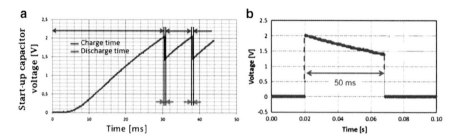

Fig. 4 Experimental results: (**a**) start-up capacitor voltage and (**b**) output load voltage during the discharge phase of the start-up capacitor

energy when it exceeds a specific value. The charge pump requires a DC voltage input to operate; this conversion is performed by the Villard circuit. The load resistance, connected to the charge pump, represents all the circuits that the electromagnetic harvester should supply; a value of 2.2 kΩ has been chosen for the output resistance [6].

Figure 4 shows the experimental results obtained with a gait cycle of 1 Hz. As shown in Fig. 4a, the first charging time of the start-up capacitor is 30.4 s. When the capacitor voltage exceeds about 2 V, the capacitor discharges through the output resistance. After the first charge and discharge, the charge time is 7.6 s. The useful time, during which the energy is provided to the load, corresponds to the capacitor discharge time and it is about 50 ms (Fig. 4b). With the start-up capacitor completely discharged, the whole power management circuit (Villard circuit and charge pump) provides an energy to the output load of 70 µJ; in this case, the experimental energy efficiency is about 7.5 %. With the start-up capacitor partially discharged, the energy efficiency is about 11 %.

5   Conclusions

In this paper, a comparison between power management circuits connected to a specific electromagnetic generator is presented. In particular, three different power management circuits have been simulated: three-step Villard circuit, three-step Cockcroft–Walton circuit, and two-step Cascode circuit. From the simulation results, the Villard circuit is the most efficient circuit for the electromagnetic harvester. Therefore, experimental measurements with this circuit have been performed. In particular, this circuit connected to a charge pump and load resistance of 2.2 kΩ has an efficiency of about 11 %. Currently, a system for the measurement of the tibiofemoral force to be connected to the described system is under development.

References

1. Luciano, V.; Sardini, E.; Serpelloni, M.; Baronio, G, Analysis of an electromechanical generator implanted in a human total knee prosthesis. Sensors Applications Symposium (SAS), 2012 IEEE, 7-9 Feb. 2012, pp. 1-5, 2012
2. Moschopoulos, G.; Jain, P.K.; "A novel single-phase soft-switched rectifier with unity power factor and minimal component count," *IEEE Transactions on Industrial Electronics*, vol. 51, no. 3, pp. 566- 576, June 2004
3. E. Arroyo, A. Badel, "Electromagnetic vibration energy harvesting device optimization by synchronous energy extraction", Sensors and Actuators A: Physical, Vol. 171, Issue 2, November 2011, pp. 266-273
4. Chung-ming Young; Ming-hui Chen; A novel single-phase ac to high voltage dc converter based on Cockcroft-Walton cascade rectifier. PEDS 2009. International Conference on Power Electronics and Drive Systems, 2009, 2-5 Nov. 2009, pp. 822-826, 2009
5. Halvorsen, T.K.; Hjortland, H.A.; Lande, T.S.; Power Harvesting Circuits in 90 nm CMOS. NORCHIP, 2008, 16-17 Nov. 2008, pp. 154-157, 2008
6. D. Crescini, E. Sardini, M. Serpelloni, "*Design and test of an autonomous sensor for force measurements in human knee implants*", Sensors and Actuators A: Physical 166(1): 1-8, 2011

Radio Frequency Energy Harvester for Remote Sensor Networks

V. Stornelli, A. Di Carlofelice, L. Pantoli, and E. Di Giampaolo

1 Introduction

Today batteries constitute the predominant source of electricity for compact low-power systems, especially for portable electronic devices. Despite the remarkable growth of their average duration, they still suffer degradation over time, therefore representing a limit to the duration of the whole system. Although the birth of low-cost batteries has encouraged the spread of mobile devices, they currently are holding back further expansion because their replacement and disposal are not possible in most applications to which the modern devices are designed. Consider, for example, a network of dozens or thousands of nodes scattered in a possibly hostile, inaccessible zone; in this case, the battery replacement becomes economically disadvantageous and physically difficult. A key role, in this sense, can be played by the energy harvesting and management technique [1, 2].

The concept of energy harvesting generally relates to the process of using ambient energy, which is converted, primarily, into electrical energy in order to power small and autonomous electronic devices [3–5].

There is a broad range of different energy domains within any typical environment, whether internal or external. Solar energy (light), radio frequency energy (RF), thermal energy (heat), and kinetic energy (motion) are examples of possible sources for harvesting electrical energy that from a typical outdoor environment is converted into an available low-voltage DC energy for autonomous system supply purposes [6] and low-power devices [7–15].

RF energy, in particular, is emitted by those sources that generate high electromagnetic fields such as TV signals, wireless radio networks, and cell phone towers and devices and is available, more or less, everywhere. In this perspective, we are

V. Stornelli (✉) • A. Di Carlofelice • L. Pantoli • E. Di Giampaolo
Department of Industrial and Information Engineering and Economics,
University of L'Aquila, L'Aquila, Italy
e-mail: vincenzo.stornelli@univaq.it

C. Di Natale et al. (eds.), *Sensors and Microsystems: Proceedings of the 17th National Conference, Brescia, Italy, 5-7 February 2013*, Lecture Notes in Electrical Engineering 268, DOI 10.1007/978-3-319-00684-0_63, © Springer International Publishing Switzerland 2014

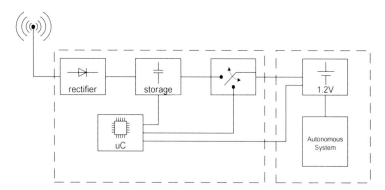

Fig. 1 Block diagram of the proposed system

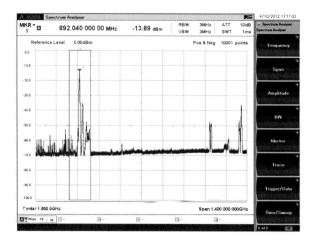

Fig. 2 Signal spectrum inside our laboratory within the university campus (600 MHz to 2 GHz)

developing a compact system architecture (see Fig. 1) able to harvest energy in the UHF region and manage it for portable sensor micro-batteries recharging purposes.

In this work, preliminary results of an RF harvesting system and power management architectural strategy for autonomous sensor network system will be presented. A particular attention has been given to the miniaturization of the antenna/rectifier structure; moreover, the measurement of the ambient RF power density in the utilized frequency range will also be presented.

In order to inspect the RF signal level of a low-density spectrum, suitable measurements have been conducted using a spectrum analyzer MS2830A and a log-periodic antenna. Figure 2 illustrates a snapshot of the experimental results obtained inside our laboratory within the university campus. From these measurements, it has been decided to optimize the antenna design, and consequently the harvester working frequency range, in the low UHF frequency band.

2   Implemented Architecture and Simulation Results

In order to convert the detected signals into a DC level, a balanced passive full-wave peak detector was designed by means of the implementation of a balanced antenna. The Thevenin equivalent schematic circuit (considering an unbalanced power source) consists of two Schottky (anode connected) diodes (it has been utilized an HSMS2850 by Avago Technologies) with a low built-in potential of approximately 0.19 V and a capacitor, as shown in Fig. 3. Schottky diodes offer low forward voltage and high switching speed and are widely considered as ideal component for RF energy harvesting.

In order to find out the best antenna impedance for the maximum power transfer into the storage capacitor, analytical and simulation experiments were conducted in the 0.5–1 GHz bandwidth at different RF input power level, and the results are presented in Table 1.

3   Conclusions

We have presented an RF harvesting system. A particular attention has been given to the miniaturization of the antenna/rectifier structure and to the full-wave rectification. The designed rectenna takes advantage from a balanced structure, allowing a full-wave rectification with only two diodes. The compactness of the used antenna allows it to be integrated in small devices area.

Fig. 3 Equivalent circuit of the antenna, T-match, and full-wave rectifier

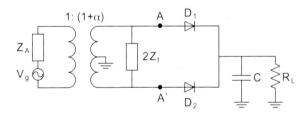

Table 1 Rectenna prototype measurement results

Distance [cm]	Output voltage with 100 kΩ load [mV]	Output voltage with 1 kΩ load [mV]
20	560	150
40	280	60
60	115	35
80	100	28
100	95	18
120	40	8
140	27	5

References

1. F.A. Mohamed, H.N. Koivo, Power management strategy for solving power dispatch problems in MicroGrid for residential applications. IEEE Energy Conference and Exhibition 2010, pp. 746–751, 2010
2. B.A. Allaf, Power system monitoring and analysis. IEEE Energy Conference and Exhibition 2010, pp. 297–301, 2010
3. T. J. Kaźmierski, Steve Beeby "Energy Harvesting Systems Principles, Modeling and Applications", Springer, New York, NY, 2011. ISBN 1441975659.
4. S. Priya and D. J. Inman, "Energy Harvesting Technologies", Springer, New York, NY, 2009. ISBN 9780387764634.
5. M. T. Penella-López and M. Gasulla-Forner, "Powering Autonomous Sensors An Integral Approach with focus on Solar and RF Energy Harvesting", Springer, New York, NY, 2011, ISBN 9789400715721.
6. S. Beeby and N. White, "Energy Harvesting for Autonomous Systems", Artech House, Norwood, MA, 2010, ISBN 9781596937185.
7. V. Stornelli. Low Voltage Low Power Fully Differential Buffer, Journal of Circuits, Systems, and Computers (JCSC), vol. 18, no. 3, pp. 497-502, 2009
8. G. Ferri, V. Stornelli and M. Fragnoli. An integrated improved CCII topology for resistive sensor application, Analog Integr. Circuits Signal Process, vol. 48, 247-250, 2006.
9. V. Stornelli, L. Pantoli, G. Leuzzi, G. Ferri. Fully differential DDA-based fifth and seventh order Bessel low pass filters and buffers for DCR radio systems, Analog Integrated Circuits and Signal Processing, May 2013, vol. 75, Issue 2, pp 305-310.
10. V. Stornelli, G. Ferri. A 0.18μm CMOS DDCCII for portable lv-lp filters, Radioengineering, Vol. 22, no. 2, p. 434, June 2013
11. P. Colucci, G. Leuzzi, L. Pantoli, "Third order integrable UHF bandpass filter using active inductors", Microwave And Optical Technology Letters, Vol. 54, no. 6 pp. 1426-1429, June 2012.
12. G. Leuzzi, V. Stornelli, S. Del Re, "A Tuneable Active Inductor With High Dynamic Range for Band-Pass Filter Applications", IEEE transactions on circuits and systems II, express briefs, vol. 58, pp. 647-651, 2011.
13. G. Ferri, V. Stornelli, A. di Simone A ccii-based high impedance input stage for biomedical applications, Journal of Circuits, Systems and Computers Vol. 20, No. 8, pp. 1441-1447, 2011.
14. G. Bucci, E. Fiorucci, C. Landi, G. Ocera. The Use of Wireless Network for Distributed Measurement Applications, Proc. IEEE Instrumentation and Measurement Technology Conference, Anchorage, USA, May 21-23, 2002
15. G. Bucci, E. Fiorucci, C. Landi, G. Ocera. Architecture of a digital wireless data communication network for distributed sensor applications, Measurement, Journal of International Measurement Confederation, vol. 35, Issue: 1, pp. 33-45, 2004.

Development of an Attitude and Heading Reference System for Motion Tracking Applications

Daniele Comotti, Michele Ermidoro, Michael Galizzi, and Andrea Vitali

1 Introduction

Inertial measurement units (IMUs) as well as magnetic, angular rate, and gravity (MARG) sensors based on low-cost MEMS are the most promising solutions for applications where an accurate and reliable orientation estimation of limbs or nodes is required, such as body motion tracking and navigation systems. MARG sensors and IMUs typically consist of a triaxial gyroscope, a triaxial accelerometer, and a triaxial magnetometer, combined together in order to provide the nine degrees of freedom needed to compute the orientation of the platform; if the processing is performed on board, then an attitude and heading reference system (AHRS) is achieved.

In motion tracking application, miniaturized and wireless AHRS can address the problem of attaching the sensor to the body and capturing the data without any impact on the body movements. These features, along with a low-power consumption for long-term usability, are also desirable for more comfortable and low-cost applications.

This work presents a novel wireless AHRS: a MARG sensor, a microcontroller, and a Bluetooth module have been assembled on the same board and have been combined with a light onboard algorithm in order to provide a low-power AHRS.

D. Comotti (✉)
Department of Electronics, Computer and Biomedical Engineering,
University of Pavia, via Ferrata 1, 27100 Pavia, BG, Italy
e-mail: daniele.comotti01@ateneopv.it

M. Ermidoro • M. Galizzi
Department of Engineering, University of Bergamo,
viale Marconi 5, 24044 Dalmine, BG, Italy

A. Vitali
STMicroelectronics, via Olivetti 2, 20864, Agrate Brianza, MB, Italy

C. Di Natale et al. (eds.), *Sensors and Microsystems: Proceedings of the 17th National Conference, Brescia, Italy, 5-7 February 2013*, Lecture Notes in Electrical Engineering 268, DOI 10.1007/978-3-319-00684-0_64, © Springer International Publishing Switzerland 2014

Even though orientation estimation of IMUs and MARG sensors has been widely investigated in the last years [1, 2], very few are the wireless and embedded full systems [3].

2 Hardware

The platform presented in this work is a wireless AHRS based on low-cost MEMS sensors and mounted on a printed circuit board (PCB). A system on board provided by STMicroelectronics, the iNemo-M1, is the sensing and processing unit with a $13 \times 13 \times 2$ mm^3 form factor. The wireless protocol is the Bluetooth V3.0 standard; during this initial development phase, the BT31 module provided by Amp'ed RF has been used. It provides a Class 1 Bluetooth V3.0 stack covering up to 100 m range in a 15×27 mm^2 form factor and comes with a wide set of commands to be used for configuration purpose. Figure 1 depicts the block diagram and a picture of the system. The power is supplied by a 3.7 V Li-ion battery with a 90 mA h charge and a $24 \times 10 \times 2$ mm^3 form factor mounted on the PCB.

3 Sensor Fusion Algorithm

The orientation of the platform is estimated using the 3D raw data provided by both the geomagnetic module and the gyroscope. Common classes of algorithms estimating the orientation are based on three steps: integration, vector observation, and Kalman filtering [4]. The integration step predicts the state by means of the gyroscope data, the vector observation step provides the state measurement using the magnetometer and the accelerometer raw data, and, finally, the Kalman filter is applied, providing the optimal estimators which minimize the error covariance [2]. Among the several methods to represent orientation, quaternions have been chosen

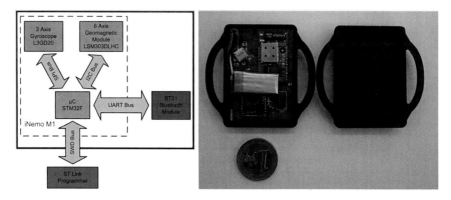

Fig. 1 Hardware architecture and picture of the presented AHRS

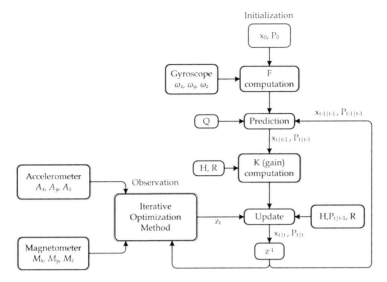

Fig. 2 Block diagram of the embedded Kalman filter estimating the orientation of the platform

because of their advantages in terms of efficiency and stability [2]. Hence, the state of the Kalman filter has been chosen as the quaternion representing the orientation:

$$x = \begin{bmatrix} q_w & q_i & q_j & q_k \end{bmatrix}^T \tag{1}$$

where q_w is the real component and q_i, q_j, and q_k are the imaginary components. The embedded sensor fusion algorithm, schematized in the block diagram of Fig. 2, consists of the following steps:

1. Prediction step, based on the angular rate acquisition. The state is predicted, $x_{t|t-1}$, and the state transition function is computed, F_t.
2. The Kalman gain, K_t, needed to perform the sensor fusion, is computed by means of the state transition function, F_t; the covariance matrix of the state prediction error, Q; and the covariance matrix of the state observation error, R.
3. The update step is performed by combining the state prediction, $x_{t|t-1}$, and the state observation, z_t. The latter component is computed by means of a Quasi-Newton method based on both the accelerometer and magnetometer raw data.

The initialization of the filter is needed during the start-up.

4 Performance

The proposed AHRS has been compared to the state-of-the-art solutions [5–7] in terms of orientation accuracy, power consumption, dimensions, maximum output data rate, and wireless capability. Concerning the orientation estimation, preliminary

Table 1 Overall performance of the presented AHRS with respect to commercial devices

	Static/dynamic RMSE (deg)						Power	ODR$_{MAX}$	Size	
	Roll	Pitch	Yaw	Roll	Pitch	Yaw	cons.	(Hz)	(mm^3)	Wireless
Xsens Mti 30	0.2	0.2	n.a.	0.5	0.5	1.0	160 mA 3.3 V	120	$57 \times 42 \times 23$	no
Xsens Mti 300	0.2	0.2	n.a.	0.3	0.3	1.0	245 mA 3.3 V	400	$57 \times 42 \times 23$	no
VN100 IMU	n.a.	n.a.	n.a.	n.a.	n.a.	n.a.	70 mA 5.0 V	300	$36 \times 33 \times 9$	no
X-io X-IMU	n.a.	n.a.	n.a.	n.a.	n.a.	n.a.	100 mA 3.6 V	512	$57 \times 38 \times 21$	BT Class 1
Proposed AHRS	0.2	0.4	0.4	1.5	1.1	0.9	50 mA 3.8 V	150	$52 \times 40 \times 15$	BT Class 1

test experiments have been performed on the sensor fusion algorithm at a 50 Hz output rate by means of a stereophotogrammetric motion analyzer (Smart Motion Capture System provided by BTS SRL, Italy) in both static and dynamic conditions. Each component of the roll, pitch, and yaw Euler angles related to the estimated quaternion has been studied separately by performing individual rotations along the x, y, and z axes with respect to an initial zero position and has been compared to the optical reference (truth reference) by computing both the RMS error (RMSE). Table 1 compares the overall performance of the proposed system with respect to commercial devices. As far as the orientation accuracy is concerned, the proposed AHRS is comparable to the state-of-the-art solutions in static conditions, whereas it is slightly less accurate during rotation. Moreover, the low-power performance, the small form factor, and the wireless capability are attractive features in AHRS designs for motion tracking applications.

5 Conclusions

In this work, the prototype of a wireless and low-power AHRS has been presented. An overview of the full development of the platform has been described, and the performance evaluation with respect to the state-of-the-art solutions has proved its good suitability for AHRS design. Even though some improvements are needed for consumer purposes, the presented AHRS is a promising solution for motion tracking applications. Ongoing activities are focused on the development of a new generation of AHRS, optimized from the standpoints of area occupancy and power consumption.

References

1. S. Madgwick, A. Harrison, and R. Vaidyanathan, Estimation of imu and marg orientation using a gradient descent algorithm. 2011 IEEE International Conference on Rehabilitation Robotics (ICORR), June 29 2011–July 1 2011, pp. 1–7, 2011
2. A. Sabatini, "Quaternion-based extended kalman filter for determining orientation by inertial and magnetic sensing," IEEE Transactions on Biomedical Engineering, vol. 53, no. 7, pp. 1346–1356, July 2006.
3. STMicroelectronics, "Xsens and stmicroelectronics demonstrate wearable wireless 3d body motion tracking," Xsens Press, Enschede, 2012
4. H. Bruckner, C. Spindeldreier, H. Blume, E. Schoonderwaldt, and E. Altenmuller, Evaluation of inertial sensor fusion algorithms in grasping tasks using real input data: Comparison of computational costs and root mean square error. International Workshop on Wearable and Implantable Body Sensor Networks, vol. 0, pp. 189–194, 2012
5. Xsens official web-site, http://www.xsens.com/en/mtw
6. VectorNav, "Vectornav vn100 product brief", http://www.vectornav.com/products/vn100-rug
7. X-IOTechnologies, "X-io technologies x-imu official-web site", http://www.x-io.co.uk/node/9

A Novel Method to Size Resistance for Biasing the POSFET Sensors in Common Drain Configuration

Arun Kumar Sinha and Daniele D. Caviglia

1 Introduction

The fabrication process of POSFETs is based on customized NMOS/CMOS technology [1, 2] and is derived from an ISFET (i.e. ion-sensitive field-effect transistor)-based process [3]. Sensor chips consist of square arrays of $N \times N$ devices, with NMOS transistors implemented in p-well on n-substrate. A 4 μm aluminium layer (i.e. bottom metal layer) is over 45 nm thick interdigitized gate consisting of SiO_2/Si_3N_4. The PVDF-TrFE (i.e. piezoelectric polymer) is spin coated on the bottom metal layer, with top metal layer implemented with an alloy of gold after which the polymer is poled in situ. The POSFET chip based on NMOS and CMOS has p-well, whose junction depth is 4.76 μm and 8 μm, respectively, and the sheet resistance is 3.5 kΩ/sq in both processes.

The specifications of POSFET chip based on the CMOS process are shown in Table 1. Additionally this POSFET chip features temperature-sensing diodes. The structural arrangement of POSFET (electrical model is shown in Fig. 1a) is similar to ISFET, as POSFET has a layer of piezoelectric polymer between metal plates on the gate of NMOS transistors, instead of ion-sensitive layer in the case of ISFET. Based on the work reported in [4], the arrangement of capacitances can be shown by Fig. 1b, where C_{PVDF} is the polymer capacitance and C_{gate} is the gate capacitance of the NMOS transistor.

A.K. Sinha (✉) • D.D. Caviglia
Department of Electrical, Electronics and Telecommunication Engineering,
University of Genova, Via opera Pia -11/A, Genoa, Italy

Department of Naval Architecture, University of Genova, Via opera Pia -11/A, Genoa, Italy
e-mail: arun.kumar.sinha@unige.it; daniele.caviglia@unige.it

C. Di Natale et al. (eds.), *Sensors and Microsystems: Proceedings of the 17th National Conference, Brescia, Italy, 5-7 February 2013*, Lecture Notes in Electrical Engineering 268, DOI 10.1007/978-3-319-00684-0_65, © Springer International Publishing Switzerland 2014

Table 1 Specifications of a type 2 CMOS-based POSFET chip used in our experiments

Specifications	Value
Dimension of PVDF-TrFE material	$W_{PVDF} = 626 \ \mu m$, $L_{PVDF} = 834 \ \mu m$
Interdigitized gate: W and L	7,268 and 12 μm
Lower and top metal electrodes	Aluminium and gold + chromium
Thickness of PVDF-TrFE material	2.5 μm
Taxel size/taxel centre to centre distance	$0.9 \times 0.9 \ mm^2/1 \ mm$
Chip size/number of POSFETs devices	$0.8 \times 1 \ cm^2/16$

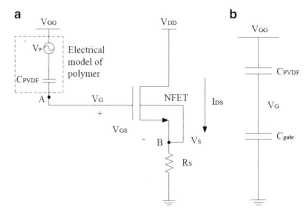

Fig. 1 (**a**) Electrical diagram of POSFET sensor along with resistance to bias the source terminal; V_{GG} is the potential at top metal layer and V_G is the potential at the gate of NMOS transistor. (**b**) Various capacitances of POSFET

2   Sizing the Resistance to Fix the Source Voltage of POSFET

Ignoring the overlap capacitances, the C_{gate} is related to the C_{OX} in weak inversion (WI) and in strong inversion (SI) as

$$C_{gate} = \left((n_n - 1)/n_n\right) \cdot C_{OX} \ (WI) \ and \ C_{gate} = (2/3) \cdot C_{OX} \ (SI). \tag{1}$$

In (1), n_n is the slope factor of the NMOS transistor. A transfer factor "A" will relate V_G to V_{GG} and hence the polymer capacitance with the gate capacitance

$$V_G = A \cdot V_{GG} = \left(C_{PVDF}/\left(C_{gate} + C_{PVDF}\right)\right) \cdot V_{GG}. \tag{2}$$

In (2), it is worth noting that "A" is a function of V_{GS}, which influences the bias point and the C_{gate}, where $C_{ox} = 7.66 \times 10^{-4} \ pF/\mu m^2$ and $C_{PVDF} = 53 \ pF$ for this sensor. The EKV equation [5] valid for long channel devices in all inversion regions of MOSFET is given by

$$I_{DSat} = I_{spec} \cdot \ln^2 \left(1 + \exp\left((V_{GS} - V_{TO})/n_n \cdot U_T\right)\right). \tag{3}$$

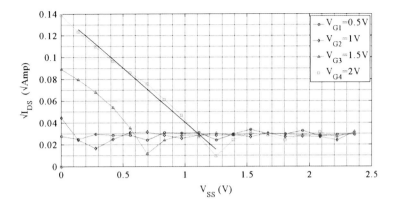

Fig. 2 Sqrt(I_{DS}) versus V_{SS} measurement of P11 showing large leakage current after the channel is pinched off

In (3), I_{spec} is the specific current and V_{TO} is the threshold voltage. Equation (3) can be expressed in terms of inversion factor (i.e. IF = I_{DSat}/I_{spec}), and after expressing in terms of V_G, we will get the source potential

$$V_S = V_G - \left(V_{TO} + 2 \cdot n_n \cdot U_T \cdot \ln \left(\exp \left(\sqrt{IF} \right) - 1 \right) \right). \tag{4}$$

In (4), V_G can be substituted by $A \cdot V_{GG}$ for exact sizing of bias resistance. From the definition of IF, we can fix the DC bias points in WI, when IF < 0.1; in moderate inversion, (MI) when 0.1 < IF < 10; and in SI, when IF > 10. The value of resistance for biasing can be calculated as

$$R_S = V_S / IF \cdot I_{spec}. \tag{5}$$

In (5), the value of I_{DSat} is the product of IF and I_{spec}, when the source voltage is fixed by resistance R_S. Also one has to ensure that the source potential should not approach the supply rail due to the large value of resistance.

3   Extraction of Parameters

Three important parameters, namely, n_n, I_{spec} and V_{TO} in (4), were extracted from SI region, because of the large subthreshold leakage current. Figure 2 shows the characteristics measurement of P11 device depicting this leakage.

The value of n_n can be extracted from the normalized transconductance

$$K_n = \left(2 \cdot L / n_n \cdot W \right) \cdot \left(-d\sqrt{I_{DS}} / dV_{SS} \right)^2, \tag{6}$$

where derivative is the slope of the curve, in the SI region of Fig. 2; a black line in the same figure shows the linear fitting with SI region of $V_{G4} = 2$ V.

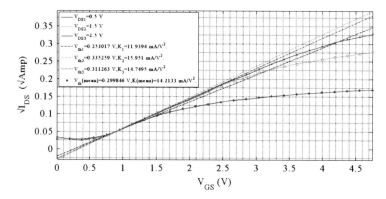

Fig. 3 Extracting transconductance and threshold voltage from the input characteristics of the P11 transistor

Table 2 Values of parameters extracted from the NMOS devices in POSFET chip

Device	K_n (W/L) mA/V^2	n_n	I_{spec} (µA)	V_{TO} (V)	A (WI)
P11	14.21	1.196	22.6	0.3	0.83
P14	14.1	1.33	25	0.25	0.76
P23	13.79	1.49	27.35	0.245	0.706
P33	14.7	1.206	23.6	0.34	0.822
P41	14.8	1.4	27.58	0.51	0.735
P44	14.28	1.183	22.49	0.64	0.836

The value of K_n can be substituted by the value of transconductance parameter, obtained from the input characteristics measurements of Fig. 3 [6], and the value of specific current will be $I_{spec} = 2 \cdot n_n \cdot K_n \cdot U_T^2$. The value of subthreshold leakage current is larger than the value of bias current ($I_B = I_{spec}/2$) to be used in the pinch-off characteristics measurement. Therefore, the measured value of the threshold voltage given by the pinch-off characteristic measurement was higher than the expected value. So we have used threshold voltage of the level 1 equation extracted from the input characteristics measurement. In EKV equation of (4), this threshold voltage can be suitably used, because the body effect is avoidable due to the connection of well with source in the NMOS transistor [7].

4 Results

The results presented in this work are of POSFETs (row, column), 11, 14, 23, 33, 41 and 44, which are sufficient to convey our experimental results. Table 2 shows the list of values extracted for six transistors; the value of "A" at SI was calculated to be 0.543 for all the devices.

Tables 3 and 4 show the value of the calculated parameters and the observed value of source potential for the known value of resistances. The supply voltage (i.e. V_{DD}) to the POSFET chip was 2.5 V, and the value of slope factor was taken,

Table 3 Calculated and observed values of bias point at the source terminal for $V_G = 1.5$ V

	P11		P14		P23		P33		P41		P44	
	V_S (V)		V_S (V)		V_S (V)		V_S (V)		V_S (V)		V_S (V)	
R_S (Ω)	Obs.	Cal.	Obs.	Cal.	Obs.	Cal.	Obs.	Cal.	Obs.	Cal.	Obs.	Cal.
220	0.52	0.57	0.55	0.56	0.56	0.51	0.54	0.52	0.411	0.39	0.33	0.36
2.2 k	0.86	0.94	0.9	0.97	0.91	0.95	0.88	0.9	0.71	0.74	0.6	0.65
22 k	1.06	1.07	1.12	1.12	0.9	1.18	1.09	1.08	0.9	0.93	0.78	0.8
0.22 M	1.2	1.23	1.25	1.28	1.26	1.3	1.23	1.17	1.04	1.03	0.92	0.89
2.2 M	1.32	1.3	1.37	1.37	1.38	1.39	1.35	1.27	1.16	1.12	1.04	0.97

Table 4 Calculated and observed values of bias point at the source terminal for $V_{GG} = 2.5$ V

	P11		P14		P23		P33		P41		P44	
	V_S (V)		V_S (V)		V_S (V)		V_S (V)		V_S (V)		V_S (V)	
R_S (Ω)	Obs.	Cal.	Obs.	Cal.	Obs.	Cal.	Obs.	Cal.	Obs.	Cal.	Obs.	Cal.
220	0.4	0.42	0.51	0.42	0.44	0.37	0.32	0.38	0.28	0.25	0.08	0.2
2.2 k	1.46	1.5	1.55	1.37	1.43	1.2	1.3	1.45	1.3	1.08	0.79	1.24
22 k	1.7	1.65	1.86	1.52	1.45	1.45	1.6	1.64	1.6	1.26	1.01	1.39
0.22 M	1.86	1.8	1.94	1.68	1.68	1.56	1.76	1.72	1.62	1.37	1.35	1.48
2.2 M	1.96	1.88	2	1.77	1.77	1.65	1.9	1.82	1.77	1.46	1.58	1.56

Table 5 Inversion factor given by different values of resistance

R_S (Ω)	P11	P14	P23	P33	P41	P44
220	105	100	93	104	68	67
2.2 k	17	16	15	17	12	12
22 k	2	2	1.5	2	1.5	1.6
0.22 M	0.24	0.23	0.2	0.34	0.17	0.18
2.2 M	0.026	0.025	0.023	0.026	0.019	0.02

same in all the regions of inversion. To calculate the bias point in MI, we have taken the "A" case of WI. In Tables 3 and 4, the observed value of source voltage was recorded using Agilent 34401A digital multimeter.

In Table 3, the gate voltage of NMOS is fixed to 1.5 V: The measured and the calculated values of the source voltages (i.e. V_S) closely agree. This demonstrates that our method of extracting the parameters is valid for this particular case of leaky transistor. It should be noted that the same value of "n_n" was taken in all the regions of inversion, because the transistors have well to source connection (shown in Fig. 1a); hence the body effect is avoidable.

In Table 4, the top metal layer of the sensor has been fixed at the supply voltage. With minor variations, the observed and calculated values of source voltages match. Again, such variations are due to the fact that potential at V_S depends on A, whose value in term depends on C_{PVDF} which is taken as constant (i.e. 53 pF) for all the devices.

Table 5 shows the value of the inversion factor realized by different values of resistance, which proves that the NMOS devices were biased in all the regions of inversion.

5   Conclusions

Thus, the following conclusions can be made regarding this sensor:

- The transistors in the POSFET chip exhibit large values of subthreshold leakage current.
- The value of "n_n" can be extracted from SI region and can be used for all the regions of inversion in the case of well to source-connected transistors. The value of V_{TO} can be substituted by the level 1 threshold voltage in this case of leaky transistors.
- The coupling factor "A" along with the technological parameters explains the reason for the bias point variation.
- Our work is an extension of cited work [8] and can be widely applicable for any electrical sensor based on POSFET configuration. This work gives better insight into the electrical behaviour of the sensor.

References

1. R. S. Dahiya *et al.*, Towards tactile sensing system on chip for robotic applications, IEEE Sensors J., **11**(12), 3216-3226, (2011).
2. R.S. Dahiya, G. Metta, M. Valle, A. Adami, and L. Lorenzelli, Piezoelectric oxide semiconductor field effect transistor touch sensing devices, Appl. Phys. Lett., **95**, 034 105-1–034 105-3, (2009).
3. S. Martinoia, N. Rosso, M. Grattarola, L. Lorenzelli, B. Margesin, and M. Zen, Development of ISFET array-based microsystems for bioelectrochemical measurements of cell populations, Biosensors and Bioelectronics, **16**(9–12), 1043-1050, (2001).
4. P. Georgiou and C. Toumazou, ISFET characteristics in CMOS and their application to weak inversion operation, Sensors and Actuators B: Chemical, **143**(1), 211–217, (2009).
5. C. C. Enz, F. Krummenacher, and E. A. Vittoz, An analytical MOS transistor model valid in all regions of operation and dedicated to low-voltage and low-current applications, Analog Integrated Circuits Signal Processing-Kluwer Academic Publishers, **8**(1), 83–114, (1995).
6. A.K. Sinha and M. Valle, A scheme for measuring and extracting level-1 parameter of FET device applied toward POSFET sensors array, the 23rd Int. Conf. on Microelectronics, Tunisia, 1-5, (2011)
7. G.A.S. Machado, C.C. Enz and M. Bucher, Estimating key parameters in the EKV MOST model for analogue design and simulation, IEEE Int. Sym. on Circuits and Systems, **3**, 1588–1591, (1995).
8. L. Barboni, M. Valle, and R. Dahiya, POSFET touch sensing devices: bias circuit design based on the g_m/I_d parameter, the 16th Annual Conf. on Sensors and Microsystems, 1–2, (2011)

A Novel Wireless Battery Recharge System for Wearable/Portable Devices

M. Galizzi, M. Caldara, V. Re, and A. Vitali

1 Introduction

In recent years, a growing interest in wireless power-charging technologies has been experienced in the market of consumer products. The developed wireless power charger system focuses its application to low-power wearable and portable devices, where a safety or galvanically insulated package is required and near field wireless transmission of electrical energy is the only way to provide enough power to recharge the internal battery.

A Wireless Power Consortium's "Qi" compliant compact prototype, composed by a full H-bridge power transmitter and a power receiver, has been developed (Fig. 1).

2 Wireless Power Transmitter Design

The wireless power transmitter complies to type A2 full H-bridge WPC topology that allows the system to work with a constant frequency. The total amount of transferred power is controlled by varying the full H-bridge supply with a PID feedback algorithm. In order to ensure an adequate power transfer control, the Wireless Power Consortium requires a 50 mV bridge supply accuracy control. A high-efficiency run-time variable step-down switching regulator (A5975D) has been used to ensure both a low-power dissipation and a proper voltage control. The resulting power supply controller ensures an efficiency up to 90 % and steps size of generated power

M. Galizzi (✉) • M. Caldara • V. Re
Department of Engineering, University of Bergamo, Viale Marconi 5, 24044 Dalmine, Italy
e-mail: michael.galizzi@unibg.it

A. Vitali
STMicroelectronics, Agrate Brianza, MB, Italy

C. Di Natale et al. (eds.), *Sensors and Microsystems: Proceedings of the 17th National Conference, Brescia, Italy, 5-7 February 2013*, Lecture Notes in Electrical Engineering 268, DOI 10.1007/978-3-319-00684-0_66, © Springer International Publishing Switzerland 2014

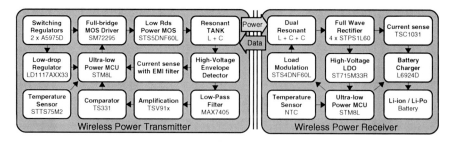

Fig. 1 Block schematic of the developed wireless power charger architecture

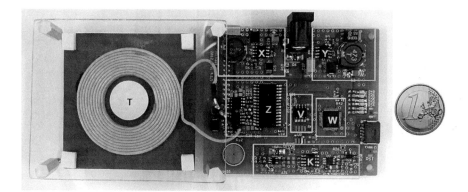

Fig. 2 The developed wireless power transmitter. The block X is the run-time variable switching regulator for full-bridge (Z) supply, the block W is the microcontroller supplied with another switching regulator (Y). The block K is the power carrier demodulator, and V is the digital temperature sensor. T is the primary coil (i.e., a 24 μH wire-wound planar inductor). The PCB can be placed under the primary coil to reduce space use

supply of 45 mV in the Qi standard range 3–12 V. The generated voltage supply is also read back with a resolution better than 3 mV by means of a 12bit microcontroller's ADC; thus, the read voltage has a much greater accuracy than required by WPC, which is 5 mV.

The feedback information is modulated on power carrier by the power receiver using the load modulation technique. The power transmitter demodulates the power carrier using a high-voltage diode-based (STPS2L40U) envelope detector of the resonant signal between primary coil and resonance capacitor. This high-voltage signal containing the envelope of the carrier is then subtracted from the high-voltage DC and is filtered with an 8th-order-switched capacitor low-pass filter (MAX7405). The filtered signal is then amplified with low-power operational amplifiers (TSV91x) and thresholded with the internal microcontroller's comparator. This analog power carrier demodulator ensures both low-power consumption and high robustness in extracting the feedback digital information from the power carrier (Fig. 2).

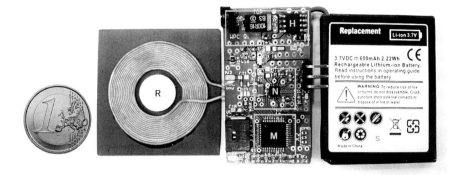

Fig. 3 The developed wireless power receiver. The block *H* is the dual resonance tank with capacitive and resistive load modulation. The block *N* is the battery charger used to charge the Li-ion battery (*S*). Block *M* the microcontroller for power control. Current sense block is on the bottom side

3 Wireless Power Receiver Design

The developed wireless power receiver makes use of an ultra-low-power down-clocked 8bit microcontroller (STM8L running down to 1 MHz) to ensure power control with low-power consumption. The microcontroller is powered directly with rectified voltage using a high-voltage LDO (ST715M33R), ensuring no consumption when not placed on power transmitter. The modulation of power carrier is made with pure capacitive loads on the AC side of power receiver using two low-Rds (on) power MOSFETs (STS4DNF60L), thus to ensure the minimum envelope of 200 mV of the power carrier, required by the standard.

A linear battery charger (L6924D) allows to charge Li-ion or Li-Po battery with a current up to 1 A, directly using the rectified power. Since the battery chargers are linear components, their power dissipation strictly depends on the voltage drop from input voltage to battery voltage. A battery voltage follower algorithm implemented into the power receiver's microcontroller ensures that the voltage drop on the battery charger (which is the rectified voltage minus the battery voltage) is closest as possible to its minimum value, providing a great reduction in power dissipation.

Single brand (STMicroelectronics) components have been chosen for the power receiver, making it suitable to be integrated in a system on chip [2]. The implemented firmware in both transmitter and receiver's microcontrollers aims to be compliant with the latest version of Wireless Power Consortium's directives [1], and it will be easily maintained up to date (Fig. 3).

4 Overall Performance

In this paper, a type A2 topology wireless power transmitter has been shown, and a simple, compliant power receiver has been illustrated. The realized prototypes have been tested with other certified wireless power systems and have proven to work as designed.

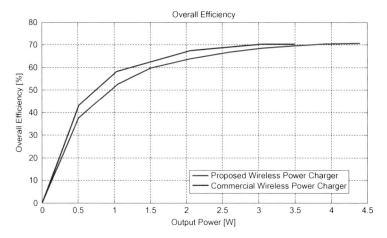

Fig. 4 Overall wireless power transfer efficiency compared with a commercial wireless power charger

Power transfer efficiency has been carefully measured, and Fig. 4 shows that the overall efficiency strictly depends on received power. An overall peak efficiency of 70 % is obtained at the maximum power, thereby respecting WPC efficiency requirements [1].

5 Conclusion

A complete solution of wireless power charger architecture has been developed with high-power transfer efficiency. The developed system is suitable for portable and mobile application in which a galvanically insulated package is required for safety reasons, such as in biomedical devices [3]. The power receiver draws no current while is not placed on a power transmitter, and the power transmitter standby consumption when grid-connected is lower than 150 mW; thus, the system is Energy Star compliant too.

References

1. Wireless Power Consortium, System description wireless power transfer Vol. 1, v1.1.1, Part I, II and III, 2010
2. V. Boscaino, F. Pellitteri, G. Capponi, R. La Rosa, A Wireless Battery Charger Architecture for Consumer Electronics. IEEE Second International Conference on Consumer Electronics, September 2012, pp. 84–88.
3. M. Caldara, C. Colleoni, M. Galizzi, E. Guido, V. Re, G. Rosace, A. Vitali, Low power textile-based wearable sensor platform for pH and temperature monitoring with wireless battery recharge. IEEE Sensors, October 2012, pp. 29–32.

Assessment of Performances of Optical Ceramics-Based Device for Monitoring of Electrical Components

L. De Maria, D. Bartalesi, and N.C. Pistoni

1 Introduction

One of the main tasks for smart grid sensors will be the real-time monitoring of electrical components of the Medium Voltage (MV) network, since failures of these components, strongly correlated to their ageing, typically represent a high percentage of outages in an MV distribution network.

RSE developed a fibre optic multisensor prototype for an early detection in MV switchboard of predischarges; these phenomena, on a long term, can give rise to breakdown and consequently to out of service [1]. One main feature of these predis-charge phenomena is their distinctive pattern correlated with both phase and polarity of the applied voltage (50 Hz). Diagnostic analysis of these events generally requires a synchronization between acquired signals and electrical phase of the MV switch-board to assess the degradation of the MV switchboard insulations. Also for the RSE prototype, it was proposed an automated temporization in order to improve its functionality in practical applications. In this work, it was proposed and investigated the feasibility of a fibre optic "trigger" sensor to be integrated into the developed prototype.

L. De Maria (✉) • D. Bartalesi
Department of Technologies for Transmission and Distribution,
RSE, via Rubattino 54, Milan, Italy
e-mail: demaria@rse-web.it

N.C. Pistoni
Optoelectronics and Fibre Optics Consultant, via del Parco 12/A, Assago, MI, Italy

C. Di Natale et al. (eds.), *Sensors and Microsystems: Proceedings of the 17th National Conference, Brescia, Italy, 5-7 February 2013*, Lecture Notes in Electrical Engineering 268, DOI 10.1007/978-3-319-00684-0_67, © Springer International Publishing Switzerland 2014

2   Fibre Optic "Trigger" Sensor

The main requirement for this application is the accurate detection of the phase of
the electric field in the MV switchboard: the device should be able to accurately
detect and to follow the electric field changes at the network frequency (50/60 Hz)
without introducing significant perturbations on the MV substation layout. It was
firstly investigated the possibility to directly detect the electrical field inside the MV
switchboard; in this configuration, the optical "trigger" probe should be located
close to one of the three electric phases (R, S, T phases) shown in Fig.1a. It was
obtained that this configuration leads to an ambiguous determination of the electric
phase. Figure 1b shows the electric field map in the XY section obtained by finite
elements nearby phase termination and insulators inside the MV compartment:
it clearly shows that the electric field of the S central phase is influenced by the
contribution of the other two phases, compromising the accurate phase detection.
As an alternative layout was proposed, the connection of the optical sensor to capacitive dividers currently mounted in MV substation [2]. In this last case, a low-cost
transparent electro-optic ceramic based on Kerr effect was proposed as optical
trigger solution [3], in which measured intensity changes are proportional to the
intensity of the applied electric field [2] (Fig. 1).

This material is generally used as a transducer element of electronically variable
optical attenuators (VOA). In the following, in view of a further adaptation of the
sensor layout as optical trigger, we started with a commercial VOA (υVOA, Bati [4]),
in order to perform a preliminary characterization of this ceramic material.

The sensor layout is based on a 1,550 nm laser source and an IR photodiode;
all these elements are telecom industry components with standard fibre pigtails to
guarantee low cost, easy assembling and high reliability.

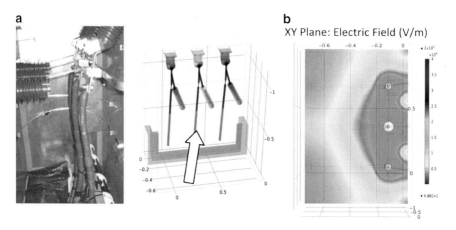

Fig. 1 (**a**) Electric phase termination inside of the MV switchboard; (**b**) Electric field mapping
(XY plane) with finite elements modelling in proximity of the central electrical phase (S, *arrow*)

3   Test Results

A preliminary characterization of the optical ceramic device was performed in laboratory under AC and DC voltage, applied by means of a voltage calibrator. Figure 2a displays the recorded optical ceramic sensor response (upper trace) to an AC applied voltage (trace below): a distortion on the optical signal for one of the applied polarity was observed. The same asymmetric behaviour was also evidenced under DC voltage (0–100 V range) values with the same amplitude and different polarities (normalized T and T_{inv} in Fig. 2b): T curve shows a first minimum at 32 V, while T_{inv} at 44 V. This unexpected sensor behaviour is probably due a piezoelectric effect (or to a not well-established fabrication process???); this intriguing behaviour could be exploited for the discrimination between positive and negative polarities of the applied 50 Hz.

The sensor response was also simulated with Jones matrix formalism, and a behaviour proportional to $\cos^2[\pi/2*(V/V\pi)^2]$ was obtained. Simulated responses of the sensor to direct and inverse DC voltage, obtained using measured attenuation values of T and T_{inv} curves as $V\pi$ values, are shown in the graphs of Fig. 2c and d, respectively, confirming the good fitting between the theoretical model and the measured behaviour in the 0–60 V voltage range.

Successive tests were carried out on a standard MV switchboard (20 kV), in which the sensor was connected to capacitive dividers. Design and parameters of the sensor were set for a low detection limit of the phase voltage equal to the 25 % of the voltage operation (20 kV). Figure 3 reports some examples of the comparison

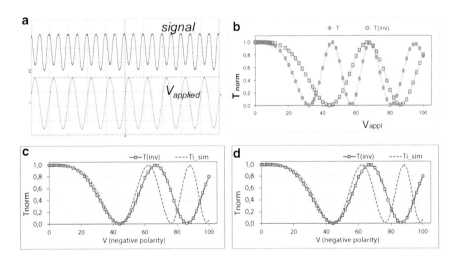

Fig. 2 Sensor response: (**a**) optical response (*upper trace*) recorded under AC voltage; (**b**) normalized sensor transmission (T and T_{inv}) with different polarity DC value (0–100 V); (**c**) measured sensor transmission and simulation curve for DC positive polarity; (**d**) measured sensor transmission and simulation curve for DC negative polarity

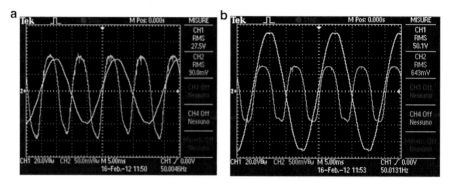

Fig. 3 Traces on scope display of 50 Hz output from the TV instrument and signal output from the optoceramic sensor: (**a**) Vappl = 5.5 kV (**b**) Vappl = 10 kV (trovare originale!!!!)

between the 50 Hz coming out from the TV and output signals from the sensors recorded for increased applied voltage values.

The optical trace wholly reproduces the phase and polarity of the applied 50 Hz; it is also able to detect distortions on the applied voltage probably due to the unloaded operation of the electrical plant. A good agreement between sensor and model results was evidenced for the range of voltage values of interest.

4 Conclusions

In this work, preliminary results of the feasibility of application of new transparent electro-optical ceramics as optical trigger for MV switchboards confirmed the capability of this transducer for an accurate detection of the phase of the electric field. The proposed optical layout also offers advantages of low cost, robustness and high immunity to electromagnetic interferences required for safe operation in HV and MV environment. Further improvements are ongoing to optimize the layout of this optical trigger and its integration in the diagnostic prototype.

Acknowledgments This work has been financed by the Research Fund for the Italian Electrical System under the Contract Agreement between RSE (formerly known as ERSE) and the Ministry of Economic Development—General Directorate for Nuclear Energy, Renewable Energy and Energy Efficiency stipulated on July 29, 2009 in compliance with the Decree of March 19, 2009. Authors also acknowledge D. Paladino for helpful discussions.

References

1. L. De Maria, D. Bartalesi "A fiber-optic multisensor system for predischarges detection on electrical equipment" IEEE Sens J 12: 207-212 (2012)
2. L. De Maria Patent Pending N. MI2012A001435.
3. H. Jiang, Y. K. Zou, Q. Chen, K. K. Li, R. Zhang, Y. Wang, Transparent Electro-Optic Ceramic and Devices. Proc. SPIE 5644, p. 380 (2005)
4. BATi, Photonics Devices *www.bostonati.com*

Characterization of a Novel Ultra-low-Power System on Chip for Biopotential Measurement

M. Caldara, V. Re, and A. Vitali

1 Introduction

Cardiovascular disease (CVD) is the major cause of death worldwide [1]. The proportion of all death due to CVD is increasing dramatically not only in developed countries but also in low-income and middle-income countries. Targeted screenings for high-risk patients are an effective approach to CVD early diagnosis, but it must cope with the present health budget reduction. In recent years, driven also from the fitness demands, microelectronics companies and research institutions started developments of high-accuracy and low-cost system on chip aimed to miniaturize electrocardiographs toward wearable solutions [2]. STMicroelectronics, in the framework of a collaboration with University of Bergamo, provided a novel ultra-low-power system on chip (SoC), named ReISC3, designed mainly for biopotential measurement and processing.

2 System Description

The miniaturized IC ReISC3 (90 nm CMOS process, 132 leads, dimensions $8 \times 8 \times 0.4$ mm^3) incorporates an ultra-low-power 32 bit microcontroller operating at 30 MHz maximum (power consumption 70 µW/MHz), embedded memories (1MByte Flash memory and 66 KByte SRAM), an extensive range of enhanced peripherals (SPI, USART, I2C, USB, four timers, two GPIO ports, one 12 bit 16 channels ADC)

M. Caldara (✉) • V. Re
Department of Engineering, University of Bergamo, Viale Marconi 5, Dalmine 24044, Italy
e-mail: michele.caldara@unibg.it

A. Vitali
STMicroelectronics, Agrate Brianza (MB), Italy

C. Di Natale et al. (eds.), *Sensors and Microsystems: Proceedings of the 17th National Conference, Brescia, Italy, 5-7 February 2013*, Lecture Notes in Electrical Engineering 268, DOI 10.1007/978-3-319-00684-0_68, © Springer International Publishing Switzerland 2014

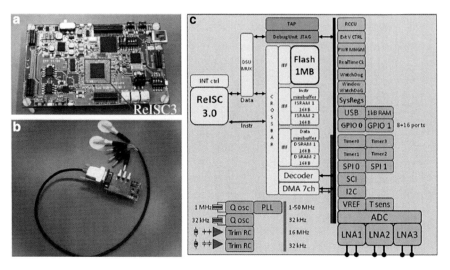

Fig. 1 ReISC3 prototype board (**a**), the setup used for characterization (**b**) and the SoC block diagram (**c**)

and low-noise analog front end (AFE) section fully programmable in terms of gain, bandwidth and offset (Fig. 1). A comprehensive set of power-saving modes allows designing low-power applications. In particular, in run mode, individual clock gating is possible for each peripheral, until interrupt occurrence, or power supply can be managed in blocks (AFE, ADC-PLL, Flash). The ReISC3 analog front end (AFE) is composed by three differential chopper amplifiers with a power consumption of only 6 µW/channel, particularly suited for low-frequency signal amplification, since the 1/f noise, dominant for low frequencies, is effectively reduced. On chip a 16-channel multiplexer dispatches the signals to a 12 bit SAR ADC; at maximum ADC clock rate (20 MHz), the sampling rate can be set between 6 kS/s and 1.25 MS/s. The prototype board is depicted in Fig. 1; all the ReISC3 pins are available on high-density connectors for tests purposes. The board includes, additionally to the SoC to be characterized, power supply stages, USB interface and crystals for clock generation. In order to perform the ReISC3 AFE characterization and the preliminary ECG measurements, a dedicated setup, shown in Fig. 1, has been developed. The AFE can be set by firmware at different gains comprised between 11.5 and 32.5 dB; its band-pass filter, having extreme cutoff frequencies of 0.05 Hz and 142 Hz, respectively, for the HP and LP, can be adjusted by firmware depending on the application.

3 Characterization

The whole electronics chain was characterized in terms of gain, bandwidth and noise. Figure 2 shows that the output noise is due to the first amplification stage. Moreover, the computed input-referred noise is about 0.5 mVrms (42 µV/$\sqrt{\text{Hz}}$),

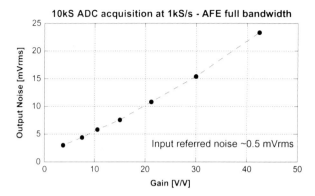

Fig. 2 Output noise of the whole electronics chain (AFE + ADC) at all settable gains, keeping the full AFE bandwidth (0.05 ÷ 142 Hz)

Fig. 3 ECG Lead I acquisition in time domain; signal is represented as raw and after processing with ANC algorithm

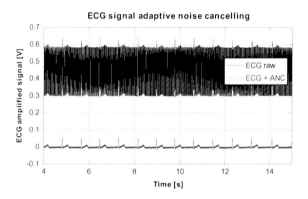

despite the declared value being $130\,\mathrm{nV}/\sqrt{\mathrm{Hz}}$. The acquired noise level is generated by the wired connection to the PC for data upload; nevertheless, it indicates that an ECG signal (with typical amplitude of a few mV), in principle, can be detected by this system. Preliminary Lead I ECG acquisitions were performed using the setup depicted in Fig. 1. The ECG signal frequency content can extend from DC up to 150 Hz; thus, it is necessary to deal with the power-line interference (PLI) issues. Figures 3 and 4 report, respectively, the acquired ECG signal in time and frequency domains, sampled at 10 kS/s; the PLI contribution is clearly visible in both graphs. In particular, on the FFT, it is evident that the body collected up to the 15th harmonic of the 50 Hz. Unfortunately this pattern, which falls in the ECG spectrum content, cannot be effectively filtered by hardware. Because of this, an adaptive noise cancellation technique (ANC) [3] has been implemented in order to track the PLI and continuously try to minimize its contribution. Figures 3 and 4 report the ANC filter results on the ECG raw signal; the filtered signal is potentially usable for heart rate determination or for diagnosis purposes.

Fig. 4 FFT of the signals depicted in Fig. 3. After ANC algorithm, the PLI harmonics are dramatically reduced

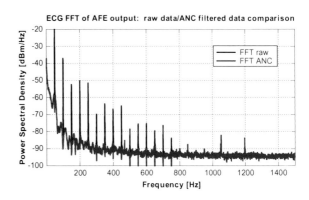

ECG FFT of AFE output: raw data/ANC filtered data comparison

4   Conclusions

The paper presented a novel SoC, named ReISC3, designed for biopotential measurement and processing. The SoC has been characterized and programmed in order to perform ECG measurements. Preliminary Lead I acquisitions have been reported, showing a major contribution of PLI, which has been effectively filtered with an adaptive noise cancelling algorithm. The system will be expanded for three-lead measurements and processing.

Acknowledgments The authors are grateful to STM for providing the hardware and the support for firmware.

References

1. "An epidemic of risk factors for cardiovascular disease", The Lancet, editorial, Vol. 377, February 12 2011.
2. M. Chan, D. Estève, J.-Y. Fourniols, C. Escriba, E. Campo, "Smart wearable systems: Current status and future challenges", Artificial Intelligence in Medicine, Available online 1 November 2012, ISSN 0933-3657, 10.1016/j.artmed.2012.09.003.
3. B. Widrow et al., "Adaptive Noise Cancelling: Principles and Applications", Proceedings of the IEEE, Vol. 63, No. 12 (1975), Pages. 1692–1716.

A Gas Multisensor Platform for Biometric and Environmental Applications

F. Armani, A. Boscolo, A. De Vecchi, and A. Palombit

1 Introduction

Gas sensing techniques can be found in a wide range of applications, from alcohol breath tester to engine exhaust emission sensor. Gases are often monitored because they represent a good precursor or indicator of events of interest, such as fire, degradation of food or nonbreathable air. The main advantage of using these sensors is the noncontact measurement. Gas sensing techniques for biometrics in particular are not well developed yet, with only few exceptions, as alcohol and oxygen breath sensors. This is because biometric applications often require disposable sensors to comply with hygiene and safety normative. Although the wide range of gas sensing materials, the most used of them presents some drawbacks in terms of integration in low power, low cost and ease of use sensors [1]. Among others, polymer-based gas sensors represent one of the most promising solutions to address these drawbacks [1, 2].

Some polymers such polypyrrole are well known in the literature to be very sensitive to a wide range of gaseous species [2, 3]. They are also of high interest for their functionalisation capabilities. When doped with the right component, the selectivity of these polymers can be highly increased, leading to very versatile and general purpose sensor platforms [4]. A single sensor board can be tuned to detect different gases just by controlling the doping species. The secondary transducer in polymer-based sensors relies on different types of interaction: electrical, thermal, optical and mechanical. Common transducing methods measure electric impedance, refractive index, viscosity, elasticity, mass or thermal capacity [5, 6].

F. Armani (✉) · A. Boscolo · A. De Vecchi · A. Palombit
APL—Department of Engineering and Architecture,
University of Trieste, via Valerio 10, Trieste 34127, Italy
e-mail: francesco.armani@phd.units.it

C. Di Natale et al. (eds.), *Sensors and Microsystems: Proceedings of the 17th National Conference, Brescia, Italy, 5-7 February 2013*, Lecture Notes in Electrical Engineering 268, DOI 10.1007/978-3-319-00684-0_69, © Springer International Publishing Switzerland 2014

The main objective of this study is to implement a novel sensor platform in compliance with the requirements of low-power consumption, low cost and high sensitivity. The platform should allow the use of different polymeric transducers, thus providing the most versatile method.

2   Principle of Operation and Sensor Structure

Multiple sensing techniques were chosen to address these requirements. In particular it was decided to combine three methods. These are differential analysis, dielectric spectroscopy and thermal capacity measurement. With this combination of interactions, a multidimensional response map of the polymer under test can be achieved. Moreover, by analysing the differential response of the functionalised polymer with a reference, interfering interactions such as temperature dependence or humidity can be reduced [5, 6].

The system presented here is comprised of an interdigit structure, an electric heater and a thermocouple. The three elements of the sensor are stacked to form a multilayer device which is able to sense electric impedance and temperature of the substrate at the same time. The multilayer device has been implemented onto a substrate of polyimide Kapton® 50 μm thick film. Metal structures were obtained via thermal vapour deposition. The heater element is a 400 Ω NiCr resistor. A third electrode positioned in the middle point of the resistor is used to compensate small variations in temperature and power between the two sensing areas. The differential interdigit structure was made with the same process. The temperature sensing layer was achieved by direct masking vapour deposition of copper and bismuth. The two thermocouples sensing elements are positioned above the two halves of the heater structure.

An electronic board for the signal conditioning and acquisition has been designed to work with the sensor and to interface with a computer for the data logging. Figure 1 shows a simplified functional diagram of the developed system. The temperature controller module uses thermocouples and heaters to maintain the temperature or thermal power at the required level. Temperature or power steps are forced to detect transitions. The thermocouples' front end is implemented with a

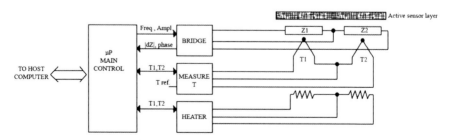

Fig. 1 Simplified diagram of the platform, on the *right* the transducer

switched-capacitor coupled acquisition system based on an MSP430 microcontroller with a 24bit DAC. The same board implements a variable frequency sine wave generator and a half bridge circuitry to measure differential impedance.

3 Sensor Characterisation

To perform initial measurements, calibration and sensitivity analysis, the multilayer transducer has been characterised using an HP4192A LF impedance analyser. A first set of measurements regard the impedance of the bare multilayer transducer. This has been characterised in air and in a saturated acetone atmosphere (~1‰) to exclude any change of electrical properties of the polyimide substrate. The sensor has then been characterised with a commonly used sensing material, methylcellulose. This polymer has been applied on the surface of the interdigit layer as an aqueous gel with a brushstroke. The polymer has been dried at room temperature and resulted as a homogeneous 100 μm thin film. A special plasma treatment has been applied to the interdigit surface before the polymer application to activate the polyimide surface and enhance methylcellulose adhesion.

The experiments have been made in a glass chamber with a controlled air feeder. Ventilation was performed after the measurements with acetone. Built-in sensor heaters were used in these experiments only to regenerate the sensitive layer and to verify the repeatability of the measurement. In this case the polymer was heated at 100 °C for 1 min while the chamber was ventilated.

The capacitance of one half of the interdigit structure was measured for frequencies in the range of 10 Hz–10 MHz. The values of this capacitance ranged from 2.6 to 3.7 pF for different interdigits exposed in air and in a 100 ppm acetone atmosphere. The responses of interdigit sensors with a methylcellulose layer in air were in the range of 3–4.5 pF as well.

Figure 2 shows the response of the interdigit structure covered with a methylcellulose film to a 1,000 ppm acetone atmosphere. In this case the capacitance variation from the background value indicates the acetone polymer absorption. At higher frequencies this value decreases to background values.

4 Conclusions

A novel gas sensor platform for biometric and environmental application has been proposed. The implementation of a novel sensor platform on a flexible and light substrate allows its use as disposable sensor for portable equipment. The ability to detect different types of interactions with the gas sensing polymer allows a more sensitive gas detection, with higher immunity to other gaseous species and to other stimuli such as humidity and temperature. Polymer ageing and degradation effect on the sensitivity should also be reduced with the use of a differential topology.

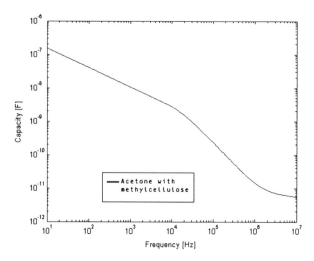

Fig. 2 Methylcellulose response to a 1,000 ppm acetone in air

References

1. I. Eisele, T. Doll, M. Burgmair, Low power gas detection with FET sensors, Sensors and Actuators B: Chemical, 78, (2001) 19–25.
2. Harold V. Shurmer, Julian W. Gardner, Odour discrimination with an electronic nose, Sensors and Actuators B: Chemical, 8, (1992), 1–11.
3. Marina Cole, Nicola Ulivieri, Jesús García-Guzmàn, Julian W. Gardner, Parametric model of a polymeric chemoresistor for use in smart sensor design and simulation, Microelectronics Journal, 34, (2003) 865–875.
4. Xiaoming Yang, Liang Li, Feng Yan, Polypyrrole/silver composite nanotubes for gas sensors, Sensors and Actuators B: Chemical, 145, (2010) 495–500.
5. Ulrich Lange, Nataliya V. Roznyatovskaya, Vladimir M. Mirsky, Conducting polymers in chemical sensors and arrays, Analytica Chimica Acta, 614, (2008) 1–26
6. A. Oprea, J. Courbat, D. Briand, N. Bârsan, U. Weimar, N.F. de Rooij, Environmental monitoring with a multisensor platform on polyimide foil, Sensors and Actuators B: Chemical, 171–172, (2012) 190–197.

A Modified De Sauty Autobalancing Bridge-Based Interface for Wide-Range Capacitive Sensor Applications

Andrea De Marcellis, Giuseppe Ferri, and Paolo Mantenuto

1 Introduction

In the last years, the development of new techniques toward miniaturization and fabrication of higher sensitivity devices has led to the production of a new generation of high-quality sensors for environmental and biological parameters detection [1, 2]. Capacitive sensors have been widely studied, thanks to the easiness of changing their electro-geometrical properties.

Bridge-based front ends are typically employed in those applications where the sensing element value shows reduced variations with respect to its baseline. Recent works on resistive sensors [3, 4] have demonstrated that high sensitivity and easy resistive value estimation features of the classical Wheatstone bridge can be improved through the development of an "autobalancing" configuration of the basic bridge, by adding a tunable resistance. The latter, set by a suitable feedback loop, spreads the equilibrium condition over a wider range.

In the literature, a number of capacitive sensor interface solutions, performing a capacitance-to-time (C–T) conversion, based on integrating [5, 6] or differentiating [7, 8] cells, have been proposed. Here, the output period is proportional to the sensor capacitance. Moreover, circuit topologies performing the capacitance-to-voltage (C–V) conversion have also been presented [9–11], but they can be employed only for low-capacitance variations.

In this paper, we propose a novel interface circuit based on a modified De Sauty bridge topology in its basic and extended configuration, for capacitive sensor estimation. Here, through a feedback loop which handles the bridge differential output voltage module and phase, voltage-controlled resistances (*VCRs*) [3] are properly

A. De Marcellis (✉) • G. Ferri • P. Mantenuto
Department of Industrial and Information Engineering and Economics,
University of L'Aquila, L'Aquila, Italy
e-mail: andrea.demarcellis@univaq.it

C. Di Natale et al. (eds.), *Sensors and Microsystems: Proceedings of the 17th National Conference, Brescia, Italy, 5-7 February 2013*, Lecture Notes in Electrical Engineering 268, DOI 10.1007/978-3-319-00684-0_70, © Springer International Publishing Switzerland 2014

tuned so to reach a new equilibrium condition, allowing a continuous capacitance estimation without the need of knowing an accurate information on the sensor, in particular its baseline.

2 Basic Capacitive Readout Circuit

The basic capacitive readout circuit is depicted in Fig. 1a. The circuit employs a bridge formed by two capacitances (the sensor under test C_{SENS} and the reference capacitance C) and two resistances (R_{VCR} and R). When the equilibrium condition $R_{VCR}/C_{SENS}=R/C$ is achieved, the sensitivity reaches its maximum value, and it is possible to detect and quantify reduced sensor variations.

Since the bridge resistors have been implemented through one *VCR* and one fixed reference resistance, thanks to the use of a feedback loop, when a bridge unbalancing occurs, a not-null differential output arises and the control signal V_{CTRL} forces the bridge to return to a steady state. Moreover, the integrator in the feedback loop ensures stability and creates a suitable V_{CTRL} value. This allows an increasing of the estimation range for the sensor capacitance.

More in particular, reminding that the circuit is excited by an AC signal, the equilibrium condition is achieved only when both the bridge differential output module and phase are nulled. Assuming that signal phasors can be ideally expressed as $\overline{\Delta V} = \Delta V \sin(\omega t + \delta)$, $\overline{V_A} = V_A \sin(\omega t + \varphi_A)$, and $\overline{V_B} = V_B \sin(\omega t + \varphi_B)$, the mixer output V_{MIX} (see Fig. 1a) is given by the following quantity:

$$V_{MIX} = K V_A \ V \sin(\delta - \varphi_A)$$ (1)

being K a constant proportional to the multiplier.

According to (1), the bridge balancing occurs when both the differential output ΔV and the sine argument are zero.

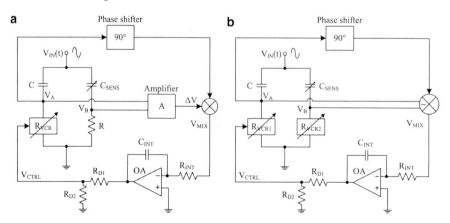

Fig. 1 Circuit schematic for the basic (**a**) and extended range (**b**) capacitance estimation version

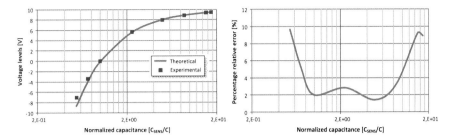

Fig. 2 Comparison between basic capacitance interface theoretical and experimental V_{CTRL} trend (*left*) and capacitance estimation percentage relative error (*right*) for the circuit in Fig. 1a

As also shown in [3], naming R_Z the internal *VCR* resistance, the emulated load can be tuned inside the range $R_{VCR} = (R_Z/2\text{-}20R_Z)$; as a consequence, a bridge balancing range width of about 1.6 decades is obtained and the capacitance estimation, performed by measuring only V_{CTRL}, is given by

$$C_{SENS} = \frac{C}{R}\left(\frac{10R_Z}{10 - V_{CTRL}}\right) \qquad (2)$$

being V_{CTRL} expressed in volt. Preliminary tests conducted on a PCB through sample capacitances have demonstrated the interface capability to correctly detect sensor values within the range 25–840 nF. In particular, as shown in Fig. 2, maintaining a percentage relative error (between measured values) lower than 3 %, an accurate capacitive estimation over about 1.2 decades is possible, where the working range can be shifted by simply changing the R_Z value.

3 Extended Capacitive Readout Circuit

Despite the previous interface good results, further optimization studies have been conducted. In order to simplify the circuit, the instrumentation amplifier has been removed and the bridge branches voltage difference has been performed directly by the mixer (see Fig. 1b) [9]. Moreover, an estimation range spreading has been performed by replacing the fixed resistance R through a *VCR*, properly tuned by an opposite phase V_{CTRL}. Thus, naming R_{Zi} (i = 1, 2) the two *VCRs* internal resistance, a bridge balancing range width of about 3.2 decades is obtained and the capacitance value can be easily estimated by

$$C_{SENS} = C_1 \frac{R_{Z1}\left(10 + V_{CTRL}\right)}{R_{Z2}\left(10 - V_{CTRL}\right)}. \qquad (3)$$

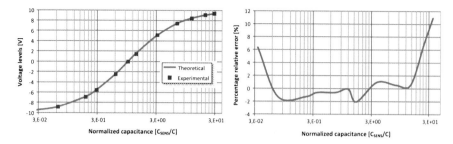

Fig. 3 Comparison between extended capacitance interface theoretical and experimental V_{CTRL} trend (*left*) and capacitance estimation percentage relative error (*right*) for the circuit in Fig. 1b

Tests conducted on a PCB through sample capacitances have demonstrated the interface capability to correctly detect sensor values within the range 900 pF–1.1 μF. The overall relative error has been reduced, and as shown in Fig. 3, maintaining a percentage relative error lower than 2 %, an accurate capacitive estimation over about 2.6 decades is possible (also in this case in a settable range).

4 Conclusions

The work proposed here has shown a novel fully analog uncalibrated interface, for wide-range capacitive sensors, and its optimized version in terms of simpler architecture and wider estimation range. Measurement tests conducted through accurate sample capacitors have shown good results in terms of accuracy and sensitivity.

References

1. A. De Marcellis, G. Ferri, Analog circuits and systems for voltage-mode and current-mode sensor interfacing applications. Springer, Netherlands (2011)
2. E. Ghafar-Zadeh, M. Sawan, D. Therriault. CMOS based capacitive sensor laboratory-on-chip: a multidisciplinary approach. Analog Integr. Cir. and Sig. Proc. **59**, 1–12 (2009).
3. P. Mantenuto, A. De Marcellis, G. Ferri. Uncalibrated analog bridge-based interface for wide-range resistive sensor estimation. IEEE Sensors Journal **12**, 1413–1414 (2012)
4. P. Mantenuto, A. De Marcellis, G. Ferri. On the sensitivity characteristics in novel automatic Wheatstone bridge-based interfaces. Proc. of Eurosensors 2012 conference, Sept. 9–12, 2012, Krakow (Poland), 261–264 (2012)
5. G. Ferri, V. Stornelli, A. De Marcellis, A. Flammini, A. Depari, Novel CMOS fully integrable interface for wide-range resistive sensor arrays with parasitic capacitance estimation. Sensors and Actuators B **130**, 207–215 (2008)
6. A. Depari, A. Flammini, D. Marioli, E. Sisinni, A. De Marcellis, G. Ferri, V. Stornelli. A new and fast-readout interface for resistive chemical sensors. IEEE Transaction on Instr. and Meas. **59**, 1276–1283 (2010)

7. A. De Marcellis, G. Ferri, P. Mantenuto, F. Valente, C. Cantalini, L. Giancaterini, CCII-based interface for capacitive/resistive sensors. Proc. IEEE Sensors 2011 conf., Limerick, 1133–1136 (2011)

8. A. De Marcellis, G. Ferri, P. Mantenuto. A novel uncalibrated read-out circuit for floating capacitive and grounded/floating resistive sensors measurement. Proc. of Eurosensors 2012 conference, Sept. 9–12, 2012, Krakow (Poland), 253–256 (2012)

9. D. Marioli, E. Sardini, A. Taroni. Measurement of small capacitance variations. CPEM '90 Digest, 22–23 (1990)

10. F.M.L. Van der Goes, G.C.M. Meijer. A novel low-cost capacitive-sensor interface. IEEE Transaction on Instr. and Meas. **45**, 536–540 (1996)

11. A.H.M. Zahirul Alam, N. Arfah, S. Khan, M. Rafiqul Islam. Design of Capacitance to Voltage converter for capacitive sensor transducer. American J. of App. Sci. **7**, 1353–1357 (2010)

A New CMOS-Integrated Analog Lock-In Amplifier for Automatic Measurement of Very Small Signals

Andrea De Marcellis, Giuseppe Ferri, and Arnaldo D'Amico

1 Introduction

In sensor applications, sometimes it is important to reveal and measure very low physical/chemical quantities, also with high accuracy and precision [1–3]. This can be achieved through the optimization/maximization of the measurement system sensitivity and resolution. In particular, in the case of small and noisy signals, coming from sensors, the lock-in technique, which is able to extract the signal from noise, can be taken into account. More in general, commercial lock-in amplifiers, as well as ad hoc solutions proposed in the literature also applied in sensor interfacing, are mainly of digital kind and, even if they show good performances and are particularly suitable for multifrequency operations, have high costs, weights, and sizes. On the contrary, in sensor applications, the analog kind of the signal to be revealed suggests, especially when the SNR is less than unity, the use of analog lock-in systems [4–9]. Unfortunately, both analog and digital traditional lock-in amplifiers have the disadvantage of knowing or fixing the operating frequency and requiring, at power on, the initial nulling of the output signal, which corresponds to the "in-quadrature" condition between input and reference signals (this should be also guaranteed continuously during the measurements). Then, the manual activation of switches provides a further 90° phase shift, allowing the reading of the output voltage, proportional to input mean value. In addition, if an operating frequency variation occurs, the system requires an additional calibration or, in worst cases, the redesign of internal blocks (e.g., band-pass filters).

A. De Marcellis (✉) • G. Ferri
Department of Industrial and Information Engineering and Economics,
University of L'Aquila, L'Aquila, Italy
e-mail: andrea.demarcellis@univaq.it

A. D'Amico
Department of Electronic Engineering, University of Roma Tor Vergata, Rome, Italy

C. Di Natale et al. (eds.), *Sensors and Microsystems: Proceedings of the 17th National Conference, Brescia, Italy, 5-7 February 2013*, Lecture Notes in Electrical Engineering 268, DOI 10.1007/978-3-319-00684-0_71, © Springer International Publishing Switzerland 2014

In this sense, here we present a new automatic analog lock-in amplifier suitable for the accurate measurements of very small signals coming from sensors, corresponding to low physical/chemical quantities. The circuit represents an advance (patented [7]) of the topology already proposed in [8] and [9]. This system operates automatically and continuously the relative phase alignment and the frequency tuning of input noisy and reference signals, both at power on and when a variation (of the phase and/or the operating frequency) occurs so allowing the correct detection of the mean value of the small input noisy signal.

2   The Proposed Automatic Lock-In Amplifier

The complete analog system, whose block scheme is shown in Fig. 1, has been implemented firstly as a prototype PCB and, successively, also designed and fabricated in a standard CMOS technology (AMS 0.35 μm) as an integrated chip, whose photo is reported in Fig. 2, with single supply voltage (1.8 V), reduced power consumption (lower than 2 mW), low costs, and small weights and sizes. The system has been optimized to operate in the working frequency range $(2.5 \div 25)$ Hz, which is suitable for different sensor applications.

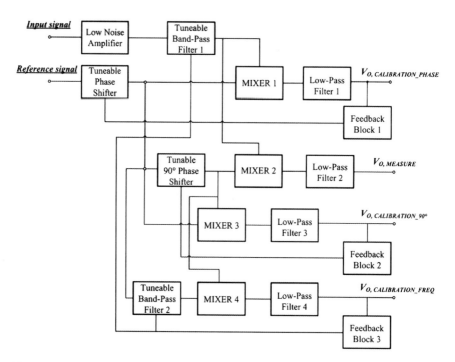

Fig. 1 Block scheme of the proposed automatic analog lock-in amplifier (when the three calibration outputs are zero, the $V_{O,MEASURE}$ value is proportional to the input signal amplitude)

Fig. 2 Photo of the fabricated chip (the lock-in amplifier is highlighted in the *white dashed box*; its area is lower than 8 mm²)

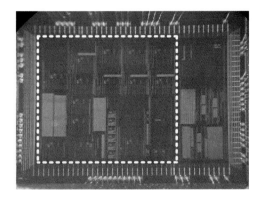

More in detail, the "in-phase" and "tuned" conditions are always constantly guaranteed by automatic operations of suitable feedbacks and control blocks which, at the same time, allow to properly extract the DC output signal, whose amplitude is proportional to that of the AC input noisy signal, as follows:

$$V_{O.MEASURE} = \frac{2AV_{Input\_signal}}{\pi},$$ (1)

being A the voltage gain given by the low noise amplifier.

3   Experimentals with the Fabricated ASIC in CMOS Technology

Figure 3 reports measurement results achieved by an electrical characterization and employment of a commercial resistive gas sensor for the ethylene glycol detection. The results have demonstrated the system validity and its satisfactory performances, confirming the circuit capability to measure both noisy signal amplitudes, down to tens of nV, and reduced ethylene glycol concentrations. In particular, with respect to a resistive gas sensor interface implemented by the simple resistive voltage divider, sensitivity and resolution improvements, given by the proposed lock-in, are of a factor over than 100, so achieving a gas theoretical resolution value of about tens of ppb.

4   Conclusions

In this paper, a new analog lock-in amplifier, fully integrated in a standard CMOS technology ASIC, for automatic detection of very small and noisy signals in sensor applications, has been proposed. Its main advantage concerns its capability to perform, continuously, an automatic phase alignment and a frequency tuning allowing

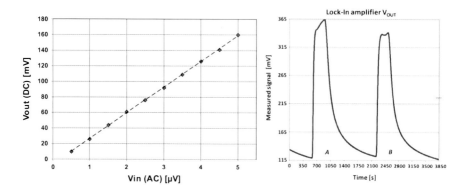

Fig. 3 Experimental results. On the *left*, measured DC output voltage (*dots*) vs. sample input AC signal amplitudes compared with theoretical calculations (*dashed line*, see (1); $f_0 = 11$ Hz). On the *right*, time response of the measured DC output voltage vs. time, for two different ethylene glycol concentrations (A = 4.8 ppm, B = 2.4 ppm) and employing a FIGARO TGS2600 as resistive gas sensor

the enhancement of the SNR and, thus, the improvement of the sensor interface sensitivity and resolution in the detection of very small physical/chemical quantities. Experimental results have shown the detection capability of very small and noisy signals, with a resolution in the order of tens of nV, making the proposed lock-in suitable for high-accuracy high-precision portable measurement systems.

Acknowledgments This work was supported by the Italian Ministry of University (MIUR) under the Program for the Development of a National Interest Research (PRIN Project 2008X744B8).

References

1. A. De Marcellis, G. Ferri. Analog circuits and systems for voltage-mode and current-mode sensor interfacing applications (Springer, Dordrecht, 2011), ISBN 978-9048198276
2. A. De Marcellis, G. Ferri, E. Palange. High sensitivity, high resolution, uncalibrated phase read-out circuit for optoelectronic detection of chemical substances. Sensors & Actuators B **179**, 328–335 (2013)
3. C. Falconi, G. Ferri, V. Stornelli, A. De Marcellis, D. Mazzieri, A. D'Amico. Current-mode high accuracy high precision CMOS amplifiers. IEEE Transactions on Circuits and Systems II **55**, 394–398 (2008)
4. M. O. Sonnaillon, F. J. Bonetto. A low-cost, high-performance, digital signal processor-based lock-in amplifier capable of measuring multiple frequency sweeps simultaneously. Review of Scientific Instruments **76**, 024703(1–7) (2005)
5. C. Azzolini, A. Magnanini, M. Tonelli, G. Chiorboli, C. Morandi. Integrated lock-in amplifier for contact-less interface to magnetically stimulated mechanical resonators. Proc. of IEEE Int. Conf. Des. Tech. Integr. Syst. Nanoscale Era, 1–6 (2008)
6. A. Gnudi, L. Colalongo, G. Baccarani. Integrated lock-in amplifier for sensor applications. Proc. of IEEE ESSCIRC, 58–61 (1999)

7. A. De Marcellis, G. Ferri, V. Stornelli, A. D'Amico, C. Di Natale, E. Martinelli, C. Falconi. Analog System Based on a Lock-In Amplifier Showing a Continuous and Automatic Phase Alignment. Patent RM2008-A194 (2008)
8. A. D'Amico, A. De Marcellis, C. Di Carlo, C. Di Natale, G. Ferri, E. Martinelli, R. Paolesse, V. Stornelli. Low-voltage low-power integrated analog lock-in amplifier for gas sensor applications. Sensors & Actuators B **144**, 400–406 (2010)
9. A. De Marcellis, G. Ferri, C. Di Natale, E. Martinelli, A. D'Amico. A Fully-Analog Lock-In Amplifier with Automatic Phase Alignment for Accurate Measurements of Very Low Gas Concentrations. IEEE Sensors Journal **12**, 1377–1383 (2012)

Impact-Enhanced Multi-beam Piezoelectric Converter for Energy Harvesting in Autonomous Sensors

F. Cerini, M. Baù, M. Ferrari, and V. Ferrari

1 Introduction

Most vibration-based generators are spring-mass-damper systems that generate maximum power when their resonant frequency matches the frequency of the ambient vibrations. Different strategies can be employed to increase the operational frequency range of vibration-based generators [1]. Among the possibilities, the exploitation of multi-element harvesters combining the outputs from multiple converters with different frequency responses into a multifrequency converter array (MFCA) or of nonlinear effects, with particular regard to bistability, were investigated [2–4]. Recently, frequency-up conversion techniques which allow to shift low-frequency mechanical vibrations towards the higher resonant frequencies of the converters were investigated using, in particular, solutions based on impact [5]. In this context we propose and experimentally validate an impact-enhanced multi-beam piezoelectric converter for energy harvesting in autonomous sensors in order to increase the overall rms output voltage and widen the equivalent bandwidth of the converter to harvest energy from broadband or random vibrations at low frequency.

2 Impact-Enhanced Multi-beam Piezoelectric Converter

The schematic of the converter is shown in Fig. 1. It is composed of a compliant driving (D) beam with a low resonant frequency (below 25 Hz) sandwiched within top (T) and bottom (B) piezoelectric parallel bimorph beams on flexible steel with resonant frequencies of 40 Hz and 60 Hz, respectively. At suitable mechanical

F. Cerini (✉) • M. Baù • M. Ferrari • V. Ferrari
Department of Information Engineering, University of Brescia, Brescia, Italy
e-mail: fabrizio.cerini@gmail.com; fabrizio.cerini@ing.unibs.it

C. Di Natale et al. (eds.), *Sensors and Microsystems: Proceedings of the 17th National Conference, Brescia, Italy, 5-7 February 2013*, Lecture Notes in Electrical Engineering 268, DOI 10.1007/978-3-319-00684-0_72, © Springer International Publishing Switzerland 2014

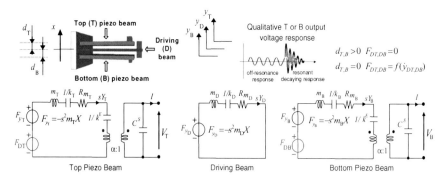

Fig. 1 Schematic diagram of the impact-enhanced multi-beam piezoelectric converter and equivalent electromechanical circuits which model the behaviour of the top and bottom piezo beams and the central driving beam

excitation conditions, the D beam impacts the piezo beams and triggers a nonlinear frequency-up conversion mechanism. The converter has been modelled by equivalent electromechanical lumped-element circuits, which have been derived for the D, T and B beams, as shown in Fig. 1. In the equivalent models, x represents the displacement of the base of the beams with respect to an external fixed frame, y_T, y_B and y_D denote the displacements of the free ends of T, B and D beams from their equilibrium positions, while d_T and d_B are the distances of the T and B piezo beams from the D beam, respectively. The impact between beams is modelled with additional impulsive generators F_{DT} and F_{TB}, representing the forces during the interaction, which act only when distances d_T and d_B are equal to zero. Qualitatively, the voltage generated by T and B piezo beams after the impact changes from a sinusoidal steady state to a resonant decaying response, as shown in Fig. 1.

3   Experimental Results

To experimentally validate the proposed architecture, two commercial piezoelectric bimorphs have been used as T and B piezo beams. They are parallel piezoelectric bimorphs on flexible steel with dimensions of $(45 \times 19 \times 0.58)\,\text{mm}^3$.

The typical electrical impedance, measured with an impedance analyzer HP4194A at 100 Hz, is a capacitance of 270 nF and a parallel resistance of 20 kΩ. The experimental setup is shown in Fig. 2a with a detailed view of the converter. The piezo beam is shown in Fig. 2b. The impact-enhanced multi-beam piezoelectric converter has been initially excited with a Brüel & Kjær 4808 electrodynamic shaker driven by sinusoidal signals with different frequencies to obtain constant peak velocity excitation and a peak acceleration $a_{peak} = 1\,g$ at 50 Hz.

The open-circuit output voltages from the T and B piezo beams have been measured with a LeCroy LT374 digital oscilloscope. Figure 3a shows the quadratic

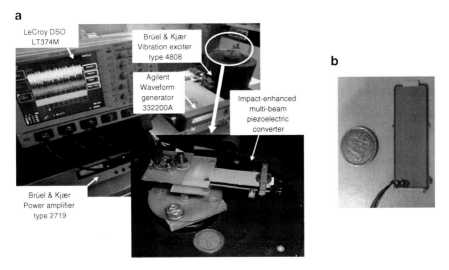

Fig. 2 (**a**) Experimental setup with a detailed view of the converter. (**b**) Piezoelectric bimorph used for the top and bottom beams

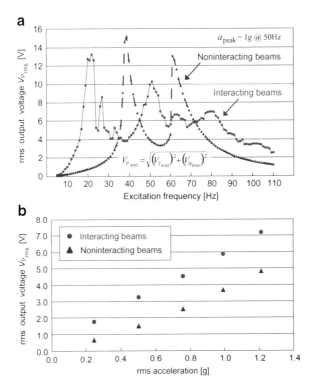

Fig. 3 V_{Prms} versus frequency for sinusoidal excitation (**a**) and V_{Prms} versus rms acceleration level for a 40-Hz low-pass filtered white noise excitation (**b**) obtained with noninteracting and interacting beams

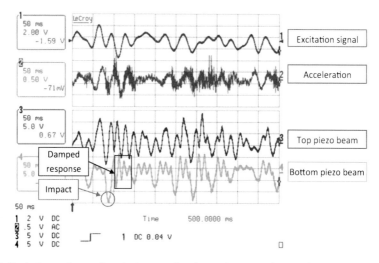

Fig. 4 Typical waveforms of excitation, acceleration and output voltages of top and bottom piezo beams for 40-Hz low-pass filtered white noise excitation

sums of the measured rms output voltages V_{Prms} for the T and B piezo beams for noninteracting and interacting conditions, i.e. with and without the D beam. As expected, a peak for both the T and B piezo beams appears around 20 Hz, corresponding to the resonant frequency of the D beam. By introducing the impact, the average of V_{Prms} is increased of up to 35 % at parity of mechanical excitation. Subsequently, the converter has been excited by a 40-Hz low-pass filtered white noise with different rms acceleration values. Figure 3b shows the quadratic sums of the measured rms output voltages V_{Prms} obtained by varying the rms amplitude of the applied acceleration, for both the interacting and the noninteracting conditions. Figure 4 shows the excitation signal fed to the shaker power amplifier, the acceleration along the vertical axis of the converter measured by an ADXL335 accelerometer and the output voltages of the top and bottom piezo beams, respectively. The impact between the piezo beams and the D beam and the subsequent resonant damped response can be identified on the output voltages of the piezo beams.

4   Conclusions

An impact-enhanced multi-beam piezoelectric converter for energy harvesting from mechanical vibrations for autonomous sensors has been presented. At parity of mechanical excitation, impact provides an increase of the overall rms output voltage V_{Prms} and widens the useful bandwidth over the no-impact condition. Experimental results show that impact increases V_{Prms} of up to 30 % and widens the bandwidth of the piezoelectric converters of up to 50 %. The harvester allows to harvest energy from broadband or random vibrations at low frequencies or mechanical movement.

Acknowledgments The work was partially carried out under the project PRIN2009-2009KFLWJA co-funded by the Italian MIUR.

References

1. D. Zhu, M.J. Tudor, S.P. Beeby. Strategies for increasing the operating frequency range of vibration energy harvesters: a review. Meas. Sci. Technol. **21**, 1–29 (2010)
2. M. Ferrari, V. Ferrari, M. Guizzetti, D. Marioli, A. Taroni. Piezoelectric multifrequency energy converter for power harvesting in autonomous Microsystems. Sens. Actuators A. **142** (1), 329–335 (2008)
3. M. Baù, M. Ferrari, V. Ferrari, M. Guizzetti. A Single-Magnet Nonlinear Piezoelectric Converter for Enhanced Energy Harvesting from Random Vibrations. Sens. Actuators A, **171** (1), 287–292 (2011)
4. F. Cottone, L. Gammaitoni, H. Vocca, M. Ferrari, V. Ferrari. Piezoelectric buckled beams for random vibrations energy harvesting. Smart. Mater. Struct. **21** (3), 035021 (11pp) (2012)
5. L. Gu. Low-frequency piezoelectric energy harvesting prototype suitable for the MEMS implementation. Microelectr. J. **42** (2), 277–282 (2011)

A Low-Cost Electronic Interface for Electrochemical and Semiconductor Gas Sensors

A. Depari, A. Flammini, and E. Sisinni

1 Introduction

Chemical sensors for gas detection are used in several applications, such as air quality, home and work safety, and food quality control. When the detection of small concentrations of substances is required, electrochemical sensors are usually employed. Conversely, when the primary concern is the low cost, semiconductor gas sensors are generally used. In the former case, the quantity to measure is the current I_s of the working electrode (*WE*) of the sensor, and in the latter, the sensor is modeled with a gas-dependent electrical resistor R_s; thus a resistance estimation is necessary. Due to the vast choice of sensors, the range of current I_s or resistance R_s to estimate is rather wide (usually 1 nA $\div$ 1 mA for I_s and 10 kΩ $\div$ 10 GΩ for R_s). Low-cost sensor interface circuits for such a wide input range are usually based on multiple-range architectures [1], with the disadvantage of having complex calibration procedures. Alternative solutions are based on current-/resistance-to-time conversion [2], the main drawback of which is the long measurement time occurring with large resistance values, making such circuits not suitable when fast sensor transients need to be acquired and analyzed. In this paper, an innovative and fast-readout electronic interface circuit, able to be used with wide-range electrochemical as well as semiconductor gas sensors, is proposed. The simple architecture together with 3.3 V single-supply and digital signal output characteristics make the presented front end particularly suitable to be replicated for the use in sensor array applications and integrated in a single-chip solution, together with the digital stages for data acquisition and processing.

A. Depari (✉) • A. Flammini • E. Sisinni
Department of Information Engineering, University of Brescia,
Via Branze 38, Brescia 25123, Italy
e-mail: alessandro.depari@ing.unibs.it

C. Di Natale et al. (eds.), *Sensors and Microsystems: Proceedings of the 17th National Conference, Brescia, Italy, 5-7 February 2013*, Lecture Notes in Electrical Engineering 268, DOI 10.1007/978-3-319-00684-0_73, © Springer International Publishing Switzerland 2014

2 The Proposed Solution

The proposed front end, based on the architecture in [3], is shown in Fig. 1a, whereas the timing diagram is reported in Fig. 1b. Connections with the sensor are displayed in Fig. 2a, b, related to an electrochemical and a resistive sensor, respectively.

The integration of the current I_s produces a ramp V_s, the slope α_s of which depends on the I_s value. A ramp V_t, with a constant slope α_t, opposite to α_s, is used to capture the ramp V_s and to generate the output signal V_o, which is also used to reset the integrators and iterate the measurement.

The *PulseGen* block of Fig. 1a, shown in Fig. 2c, is a monostable circuit to generate, during the low-to-high commutation of V_c, a positive pulse of V_o long enough to assure a complete reset of the integrators Int_s and Int_t as in Fig. 1b.

The time T_{meas} is related to the unknown quantities I_s or R_s by means of the relation in (1). The use of a moving threshold V_t allows the measurement time T_{meas} to

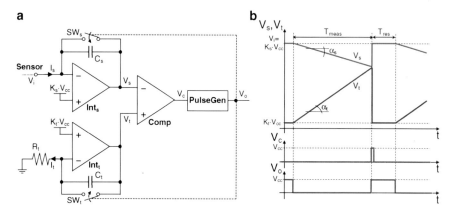

Fig. 1 (a) Scheme of the proposed interface circuit. (b) Time diagram of the circuit signals

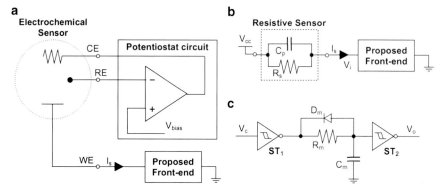

Fig. 2 The connection of the sensor to the front end in Fig. 1a in case of (a) electrochemical sensor and (b) resistive sensor. (c) The *PulseGen* block of Fig. 1a for the creation of the reset/output signal V_o

be limited, particularly when small current or large resistance values (almost flat V_s ramp) are under examination [4]. In this case, the maximum measurement time $T_{meas,MAX}$ is given by the relation in (2):

$$T_{meas} = V_{cc} \frac{K_s - K_t}{\dfrac{I_s}{C_s} + V_{cc} \dfrac{K_t}{R_t \cdot C_t}} = \frac{K_s - K_t}{\dfrac{(1 - K_s)}{R_s \cdot C_s} + \dfrac{K_t}{R_t \cdot C_t}} \quad with \quad K_t < K_s < 1 \qquad (1)$$

$$T_{meas,MAX} = R_t \cdot C_t \left(\frac{K_s}{K_t} - 1 \right) \quad with \quad K_t < K_s < 1 \qquad (2)$$

3 Experimental Results and Conclusions

The experimental validation of the proposed approach has been conducted by means of a discrete component prototype. Sample resistors from 1 kΩ to 10 GΩ in the configuration of Fig. 2a have been used to emulate resistive sensors; by assuring a resistor voltage drop of 1 V (referring to Figs. 1a and 2a, $V_{cc} = 3.3$ V and $V_i = K_s \cdot V_{cc} = 2.3$ V), a current from 100 pA to 1 mA flows in the test resistor, thus emulating the electrochemical sensor output current. A digital counter (Agilent 53230A) has been employed for the measurement of T_{meas}, and estimation of I_s and R_s has been computed by inverting, respectively, the second and third terms of the relation in (1). Results obtained with the aforementioned experimental setup are shown in Table 1, in terms of relative standard deviation σ_{Rel} as well as of relative linearity error $\varepsilon_{L,Rel}$ (evaluated with the weighted least mean square line). The experimental results show a relative standard deviation less than 0.1 % and a relative linearity error below 5 % in the whole considered range (seven decades) for both I_s and R_s estimations [5].

The measuring time T_{meas} spans across five decades (from hundreds of nanoseconds to about 25 ms), thus showing a time compression behavior with large sensor

Table 1 Experimental results obtained with the discrete component prototype and sample resistors emulating the sensor

R_s (MΩ)	I_s (μA)	T_{meas} (μs)	R_s estimation		I_s estimation	
			σ_{Rel} (%)	$\varepsilon_{L,Rel}$ (%)	σ_{Rel} (%)	$\varepsilon_{L,Rel}$ (%)
1.00E−03	1.00E+03	2.91E−01	0.03	−4.14	0.03	4.32
1.00E−02	1.00E+02	2.22E+00	0.02	−0.58	0.02	0.58
1.00E−01	1.00E+01	2.15E+01	0.01	−0.16	0.01	0.16
1.00E+00	1.00E+00	2.14E+02	0.02	−0.52	0.02	0.52
1.00E+01	1.00E−01	2.00E+03	0.01	0.00	0.01	0.00
1.00E+02	1.00E−02	1.22E+04	0.03	1.30	0.03	−1.28
1.00E+03	1.00E−03	2.51E+04	0.04	4.16	0.04	−3.99
1.00E+04	1.00E−04	2.83E+04	0.12	−0.43	0.12	0.43

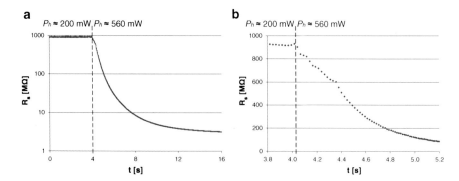

Fig. 3 (**a**) Fast thermal transient of a MOX sensor acquired of the proposed front end. (**b**) Detail of the sensor transient around the heater power variation

resistance (small sensor current) values which keeps the measurement time short. The power dissipation of the front end is less than 30 mW (at 3.3 V), and the cost of the realized prototype is less than 10 EUR, making it suitable for the use in low-cost and low-power gas detection systems.

A titanium dioxide MOX sensor has been used to test the capability of the proposed front end of acquiring fast sensor transients. To force such a sensor behavior, the heater voltage V_h has been quickly changed from $V_h = 2$ V (heater power $P_h \approx 200$ mW which corresponds to a sensor temperature of about 215 °C) to $V_h = 4$ V ($P_h \approx 560$ mW which corresponds to about 440 °C). As visible in Fig. 3a, the sensor resistance has a fast drop of almost three decades in about 15 s. The transient detail in Fig. 3b shows how the presented front end has been able to accurately track the resistance variation, even in the presence of a large resistance value (sensor baseline around 1 GΩ) and fast resistance variation, thus demonstrating the effectiveness of the proposed approach.

Acknowledgments The authors would like to thank Prof. Giorgio Sberveglieri and his staff for the technical and equipment support during the experimental tests.

References

1. M. Baroncini, P. Placidi, G. C. Cardinali, A. Scorzoni. A simple interface circuit for micromachined gas sensors. Sens. Actuators A **109** (1–2), 131–136 (2003)
2. G. Ferri, C. Di Carlo, V. Stornelli, A. De Marcellis, A. Flammini, A. Depari, N. Jand. A single-chip integrated interfacing circuit for wide-range resistive gas sensor arrays. Sens. Actuators B **143** (1), 218–225 (2009).
3. A. Depari, A. Flammini, D. Marioli, S. Rosa, A. Taroni. A low-cost circuit for high-value resistive sensors varying over a wide range. IOP Meas. Sci. Tech. **17** (2), 353–358 (2006).
4. A. Depari, A. Flammini, D. Marioli, E. Sisinni, A. De Marcellis, G. Ferri, V. Stornelli. A New and Fast-Readout Interface for Resistive Chemical Sensors. IEEE Trans. Instr. Meas. **59** (5), 1276–1283 (2010).
5. A. Depari, A. Flammini. Flexible and low-cost interface circuit for electrochemical and resistive gas sensors. Proc. Eng. **47**, 148–151 (2012).

Modeling Hysteresis Losses in Magnetic Core Inductors for DC–DC Conversion

G. Calabrese, M. Granato, G. Frattini, and L. Capineri

1 Introduction

Size reduction of electronic DC–DC converters is a topic of major interest for power electronics. The design goal is usually the achievement of size shrinking without scarifying the quantity of managed power. This means an increase in power density, expressed as managed power over converter volume. Increasing the switching frequency of the converter has often a key role in this direction, aiming at the size reduction of the passive components. The switching frequencies with the potential of size reduction with respect to current converters are in the range of 1–100 MHz [1]. Below 100 MHz it has been shown that inductors which present magnetic cores are still capable of obtaining the same inductance values of an air-core inductor on a smaller volume (higher inductance density) [2], and they can also provide advantages in terms of EMI (magnetic flux confinement).

For these reasons it is important to investigate magnetic core losses affecting the overall inductor, and thus converter, efficiency.

G. Calabrese (✉)
Dipartimento di Ingegneria dell'Informazione, Università degli Studi di Firenze,
Firenze 50139, Italy

Kilby Europe Labs, Texas Instruments, Rozzano, MI 20089, Italy
e-mail: giacomo.calabrese@unifi.it

M. Granato • G. Frattini
Kilby Europe Labs, Texas Instruments, Rozzano, MI 20089, Italy

L. Capineri
Dipartimento di Ingegneria dell'Informazione, Università degli Studi di Firenze,
Firenze 50139, Italy

C. Di Natale et al. (eds.), *Sensors and Microsystems: Proceedings of the 17th National Conference, Brescia, Italy, 5-7 February 2013*, Lecture Notes in Electrical Engineering 268, DOI 10.1007/978-3-319-00684-0_74, © Springer International Publishing Switzerland 2014

2 Magnetic Core Inductor Losses

Magnetic core inductors contribute to the converter power losses, mainly with conductor losses and core losses. Conductor losses are related to the resistivity of the windings, skin, and proximity effects. Core losses are mainly addressable to two terms: hysteresis losses and eddy currents losses. Magnetic materials are usually designed to present high resistivity in the operating frequency range. As an example, ferrite materials present eddy currents losses few orders of magnitude below the hysteresis ones, leaving the latter as the dominant contribution.

Hysteresis losses are proportional to the area of the B–H characteristic loop. The loop area directly depends on the shape of the current signal crossing the inductor. In power converters, the shape of the current crossing the inductors usually belongs to one of these three categories: sinusoidal, triangular, or triangular plus a DC bias, depending on the inductor function; this ultimately affects the power losses estimation.

The common practice used in power converters design is to estimate power losses with parameters or magnetic characterization curves provided by the core or inductor manufacturer (e.g., Q factor, complex permeability, loss tangent, empirical equations). However, these parameters are defined for specific inductor operating conditions (e.g., sinusoidal currents, small-signal analysis).

A methodology which estimates the losses for an arbitrary periodic current waveform would be preferable, especially in terms of magnetic material choice for custom inductors design. Using magnetization hysteresis models based on differential equations, it is possible to provide an estimation which depends on an arbitrary current waveform.

3 Model Implementation

The proposed model for hysteresis power losses estimation has been implemented following the scheme shown in Fig. 1.

It is based on Jiles-Atherton hysteresis model [3] which has been chosen due to the availability of modeling parameters for several magnetic materials and parameters extraction techniques for materials not yet characterized [4]. The model receives as inputs the magnetic field H(t) waveform (obtainable from inductor current once the geometry is defined) and the parameters which describe the magnetic

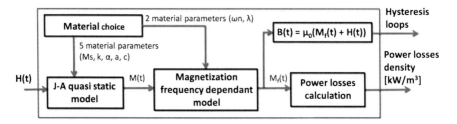

Fig. 1 Model structure

material hysteresis [5]. Using H(t) as stimulus, the differential equation based on the Jiles-Atherton model for magnetic hysteresis is solved. The output is the quasi-static magnetization M(t). A second processing stage applies a frequency-dependent model [6] to obtain the magnetization behavior $M_f(t)$ at the operating frequency of the input signal. At this point the graph of the B–H hysteresis loop is extracted and power losses are evaluated integrating its area.

4   Hysteresis Loop Plots

Figure 2 shows the hysteresis loops obtained with the described model using 3F4 material [7] in three different operating conditions which simulate different inductor currents in power converter applications. It is possible to see the difference between the loops areas as the inductor operating condition varies.

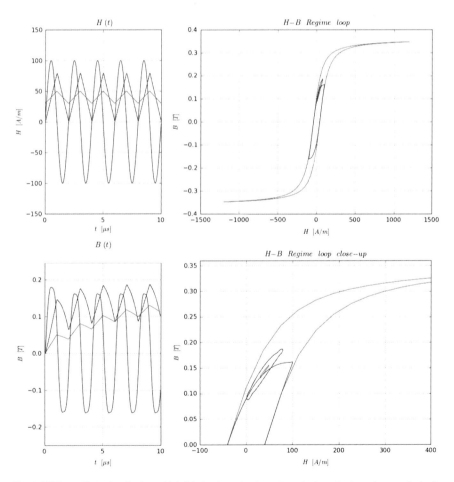

Fig. 2 Different H(t) stimuli: sinusoidal (*blue*), triangular (*green*), and triangular boundary mode (*red*); the calculated B(t) and their B–H loops compared with full saturation hysteresis (*black*) for 3F4 material

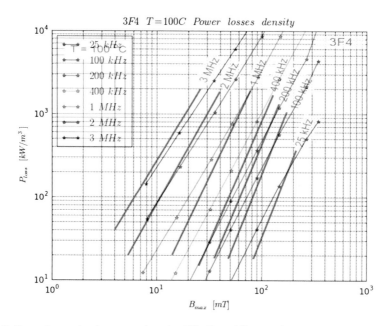

Fig. 3 Power losses density comparison for 3F4. *Dotted lines*: estimated values at 25 °C. *Bold black lines*: material datasheet [7] loss densities measured at 25 °C

5   Model Validation

The first validation step of the proposed model has been carried out comparing the power losses densities obtained from the model loaded with 3F4 parameters at 25 °C and material datasheet graphs (Fig. 3). In this case the eddy current losses contribution has been included. The estimation of discrepancies requires further investigation and modeling improvement. The first is related to the frequency dependency model. In fact, the frequency dependency is modeled as a second order system with parameters obtained from [6]. A better estimation could be done measuring the frequency dependency of magnetization on a test sample and modeling its transfer function with a higher-order system.

6   Conclusions and Future Works

A modeling technique to estimate hysteresis power losses in magnetic materials has been implemented and preliminarily verified. A complete model capable of estimating the complete inductor losses adding the geometry-dependent contributions (conductor losses and eddy current losses) has been implemented.

Future work involves a measurement campaign on different magnetic materials to further calibrate and validate the different model sections.

This tool allows for accurate power losses estimation, and so it will enable power density maximization of inductors for DC–DC power conversion, taking also into account their specific operating mode.

References

1. C. O'Mathúna, W. Ningning, S. Kulkarni, S. Roy, "Review of Integrated Magnetics for Power Supply on Chip (PwrSoC)," *Power Electronics, IEEE Transactions on*, vol. 27, no. 11 (2012), pp. 4799–4816
2. Y. Han, G. Cheung, L. An, C. Sullivan, D. Perreault, "Evaluation of magnetic materials for very high frequency power applications," *Power Electronics Specialists Conference, 2008. PESC 2008. IEEE* (2008), pp. 4270–4276
3. D.C. Jiles and D.L. Atherton, "Theory of Ferromagnetic Hysteresis", Journal of Magnetics and Magnetic Materials, Vol. 61 (1986), pp. 48–60
4. X. Wang, D.W.P. Thomas, M. Sumner, J. Paul, S.H.L. Cabral, "Numerical determination of Jiles-Atherton model parameters", COMPEL: The International Journal for Computation and Mathematics in Electrical and Electronic Engineering, Vol. 28, Issue: 2 (2009), pp. 493–503
5. D. Jiles, Z. Gao, "Modeling the magnetic properties of materials for circuit simulator applications" in Nonlinear electromagnetic systems: proceedings of the International ISEM Symposium on Nonlinear Electromagnetic Systems, Cardiff, Wales, UK, (1995)
6. D.C. Jiles, "Frequency dependence of hysteresis curves in 'non-conducting' magnetic materials," *Magnetics, IEEE Transactions on*, vol. 29, no. 6 (1993), pp. 3490–3492
7. Ferroxcube – 3F4 Datasheet (2008) URL: http://www.elnamagnetics.com/wp-content/uploads/library/Ferroxcube-Materials/3F4_Material_Specification.pdf

Nonlinear Multi-frequency Converter Array for Energy Harvesting from Broadband Low-Frequency Vibrations

D. Alghisi, M. Baù, M. Ferrari, and V. Ferrari

1 Introduction

Piezoelectric energy harvesters from vibrations and motion can be based on linear resonant converter systems which achieve the best harvesting effectiveness when operated at mechanical resonance, but they are suboptimal with frequency-varying and broadband low-frequency vibrations. The exploitation of multi-element harvesters combining the outputs from multiple converters with different frequency responses into a multi-frequency converter array (MFCA) was investigated [1]. However, for a given excitation frequency, mostly one single converter at a time contributes to the output power, with a corresponding limitation in the whole power density.

To widen the bandwidth of the converter, the exploitation of nonlinear effects [2] and in particular of bistable systems was introduced, creating external nonlinear forces by means of magnets [3]. The presence of bistability makes the system capable to rapidly switch between the stable states, thereby increasing the converted power and widening the bandwidth of the harvester, but the conversion effectiveness is dependent on the vibration amplitude [4, 5].

To overcome these limitations, the combination of the multi-frequency and nonlinear approaches into an innovative converter array is proposed.

D. Alghisi (✉) • M. Baù • M. Ferrari • V. Ferrari
Department of Information Engineering, University of Brescia, Via Branze 38,
Brescia 25123, Italy
e-mail: davide.alghisi@ing.unibs.it

C. Di Natale et al. (eds.), *Sensors and Microsystems: Proceedings of the 17th National Conference, Brescia, Italy, 5-7 February 2013*, Lecture Notes in Electrical Engineering 268, DOI 10.1007/978-3-319-00684-0_75, © Springer International Publishing Switzerland 2014

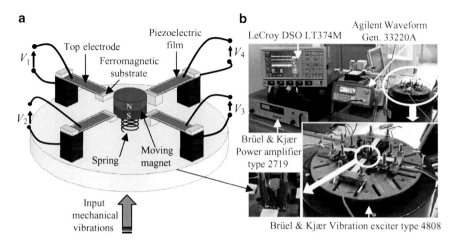

Fig. 1 Nonlinear multi-frequency converter array. Schematic diagram (**a**) and the prototype (**b**) realized with four piezoelectric cantilevers on ferromagnetic substrates and a central permanent moving magnet

2 Nonlinear Multi-frequency Converter Array

The converter array is composed of four piezoelectric cantilevers made by ferromagnetic stainless steel substrates with screen-printed PZT layers, coupled with a permanent moving magnet spring suspended at the center of the array base, as shown in Fig. 1a.

The piezoelectric cantilevers were fabricated by screen-printing PZT films starting from a paste. The PZT paste is composed of commercial powders (Piezokeramica APC 856) and a low-curing-temperature polymeric binder. The printed films were cured at 150 °C for 10 min and then poled at 300 V at the same temperature for 10 min. The steel substrates have planar dimensions of 40 mm × 5 mm and thickness between 100 and 200 μm, while the PZT layer thickness is about 65 μm.

Differences in steel thickness and tip masses among cantilevers determine different resonant frequencies. By adjusting the distance and vertical alignment from the central magnet, each cantilever presents a different potential energy function and frequency response. The spring-suspended moving magnet allows to trigger the nonlinearity and bistability by mutual interactions among the cantilevers which therefore become strongly coupled.

3 Experimental Results

The converter array was excited by an electrodynamic shaker with a band-pass filtered white noise acceleration with different peak values up to $2.2\,g$. Both the linear, i.e., without the central magnet, and nonlinear conditions with fixed and

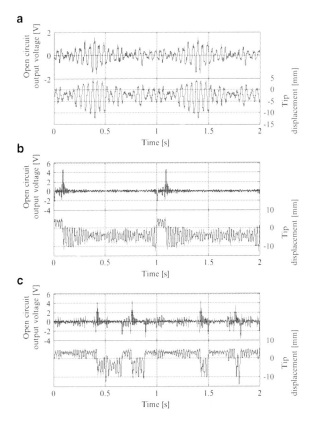

Fig. 2 Measured tip displacement and open-circuit output voltage of a piezoelectric cantilever for linear (**a**) and nonlinear conditions with the permanent magnet held in a fixed position (**b**) and spring suspended (**c**) on the converter base, obtained with band-pass filtered white noise acceleration ($a_{RMS} = 1.5\,g$)

spring-suspended magnet were investigated. Figure 1b shows the experimental setup. Custom-made single-point optical triangulators, based on the reflective sensor Vishay TCRT1000, were clamped to the array base and used to measure the tip displacement of each converter.

Figure 2 compares the measured tip displacement and open-circuit output voltage for one of the piezoelectric cantilever in the linear (a) and nonlinear conditions with the permanent magnet held in a fixed position (b) and spring suspended (c) on the converter array base. In the latter case, the bounces between the two equilibrium points are favored by the moving magnet, therefore increasing the output voltage.

Figure 3 shows that at parity of mechanical excitation, the system with the spring-suspended magnet provides an increase in the RMS output voltage over both the linear system and the nonlinear system with the fixed magnet [6].

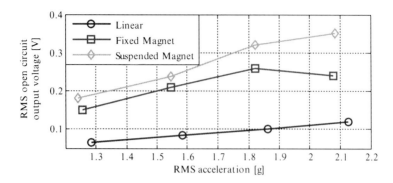

Fig. 3 Measured RMS output voltage of a piezoelectric cantilever for linear and nonlinear conditions with the permanent magnet held in a fixed position and spring suspended on the converter base for different acceleration RMS values

4   Conclusions

A multi-frequency nonlinear piezoelectric energy converter array for energy harvesting from mechanical vibrations was presented. The combination of the nonlinear and multi-frequency approaches allows to increase the converter equivalent bandwidth and shifting it towards lower frequencies without worsening the peak response. The possibility to fabricate cantilevers with different potential curves can be useful to obtain a collective bistable behavior with a reduced dependency on the sensitivity of each converter to the amplitude of the mechanical excitation, therefore increasing the overall effectiveness of the converter array.

Acknowledgments The work was partially carried out under the project PRIN2009-2009KFLWJA co-funded by the Italian MIUR.

References

1. M. Ferrari, V. Ferrari, M. Guizzetti, D. Marioli, A. Taroni. Piezoelectric multifrequency energy converter for power harvesting in autonomous Microsystems. Sens. Actuators A **142** (1), 329–335 (2008)
2. M. Ferrari, V. Ferrari, M. Guizzetti, B. Andò, S. Baglio, C. Trigona. Improved Energy Harvesting from Wideband Vibrations by Nonlinear Piezoelectric Converters. Sens. Actuators A **162** (2), 425–431 (2010)
3. J. Qiu, J.H. Lang, A.H. Slocum. A curved-beam bistable mechanism. IEEE J. Microelectromech. Syst. **13** (2), 137–146 (2004)
4. R. Ramlan, M.J. Brennan, B.R. Mace, I. Kovacic. Potential Benefits of a Non-linear Stiffness in an Energy Harvesting Device. Nonlinear Dyn. **59** (4), 545–558 (2010)
5. S.C. Stanton, C.C. McGehee, B.P. Mann. Nonlinear dynamics for broadband energy harvesting: Investigation of a bistable piezoelectric inertial generator. Phys. D. **239** (10), 640–653 (2010)
6. M. Ferrari, D. Alghisi, M. Baù, V. Ferrari. Nonlinear Multi-Frequency Converter Array for Vibration Energy Harvesting in Autonomous Sensors. Proc. Eng. **47**, 410–413 (2012)

A Wireless Sensor Network Architecture for Structural Health Monitoring

D. Tignola, Saverio De Vito, Grazia Fattoruso, Francesca D'Aversa, and Girolamo Di Francia

1 Introduction

Structural health monitoring (SHM) is a growing technological field based on sensing technologies and algorithms for monitoring the structural system state, diagnosing the structure current conditions, performing a prognosis of expected performance and supporting maintenance, safety and emergency actions.

Advances in Micro-Electro-Mechanical Systems (MEMS) and Wireless Sensor Network (WSN) technologies provide opportunities for sensing, wireless communication and distributed data processing for a variety of new SHM applications. WSN have become an attractive alternative to traditional wired sensor systems in order to reduce implementation costs of SHM systems. Wireless sensors equipped with local computation and communication capabilities have successfully been used in several SHM applications.

Several software platforms for the control of WSN are available. Many of these are designed for the management of WSN based on single technology. These platforms balance their low flexibility with being highly integrated with the used WSN technology [1, 2]. A limited number of software platforms allow to integrate and manage heterogeneous WSN, but they are generally oriented for monitoring global parameters such as autonomy, consumption and position of single sensor node [3, 4] and have never been applied to the SHM scenario.

D. Tignola (✉)
Consorzio T.R.E., Via Privata D.Giustino, 3/A, Naples, NA 80125, Italy
e-mail: diego.tignola@gmail.com

S. De Vito • G. Fattoruso • G. Di Francia
UTTP/MDB, ENEA Portici Research Centre, P.le E. Fermi, 1, Portici, NA 80055, Italy

F. D'Aversa
Consorzio T.R.E., Via Privata D.Giustino, 3/A, Naples, NA 80125, Italy

UTTP/MDB, ENEA Portici Research Centre, P.le E. Fermi, 1, Portici, NA 80055, Italy

C. Di Natale et al. (eds.), *Sensors and Microsystems: Proceedings of the 17th National Conference, Brescia, Italy, 5-7 February 2013*, Lecture Notes in Electrical Engineering 268, DOI 10.1007/978-3-319-00684-0_76, © Springer International Publishing Switzerland 2014

2   The PROVACI Project

PROVACI is a cooperative research project funded by Italian Ministry of Scientific Research targeted to the development of new technologies for the management of cultural heritage buildings. In the framework of this project, ENEA, together with Consorzio T.R.E., is involved in the development of WSN technologies for SHM and particularly in the development of a harmonization framework capable to cope with different WSN solutions, hence proposing an innovative approach compared to the state of the art. The purpose of this paper is, in fact, to describe the design and development of a hardware/software architecture (Fig. 1) for the integration and fusion of data gathered by multiple WSN for SHM purposes.

3   Architectural Design

In order to address WSN heterogeneity introduced by multiple platforms and technologies, a multilayer architecture is proposed for allowing the logical separation of the different functions hiding their implementation details to the other layers. From an architectural point of view, the designed system is composed of three functional levels. The first one includes the WSNs that send the gathered data to their gateway. The second level, implemented on a specialized server machine, is the core of the architecture. It is composed of several subsystems (see Fig. 1): (1) a set of database for storing the raw data coming from different WSNs; (2) a set of adapters which, acting as middleware, are able to enable consistency and homogeneity in the data coming from previous subsystem; (3) a database for collecting data from the adapters; (4) a data fusion system for building a global knowledge base represented in (5); and, finally, (6) a web application for clients data mining.

At the third level, the end users, typically by means of client apps, can enforce the application of data fusion techniques obtaining a multidimensional reconstruction based on preloaded models of the monitored buildings. In particular localization, acceleration and displacement data will be used for the superimposition of colour-coded warning signals on the 3D building model. The main structural components subjected to significant displacement of acceleration will be linked to colours related with high danger by using statistical decision support systems. Reporting features including analysis of historical data will be also allowed.

To further investigate the basic concept of the architecture, some aspects have to be highlighted particularly in the software business model. A registration is actually necessary before any WSNs send data to the server, in order to make the system flexible and facilitate the addition of new types of WSN. During this phase, WSN characteristics are defined such as type of measured data and network proprieties (e.g. number of motes and gateway type). Another crucial aspect is the adapter's role, which processes data that could be stored in different ways (e.g. relational database and text file) or have different syntax and semantics (e.g. different

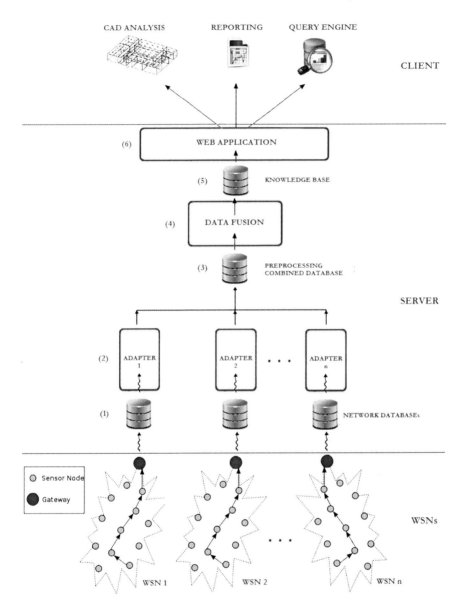

Fig. 1 Multilayer hardware/software architecture for the integration and data fusion of heterogeneous WSN for structural health monitoring

measurement units and different sampling frequency). Adapters are being developed based on Pentaho. The proposed architecture effectively shield the storage systems and data fusion systems from the differences in the underlying sensing systems, making the overall service independent from the technological platforms.

4   Conclusions

Though the described architecture is currently in a (advanced) design phase, it will be deployed in the field in the next months and specifically in the framework of a two-building complex, one of which has been damaged during the earthquake of the Aquila. Two structural monitoring networks, based on different technologies and produced by two different project partners, are in fact in their deployment phase. They will represent the basic data production frameworks that will be harmonized by the proposed architecture.

Acknowledgments The research work has been funded by PROVACI Project—PON "Ricerca e Competitività 2007–2013".

References

1. R. Jurdak, A. G. Ruzzelli, A. Barbirato, S. Boivineau, "Octopus: monitoring, visualization, and control of sensor networks", Wireless communications and Mobile computing, **11**(8), 1073–1091, 2011.
2. M. Turon, "MOTE-VIEW: A sensor network monitoring and management tool", Embedded Networked Sensors, 2005. EmNetS-II. The Second IEEE Workshop on, pp. 11–18, 30–31 May 2005.
3. I. Chatzigiannakis, G. Mylonas, S. Nikoletseas, "jWebDust: A Java-Based Generic Application Environment for Wireless Sensor Networks", IEEE International Conference on Distributed Computing in Sensor Networks (DCOSS), Lecture Notes in Computer Science (LNCS), vol. 3560, pp. 467, 2005.
4. M. Navarro, D. Bhatnagar, Y. Liang, "An Integrated Network and Data Management System for Heterogeneous WSNs", Proceedings of the 8th IEEE International Conference on Mobile Ad-hoc and Sensor Systems ('MASS 11), Valencia, Spain, October 2011.

Characterization of Multi-sensor System for Noninvasive Measurement of Vital Parameters

C.M. De Dominicis, A. Depari, A. Flammini, S. Rinaldi, and A. Vezzoli

1 Motivation

In the last years, the growing availability of performing personal computing devices (such as smartphones) opens new scenarios to the so-called body area network, i.e., sensor networks used to monitor physical conditions of the user. As a matter of fact, the commercially available sensors, adopted to detect physical parameters, like heartbeat or exertion level, are not suitable for run-time measurements. However, the monitoring of vital parameters, specifically during sport activities, can help the athlete by providing useful information about the response of the body to the fatigue. If the information is given in real time, the athlete can suitably adjust his activity to improve his performance or to avoid serious health risks. Such systems should be compact and portable, noninvasive, and low power in order to be battery operated during outdoor activities. The market of wearable sensors is quickly growing, and the interest of the research about that is increasing, not only for singular wearable sensor applications, but also for the so-called wearable sensor system [1]. In this paper a simple wearable sensor system is proposed.

2 The Proposed System

The proposed system consists of an optical PPG sensor and two electrodes for tissue impedance measurement [2]. The PPG sensor is composed of a phototransistor and a photodiode [3], which need to be positioned on both sides of the target tissue, one in front of the other. The proposed PPG sensor adopts the transmission technique,

C.M. De Dominicis • A. Depari (✉) • A. Flammini • S. Rinaldi • A. Vezzoli
Department of Information Engineering, University of Brescia, Via Branze 38,
Brescia 25123, Italy
e-mail: alessandro.depari@ing.unibs.it

C. Di Natale et al. (eds.), *Sensors and Microsystems: Proceedings of the 17th National Conference, Brescia, Italy, 5-7 February 2013*, Lecture Notes in Electrical Engineering 268, DOI 10.1007/978-3-319-00684-0_77, © Springer International Publishing Switzerland 2014

Fig. 1 The realized
sensor components used
for the experimental
characterization: (**a**) light
emitter, (**b**) phototransistor

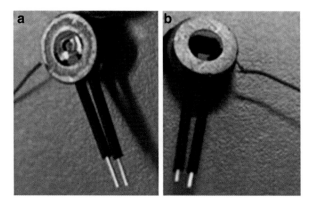

Fig. 2 The equivalent model
of the impedance sensor

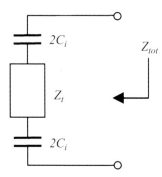

and thus it works only in areas without bone obstructions, such as earlobe [4]. The
PPG sensor components have been placed inside two metallic rings, conveniently
carved in order to hold the PPG sensor parts. The rings also act as the electrodes for
the impedance measurement, as shown in Fig. 1.

The realized prototype is shown in Fig. 1: at the stage of the project, the electrodes are in direct contact with the skin; covering the electrodes with an insulating
film is advisable, but in this case the estimated impedance Z_{tot} would be the combination of the target impedance Z_t and the capacitance effects $2C_i$ due to the two
interfaces of electrode-insulator film, as shown in Fig. 2.

The relationship between Z_{tot} and Z_t with the dependence on C_i and operative
frequency f is shown in (1):

$$Z_{tot} = Z_t + Z_i = \mathrm{Re}(Z_t) + (j\mathrm{Im}(Z_t) + \mathrm{Im}(Z_i)) = \mathrm{Re}(Z_t) + \left(j\mathrm{Im}(Z_t) + \frac{1}{j2\pi fC_i} \right) \quad (1)$$

It is important to accurately select the insulator film in order not to lose sensitivity
in the measurement of the target impedance.

3 Experimental Results

As previously stated, the PPG sensor is used to monitor change in volume of blood in the arteries [5] and thus to obtain information about heart beating and possible arrhythmia. The use of an additional PPG sensor working with two different wavelengths would allow the estimation of other parameters, like the saturation of oxygen in blood, to be performed [6, 7]. The frequency spectrum and the time domain behavior of the PPG output signal applied on the earlobe are shown in Fig. 3a, b, respectively. Data have been acquired with a data acquisition board, placed after a suitable filtering and amplifying stage [8].

The signal shown in Fig. 3a is useful both for the heart rate estimation and for PPG shape analysis. Information of Fig. 3 can be exploited to evaluate the occurrence of heart diseases, such as arrhythmias or arterial occlusions, by examination of the distances between the heartbeats or the steepness of the systoles' phase (the first phase of the heartbeat). All these elaborations could be performed by portable devices that nowadays have relatively high computing features, such as tablets and smartphones.

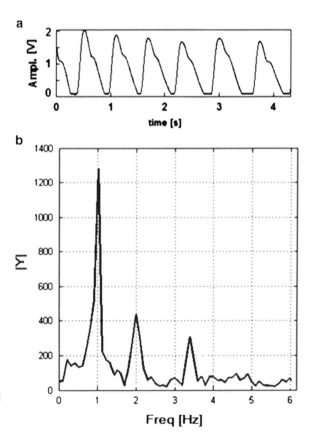

Fig. 3 Filtered and amplified signal (**a**) of the PPG sensor and its relative spectrum (**b**)

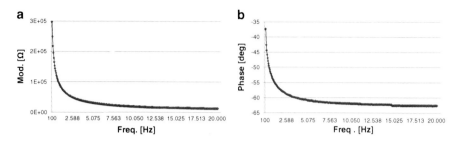

Fig. 4 Module (**a**) and phase (**b**) of the impedance measurement of the earlobe of a person in normal conditions

Additional physical parameters could be estimated, according to the specific application, thanks to measurement of the earlobe impedance performed by the sensor electrodes [9]. In Fig. 4, the module (a) and the phase (b) of the earlobe impedance, measured with an impedance analyzer and by using the electrodes without the insulating film, are shown. The impedance estimation system could be extremely simplified under the hypothesis of using operating frequencies at which the earlobe impedance resembles a pure capacitor. According to Fig. 4, for frequencies greater than 20 kHz, the earlobe impedance behaves mainly as capacitive component; enhancement of this capacitive behavior can be obtained by applying the insulator film to the electrodes.

4 Conclusions

In conclusion, noninvasive measurement of physical parameters requires simple, portable, and low-cost systems, able to collect the information in real time, acquiring data from different sensors. A system composed of different sensors, which could be easily integrated in a commercial earphone, has been proposed. The feasibility of the proposed multi-sensor approach has been experimentally demonstrated in the paper, collecting proper information about the cardiac signal (provided by a PPG sensor) and impendence of human earlobe tissue.

References

1. W.H. Wu, A.A.T. Bui, M.A. Batalin, D. Liu, W.J. Kaiser. Incremental Diagnosis Method for Intelligent Wearable Sensor Systems. IEEE Transactions on Information Technology in Biomedicine, vol. 11, no. 5, pp. 553–562, september 2007
2. A. Depari, A. Flammini, S. Rinaldi, A. Vezzoli. Multi-sensor system with Bluetooth connectivity for non-invasive measurements of human body physical parameters. Sensors and Actuators A: Physical, In press, DOI: 10.1016/j.sna.2013.05.001
3. Vishay Semiconductor, Available on line at: www.vishay.com

4. J.A.C. Patterson, D.G. McIlwraith, Yang Guang-Zhong. A Flexible, Low Noise Reflective PPG Sensor Platform for Ear-Worn Heart Rate Monitoring. Wearable and Implantable Body Sensor Networks, 2009. BSN 2009. Sixth International Workshop on, vol., no., pp. 286–291, 3–5 June 2009
5. A. L. Lee, A.J. Tahmoush, J. R. Jennings. An LED-Transistor Photoplethysmograph. IEEE Transactions on Biomedical Engineering, 22 (1975), 248–250
6. Y. Mendelson, B. D. Ochs. Noninvasive pulse oximetry utilizing skin reflectance photoplethysmography. IEEE Transactions on Biomedical Engineering, 35, (1998), 798–805
7. Medical Instrumentation: Application and Design, John G. Webster Editor
8. C. M. De Dominicis, D. Mazzotti, M. Piccinelli, S. Rinaldi, A. Vezzoli, A. Depari. Evaluation of Bluetooth Hands-Free Profile for Sensors Applications in Smartphone Platforms. Proc. of IEEE Sensors Applications Symposium (SAS 2011), 120–125
9. D. Miklavčič, N. Pavšelj, F. X. Hart. Electric Properties of Tissues. Wiley Encyclopedia of Biomedical Engineering, (2006), 1–12

Wireless Sensor Network Based on wM-Bus for Leakage Detection in Gas and Water Pipes

P. Ferrari, A. Flammini, S. Rinaldi, and A. Vezzoli

1 The Proposed Approach

Several technologies for the detection of the leakages in pipes carrying gas, oil, or water are available in literature. Some of them allow also the localization of the losses over the pipe. In the following, the so-called negative pressure wave (NPW) technique for detection and localization of leakages in water and gas distribution grid has been considered. The system estimates the distance between a measurement station and the leakage by means of the NPW generated by the leakage itself. The time delay (Δt) between the receptions of the same NPW by two stations placed on the same pipe segment is estimated. Then, it is possible to evaluate the leakage position knowing the speed of the NPW. In (1), the equation for the estimation of the distance of the leakage (L_1) from the pressure station 1 (see Fig. 1) is showed:

$$L_1 = \frac{L + v_w \Delta t}{2} \tag{1}$$

where L is the distance between the two stations, v_w is the speed of the wave, and Δt is the time delay between the detection of the NPW by the two stations.

Nowadays, WSN (wireless sensor network) is a mature technology and adopted in a great number of application fields, from health care [1] to industrial applications [2]. Solutions based on WSN are suitable for the data exchange and synchronization requirements involved in NPW system. The wireless communication technologies used in smart metering system, which performs distributed measurements of water and gas, can be also applied to identify and to localize leakages on pipe. The new

P. Ferrari • A. Flammini • S. Rinaldi (✉) • A. Vezzoli
Department of Information Engineering, University of Brescia, via Branze 38,
Brescia 25123, Italy
e-mail: stefano.rinaldi@ing.unibs.it

C. Di Natale et al. (eds.), *Sensors and Microsystems: Proceedings of the 17th National Conference, Brescia, Italy, 5-7 February 2013*, Lecture Notes in Electrical Engineering 268, DOI 10.1007/978-3-319-00684-0_78, © Springer International Publishing Switzerland 2014

Fig. 1 The operating
principle of NPW leakage
localization system

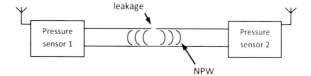

Fig. 2 The architecture of
the monitoring WSN based
on wM-bus

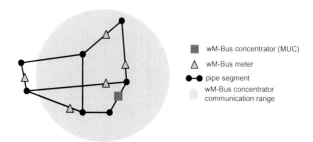

regulation about gas and water metering stimulates the producers of metering devices to focus on distributed monitoring, exploiting for data transmission the wM-bus communication [3]. This communication technology provides a wide data communication range and long operational life for battery-operated devices. The monitoring of an urban water or gas distribution grid requires a complex WSN, in order to monitor a wide area with several measurement points. An example of WSN network over a pipe grid, where a wM-Bus master and several measurement points are distributed through the whole grid, is shown in Fig. 2.

In the proposed system, the distributed measurement of the NPW events requires an accurate time synchronization, to be able to provide an accurate estimation of spatial localization. The speed of the NPW depends on many factors but is easily down to 1,000 m/s [4]; thus, a time synchronization uncertainty below 1 ms is required to obtain an estimation of the position with an uncertainty on the order of 1 m.

2 The Results

As mentioned in the previous section, accurate time synchronization is required to satisfy the requirements of the adopted method. The time synchronization of the nodes is performed by a broadcast message sent by the wM-Bus master (concentrator) to the meters (Fig. 3). The time data format provided by the wM-Bus standard has a time resolution on the order of 1 s, not enough to satisfy the requirements of the measurement methodology adopted. Nevertheless, the M-bus protocol supports COSEM time format, which has a time resolution below 1 ms.

The main sources of uncertainty in wireless synchronization depend on the management of the messages exchanged by wireless devices of the WSN and on the

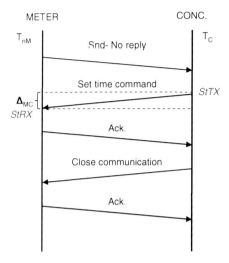

Fig. 3 The messages exchanged during the wM-bus synchronization

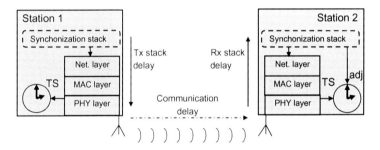

Fig. 4 Sources of uncertainty in wM-Bus communication

protocol synchronization stack adopted, as well described in [5]. For this reason, a commercial wM-Bus device has been characterized, in order to evaluate its contribution to the synchronization uncertainty. Then, this information can be used to validate the feasibility of the proposed approach. The source of synchronization uncertainty involved in a simple message transmission from station 1 to station 2 [6] is shown in Fig. 4.

The synchronization obtained using broadcast messages is mainly affected by the uncertainty of the reception stage of the wireless nodes of the network. The distribution of the time difference (delay) of the receptions (at hardware level) of the same message between the RX stages of two wM-Bus devices (Fig. 4) is shown in Fig. 5. The jitter, i.e., the maximum time difference variation, is well below 1 μs; thus, a synchronization uncertainty on the order of few tens of microseconds is expected.

Fig. 5 Normalized delay
between two wM-Bus
receivers

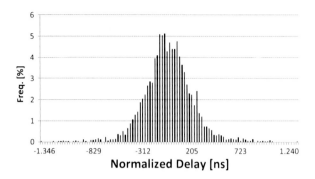

Normalized Delay [ns]

3   Conclusions

The growing demand for an efficient management of the resources, like gas and
water, requires a careful management and monitoring of the distribution grid. In this
paper, a system for leakage detection, identification, and localization in water and
gas pipe has been proposed. The system requires the installation of sensors on the
distribution gas and water grids. A WSN, based on wM-Bus technology, is adopted
to provide the required communication and synchronization services. In particular,
the distributed measurement system requires a synchronization of the time reference of each acquisition point below the millisecond. The preliminary characterization of wM-Bus devices proved the feasibility of the proposed approach.

References

1. A. Depari, A. Flammini, S. Rinaldi, A. Vezzoli, "A Portable Multi-sensor System for Non-
   invasive Measurement of Biometrical Data", Procedia Engineering - 26th European Conference
   on Solid-State Transducers, EUROSENSOR 2012, Krakow, Poland, September 9–12, 2012,
   Vol. 47, pp. 1323–1326.
2. P. Ferrari, A. Flammini, D. Marioli, E. Sisinni, A. Taroni, "Coexistence of Wireless Sensor
   Networks in Factory Automation Scenarios", Sensors & Transducers Journal, April, 2008, Vol.
   90, No. 4, pp. 48–60.
3. "Communication systems for meters and remote reading of meters, part 4: wireless meter read-
   out (Radio meter reading for operation in the 868 MHz SRD band)", EN 13757-4, 2005.
4. S. Srirangarajan, M. Allen, A. Preis, M. Iqbal, H. B. Lim, and A. J. Whittle, "Wavelet-Based
   Burst Event Detection and Localization in water distribution system", Journal of Signal
   Processing Systems, DOI: 10.1007/s11265-012-0690-6, Sep. 2012.
5. P. Ferrari, A. Flammini, S. Rinaldi, A. Bondavalli, F. Brancati, "Experimental Characterization
   of Uncertainty Sources in a Software-Only Synchronization System", IEEE Trans.
   Instrumentation and Measurement, May, 2012, Vol. 61, No. 5, pp. 1512–1521.
6. A. Flammini, S. Rinaldi, A. Vezzoli, "The Sense of Time in Open Metering System", 2011
   IEEE International Conference on Smart Measurements for Future Grids (SMFG), Bologna,
   Italy, November 14–16, 2011, pp. 22–27.monitoring WSN based on wM-bus.

Arduino-Based Shield for Resistive Gas Sensor Array Characterization Under UV Light Exposure

D. Aloisio, N. Donato, G. Neri, M. Latino, T. Wagner, M. Tiemann, and P.P. Capra

1 Introduction

The development of nanostructured sensing material and its integration in micro-electronics devices are the key for the actual increasing of sensors market. Nanostructured sensing material optimization can be considered as the turning point for the development of devices able to achieve the right balance between low power consumption requirements and sensing performance. The low power consumption requirements can be fulfilled by developing sensing materials able to operate itself at conditions close to room temperature or by UV "activation." As an example of this concept, mesoporous In_2O_3-based material can be irradiated with UV light to enhance the sensing properties to decrease the operating temperature value [1]. Therefore, new applications and services are arising, supported by the improvement of new portable and low-cost sensing systems. In such task, here is reported the development of a characterization system for metal oxide (MOX)-based gas sensors with Arduino UNO platform. The developed shield is able to measure the response of up to six sensors under UV radiation by means of LED devices. In the shield

D. Aloisio (✉) • N. Donato • G. Neri
Department of Electronic Engineering, Chemistry and Industrial Engineering,
University of Messina, Contrada di Dio, Messina 98166, Italy
e-mail: daloisio@unime.it

M. Latino
Consiglio Nazionale Delle Ricerche, Istituto per la Microelettronica e Microsistemi,
Zona Industriale VIII Strada 5, Catania 95121, Italy

T. Wagner • M. Tiemann
Naturwissenschaftliche Fakultät, Department Chemie, Universität Paderborn,
Warburger Straße 100, Paderborn 33098, Germany

P.P. Capra
Istituto Nazionale di Ricerca Metrologica, Torino 10135, Italy

C. Di Natale et al. (eds.), *Sensors and Microsystems: Proceedings of the 17th National Conference, Brescia, Italy, 5-7 February 2013*, Lecture Notes in Electrical Engineering 268, DOI 10.1007/978-3-319-00684-0_79, © Springer International Publishing Switzerland 2014

takes place the board with six inverter-based oscillators where the RC tank of every circuit is composed of the resistance of the sensor (R) and an external capacitance (C). By properly choosing the capacitance value and by measuring the period (frequency) of the oscillator, the resistance response of each sensor can be evaluated.

2 Experiments

The mesoporous sensing material development and characterization were reported in previous papers [1, 2]. Here are reported the experimental activities regarding the board for hardware interfacing. A good alternative to realize real-time resistance measurements is to use an oscillator system whose frequency is resistance value dependent [3]. This approach avoids the use of A/D converters and simplifies the interfacing circuit of the sensor. The measurement system core is based on an oscillator circuit in which frequency depends on the resistor value present in the RC tank of the resonant ring. It is based on two inverter gates and a Schmidt trigger, essential for a proper RC charge and discharge, all connected in a loop for triple inversion of feedback signal. In this configuration choosing proper capacitance value is possible to adjust the range of measurement; see block diagram in Fig. 1. The single sensor cell was then replicated to handle an array configuration by developing a sensor shield for Arduino environment, fully compatible and able to be assembled with other ones in a modular structure (see Fig. 2).

The frequency/period measurement is performed with an Arduino UNO platform. This choice is justified by a series of factors as low cost of the system, possibility to change firmware on-fly and overall great flexibility of the platform to control more than one sensor and/or other systems. The peripherals of ATmega328, mounted over this board, ensure to control contemporary up to six sensor and UV exposition of the sample using UV LEDs. The frequency range is a balance related to the number of the sensors to characterize. For a single sensor, a wider frequency range can be achieved spanning from 1 kHz to 5 MHz.

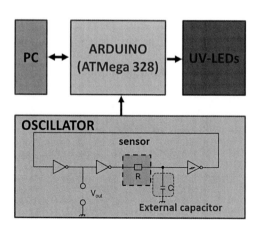

Fig. 1 Measurement system scheme with single-cell oscillator

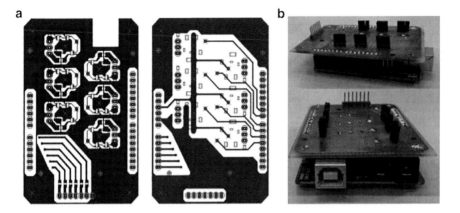

Fig. 2 (**a**) Shield top and bottom layers, (**b**) assembled shields

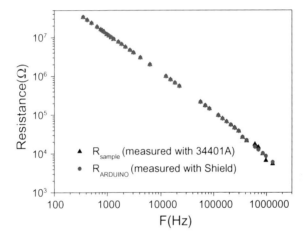

Fig. 3 Calibration diagram

Arduino board programmed with a proper developed library is able to measure the frequencies, by using internal interrupt events, and to send data by means of serial port to a personal computer. The GUI, developed in MATLAB environment, is able to control the communication protocol, the measurement procedures, as well as the LED switching. Two libraries were used to realize the measurements: the first one for one channel use with a greater range and the second one for multichannel use. Figure 3 reports the calibration diagram, obtained by comparing the measurement data of commercial resistive samples measured both with Agilent 34401A multimeter and the developed system, respectively.

3   Results

Gas sensing tests were carried out inside a Teflon chamber equipped with 400 nm UV LEDs under controlled atmosphere. Mass flow controllers were used to adjust desired concentrations of target gas in dry air. The sensors were tested on a concentration range of NO_2 spanning from 0.3 to 5 ppm. The experimental data show the UV radiation influence on the resistance value. In Fig. 4a the resistive behavior of a mesoporous In_2O_3-based sensor exposed in dry nitrogen atmosphere and dark/UV light conditions is reported. It can be seen that UV light exposure leads to a sudden resistance decrease. After turning the UV light off, the resistance value slowly recovers.

Figure 4b reports the response of the sensor towards NO_2 at room temperature (RT) and under UV irradiation. Because of the high response, the sensor devices are promising candidates for low power sensing applications.

4   Conclusions

We report about the development of an Arduino shield for the characterization of resistive sensors. The shield is able to handle up to six sensors and six LEDs for UV irradiation. The system was calibrated and then validated in the characterization of mesoporous In_2O_3 sensing films towards NO_2 as target gas. Further investigations are in progress in order to characterize several sensing materials and to improve the shield by integrating a temperature monitor/control for the sensing devices.

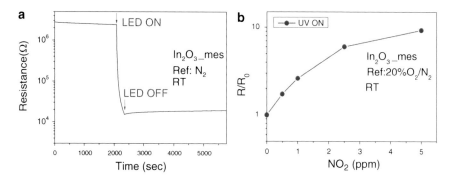

Fig. 4 (a) Sample resistance under dry nitrogen and dark/radiation conditions, (b) calibration curve towards NO_2 under UV irradiation at room temperature

References

1. T. Wagner, C.-D. Kohl, S. Morandi, C. Malagú, N. Donato, M. Latino, G. Neri, M. Tiemann, Photoreduction of mesoporous In2O3: Mechanistic model and utility in gas sensing, Chemistry - A European Journal, Vol. 18, Issue 26, 25 June 2012, pp. 8216–8223.
2. T. Wagner, C.-D. Kohl, C. Malagù, N. Donato, M. Latino, G. Neri, M. Tiemann, UV light-enhanced NO_2 sensing by mesoporous In_2O_3: Interpretation of results by a new sensing model, Sensors and Actuators B, 187, pp. 488–494, (2013) http://dx.doi.org/10.1016/j.snb.2013.02.025
3. J. L. Merino, S. A. Bota, R. Casanova, A. Diéguez, C. Cané, J. Samitier, "A Reusable Smart Interface for Gas Sensor Resistance Measurement", IEEE TRANSACTIONS ON INSTRUMENTATION AND MEASUREMENT, vol. 53, no. 4, pp. 1173–1178, 2004.

Piezoelectric Energy Harvesting from von Karman Vortices

M. Demori, V. Ferrari, S. Farisè, and P. Poesio

1 Introduction

The recovery of energy from von Karman vortices can be effective since they are common behind obstacles immersed in a flow (at Reynolds numbers Re 300–600) and they represent an ideal forcing for a vibrating energy harvester due to the conversion of a unidirectional (DC) flow into an alternated (AC) periodic pressure field [1].

In this work an energy-harvesting system where von Karman vortices excite an oscillating beam is proposed. The vortices are generated by a bluff body immersed in an airflow, and they are collected by placing the beam in the vortices region [2]. A piezoelectric converter has been embedded in the beam to harvest energy from the generated oscillations. The system composed of the bluff body and the beam has been placed in a wind tunnel with the possibility to direct the beam with different orientation angles with respect to the flow direction and to generate different flow velocities. In this way the harvesting effectiveness can be investigated as a function of these two different parameters. In fact, the orientation angle affects the interaction between the beam and the forcing pressure field. On the other hand, the repetition frequency of the vortices and the coupling with the mechanical beam resonance depend on the flow velocity [3].

M. Demori (✉) • V. Ferrari
Department of Information Engineering, University of Brescia, Via Branze 38,
Brescia 25123, Italy
e-mail: marco.demori@ing.unibs.it

S. Farisè • P. Poesio
Department of Mechanical and Industrial Engineering, University of Brescia,
Via Branze 38, Brescia 25123, Italy

C. Di Natale et al. (eds.), *Sensors and Microsystems: Proceedings of the 17th National Conference, Brescia, Italy, 5-7 February 2013*, Lecture Notes in Electrical Engineering 268, DOI 10.1007/978-3-319-00684-0_80, © Springer International Publishing Switzerland 2014

Fig. 1 Schematization of the
system: the bluff body is
placed perpendicular to the
flow direction, and it is
20 mm thick and 55 mm
wide; the beam is composed
of the piezoelectric converter,
with 45 mm length and
20 mm width, and the blade,
with 70 mm length and
30 mm width

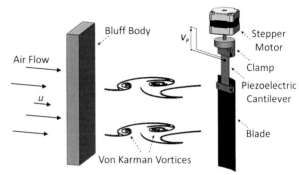

2 System Description

A parallelepiped-shaped obstacle has been placed in a wind tunnel, and it acts as
bluff body for the generation of the vortices. As shown in Fig. 1, the oscillating
beam system has been placed behind the bluff body. The beam is composed of the
piezoelectric converter and a blade profile. The converter is a bimorph piezoelectric
element, WAC3X/18, with internal impedance composed of a capacitance
$C_P \approx 270$ nF and a parallel resistance $R_P \approx 20$ kΩ measured at a frequency of 100 Hz.
The converter is based on a steel slat, which has been clamped at one end to obtain
a cantilever structure. The blade has been connected to the free end of the converter
to collect the alternate pressure field of the vortices as the forcing for the oscillation.
A tailored clamp to connect the beam to the shaft of a stepper motor has been manufactured. In this way the stepper motor allows to vary the orientation of the beam
with respect to the flow direction.

3 Experimental Results

In the experimental tests, different airflows have been generated in the wind tunnel,
while the open-circuit converter voltage V_P, due to the beam oscillations, has been
measured for different beam orientations. In Fig. 2a the indication of the blade position and the orientation angle θ are depicted on an example of smoke visualization
of the air streamline in the vortex region. The whole range of orientations has been
divided in 200 steps, and for each step, the voltage V_P has been measured for 20 s.
In Fig. 2b the rms values of the V_P measurements are shown as a function of θ for
three different flow velocities, $u_1 = 1.5$ m/s, $u_2 = 3.6$ m/s, and $u_3 = 5$ m/s. As it can be
seen, at the intermediate velocity, higher rms values, up to about 30 V, have been
achieved. On the other hand, the values obtained for the other velocities are below
10 V. In addition, it can be observed that higher rms values have been obtained at
orientation angles around $\theta = \pm 60°$ and $\theta = \pm 120°$ where the oscillating beam presents a better coupling with the forcing pressure field.

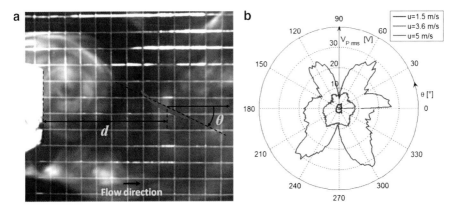

Fig. 2 (**a**) Top view picture of the air streamline obtained with smoke visualization: the position of the blade, at distance $d = 80$ mm from the bluff body, and the orientation angle θ are shown. (**b**) Polar visualization of the rms value of the open-circuit voltage V_P measurements as a function of the orientation angle θ

The larger oscillations and therefore the higher value of V_P obtained at the velocity u_2, compared with those obtained at other velocities, can be associated to a forcing frequency close to the mechanical resonance frequency of the oscillating beam, which is of about 14 Hz. In Fig. 3 the time behavior and the frequency spectrum of the voltage V_P have been compared for the three different flow velocities at the same angle θ. The main components in the spectrum of V_P at u_1 and u_3 are in the region of the beam resonance frequency and near the frequencies that can be related to the forcing vortex. On the other hand, in the spectrum at u_2, the main components are concentrated near the beam resonance frequency with larger values due to the resonance effect.

The converter has been connected to a resistive load to evaluate the power that can be harvested from the system. A $R_L = 15$ kΩ has been used as the optimal load at the obtained oscillation frequencies. The measured powers as a function of the flow velocity are shown in Fig. 4. Higher values, up to 1.2 mW, are obtained for intermediate velocities where the frequency of the vortices matches the resonance frequency of the beam.

4   Conclusions

In this work the possibility to use a piezoelectric converter to recover energy from an airflow has been demonstrated. The alternate forcing for the vibrations is provided by von Karman vortices generated behind a bluff body placed in the airflow. In is shown that the harvesting effectiveness can be improved by matching the frequencies of the beam resonance and the vortex repetition.

Fig. 3 Time behavior (**a**) and frequency spectrum (**b**) of the open-circuit voltage V_P measured at different flow velocities for $\theta = 300°$

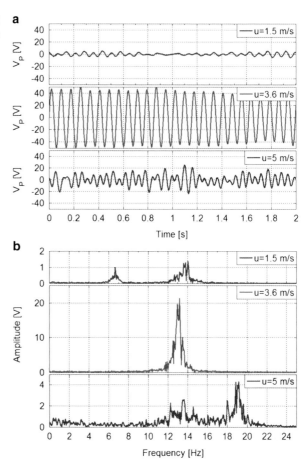

Fig. 4 Measured power on a resistive load $R_L = 15$ kΩ as a function of the flow velocity

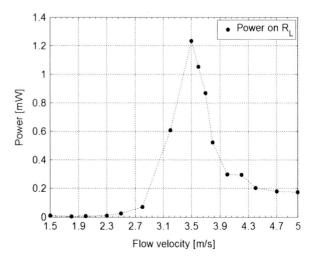

References

1. A. Erturk, W.G. R. Vieira, C. De Marqui Jr., D. J. Inman. On the energy harvesting potential of piezoaeroelastic systems. Appl. Phys. Lett, **96**, 184103 (2010)
2. S Shi, T.H. New, Y. Liu. Flapping dynamics of a low aspect-ratio energy-harvesting membrane immersed in a square cylinder wake. Exp. Therm. Fluid Sci. **46**, 151–161 (2013)
3. X. Gao, W. Shih, W.Y. Shih. Flow Energy Harvesting Using Piezoelectric Cantilevers With Cylindrical Extension. IEEE Transactions on Industrial Electronics **60** (3), 1116–1118 (2013)

Mobile System for Air Pollution Evaluation

A. Bernieri, G. Betta, L. Ferrigno, and M. Laracca

1 Introduction

In recent years, the pollution monitoring in urban areas has become one of the most critical issues for local public authorities, in order to verify that pollution levels not exceed limits considered unsafe or that are regulated by local laws. Generally, the pollution monitoring is performed by using measurement stations located in few points of the region of interest, since these stations are generally characterized by high costs, weights, and dimensions. Then, the pollution levels over the remaining area are predicted by means of suitable interpolation models. Due to the great variety of urban scenarios, it becomes very difficult to obtain reliable pollution levels in area in which the measurements have not been directly taken but only predicted. Consequently, the pollution monitoring can suffer a lack of reliable information indispensable for the actuation of proper environmental management policies. In this scenario, it becomes of interest to use a number of low-cost measurement system installed on public vehicles in urban service, in order to measure the actual pollution level in all parts of the territory in which these vehicles perform their paths. In this way, a punctual, reliable, and real-time pollution data can be obtained.

A. Bernieri • G. Betta (✉) • L. Ferrigno • M. Laracca
Department of Electric and Information Engineering, University of Cassino and South Lazio, Cassino, Italy
e-mail: bernieri@unicas.it; betta@unicas.it

C. Di Natale et al. (eds.), *Sensors and Microsystems: Proceedings of the 17th National Conference, Brescia, Italy, 5-7 February 2013*, Lecture Notes in Electrical Engineering 268, DOI 10.1007/978-3-319-00684-0_81, © Springer International Publishing Switzerland 2014

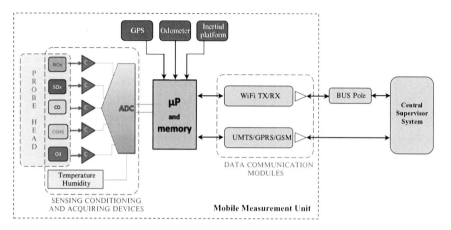

Fig. 1 Architecture of the proposed system

2 The Proposed Measurement System

Thanks to previous experience in the field of wireless measurement systems [1], sensing devices [2–4], wireless sensor networks [5, 6], and uncertainty estimation [7]; in this paper the authors propose an effective mobile measurement system for the real-time monitoring of environmental pollutions in urban areas [8, 9]. Key features of the proposed system are the significant low cost, the high portability of the measurement units, the modularity, and the autonomous power supply. The architecture of the proposed pollution measurement system is shown in Fig. 1. It is constituted by two main classes of devices:

- Mobile Measurement Units (MMU): designed to be hosted on board of vehicles as urban service buses or service vehicles of the local authority; these units are equipped with useful sensors to measure the concentration of relevant pollutants (CO, NO_2, SO_2, O_3, C_6H_6, and so on), a GPS/odometer to determine the unit position, and a WiFi and/or UMTS/GPRS/GSM communication system to send the measured data to the Central Supervisor System.
- Central Supervisor System (CSS): designed to be located in a fixed position and devoted to remotely retrieve the measured data; in practice, the CSS is a workstation equipped with the necessary devices and software for retrieving, storing, and viewing measurement data collected by MMUs, as well as for managing the data elaboration and for supervising the MMUs status and operations.

3 Experimental Results

More research activities are achieved in order to verify the metrological performance of the measurement system and to realize an efficient and useful data interpolation/representation technique. As regards the first point, a comparison between

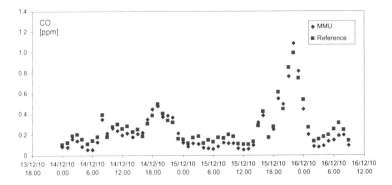

Fig. 2 MMU validation test results

the results obtained from the MMU and those obtained from a calibrated traditional measurement instrumentation, positioned at the same location, was performed.

Figure 2 shows the first result obtained with reference to the CO pollutant during a test phase of about 60 h, highlighting the compatibility of the obtained measurement with the adopted CO sensor specification and with the actual standard limits [10].

In order to verify the measurement performance in a real application, a series of measurement was performed during predefined paths in the city of Bologna using a service vehicle on which the MMU was installed; the measurements were then elaborated by means of suitable procedures, running on the CSS and performing the elaboration of Voronoi diagrams and Kriging variograms as the first representation techniques [11, 12].

Figure 3 shows the CO-pollutant distribution in the considered urban area. In the figure, the circles indicate the punctual measurement data acquired by means of the MMU and georeferenced using the Google map of the city. The results showed are computed as mean of seven different paths, performed in a period of 1 week of stable weather conditions (absence of wind and of rain). In the figure, the red triangles indicate the positions of two traditional fixed measurement units. It is evident that the use of the proposed mobile measurement system is capable to highlight some urban areas in which the pollutant concentration is very high. Using the traditional approach, in these areas the pollution information may be neglected or obtained by means of prediction models which could not assure the same accuracy of the punctual measurements.

4   Conclusions

The paper proposes an efficient architecture for mobile systems able to perform the urban pollution monitoring. The modular architecture of the proposed measurement system assures good feasibility and efficiency. In addition, the described solution is very attractive because it allows exploiting the spatial and time capillarity of a typical public transportation system to achieve in an easy way a high amount of data

Fig. 3 Voronoi diagrams and Kriging variograms of the considered pollutant (the *red triangles* indicate the position of traditional fixed measurement systems)

(from the spatial and time points of view) useful also for improving the prediction models. With reference to the employed data elaboration techniques (Voronoi diagrams and ordinary Kriging variograms), the achieved results show a good agreement, thanks to the high number of the punctual measurements carried out. Moreover, the use of many mobile measurement systems installed on different vehicles, which perform the same or different urban paths, can improve the whole measurement uncertainty using suitable data fusion techniques.

References

1. G. Betta, D. Capriglione, L. Ferrigno, G. Miele, Influence of WiFi Computer Interfaces On Measurement Apparatuses. IEEE Transactions on Instrumentation and Measurement, **59** (12), 3244–3252 (2010).
2. A. Bernieri, L. Ferrigno, M. Laracca, A. Tamburrino. Improving GMR Magnetometer Sensor Uncertainty by Implementing an Automatic Procedure for Calibration and Adjustment. Proc. of the XIX IEEE IMTC/07, May 2007, Warsaw (PL), 1–6 (2007).
3. D. Amicone, A. Bernieri, L. Ferrigno, M. Laracca. A Smart add-on Device for the Remote Calibration of Electrical Energy Meters. Proc. of the IEEE I2MTC/2009, May 2009, Suntec (SGP), 1599–1604 (2009).
4. G. Betta, S. Esposito, M. Laracca, M. Pansini. A novel Sol-Gel-based sensor for humidity detection. Proc. of the 16th IMEKO TC4 Symposium, September 2008, Florence (I), (2008).
5. A. Bernieri, L. Ferrigno, M. Laracca, P. Verde. Wireless sensor network for power quality monitoring according to IEC 61000-4-30. Proc. of the XIII IMEKO-TC4, September 2004, Athens (GR), **2**, 752–758 (2004).
6. L. Ferrigno, V. Paciello, A. Pietrosanto. Low-cost visual sensor node for Bluetooth-based measurement networks. IEEE Transactions on Instrumentation and Measurement, **55** (2), 521–527 (2006).
7. G. Betta, D. Capriglione, L. Ferrigno, G. Miele. Experimental Investigation of the Electromagnetic Interference of Zigbee Transmitters on Measurement Instruments. IEEE Transactions on Instrumentation and Measurement. **57** (10), 2118–2127 (2008).
8. A. Bernieri, D. Capriglione, L. Ferrigno, M. Laracca. A mobile measurement system for urban pollution monitoring. Proc. of the XIII TC-4 IMEKO Symposium and IX Semetro, September 2011, Natal (BR), 1–6 (2011).
9. A. Bernieri, D. Capriglione, L. Ferrigno, M. Laracca. Design of an efficient mobile measurement system for urban pollution monitoring. ACTA IMEKO, **1**, 77–84 (2012).
10. Directive 1999/30/EC of 22 April 1999 and Directive 2000/69/EC of 16 November 2000.
11. B. Denby, J. Horálek, S.E. Walker, K. Eben, and J. Fiala. "Interpolation and assimilation methods for European scale air quality assessment and mapping. Part I: Review and recommendations", ETC/ACC Technical Paper 2005/7.
12. J. Horálek, P. Kurfürst, B. Denby, P. de Smet, F. de Leeuw, M. Brabec, J. Fiala,. "Interpolation and assimilation methods for European scale air quality assessment and map-ping. Part II: Development and testing new methodologies", ETC/ACC Technical Paper 2005/7.

Part VII
Applications

A Wearable Sweat pH and Body Temperature Sensor Platform for Health, Fitness, and Wellness Applications

M. Caldara, C. Colleoni, E. Guido, V. Re, G. Rosace, and A. Vitali

1 Introduction

Wearable sensors development and diffusion on the market are in rapid expansion [1]. The medical community is particularly interested in the remote monitoring, as a result of budget reduction together with an increasing number of elderly and high-risk patients. Concerning fitness applications, the focus is on real-time monitoring of athletes' performances, in particular hydration and the sweat quality. The wellness field finally demands real-time data on personal health and environment. The wearable platform presented in this study (Fig. 1) is aimed to monitor the sweat pH and the skin temperature, by using novel high-resolution and accuracy techniques with ultralow power consumption. The sweat pH varies due to various endogenous and exogenous factors [2]. Skin temperature is linked to body thermoregulation [3], and its trend can provide information on the metabolic activity.

2 System Description

The developed pH sensor is a combination of a halochromic smart textile (able to change its color depending on the pH value) and a high-sensitivity and low-power color sensor. The main features of the smart textile are the nontoxicity of the pH-sensitive molecule, its pH range of color variations compatible with sweat pH typical values, and the solgel technique, which has been used in order to achieve dye

M. Caldara (✉) • C. Colleoni • E. Guido • V. Re • G. Rosace
Department of Engineering, University of Bergamo, Viale Marconi 5, 24044 Dalmine, Italy
e-mail: michele.caldara@unibg.it

A. Vitali
STMicroelectronics, Agrate Brianza, MB, Italy

C. Di Natale et al. (eds.), *Sensors and Microsystems: Proceedings of the 17th National Conference, Brescia, Italy, 5-7 February 2013*, Lecture Notes in Electrical Engineering 268, DOI 10.1007/978-3-319-00684-0_82, © Springer International Publishing Switzerland 2014

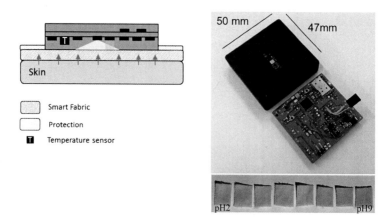

Fig. 1 pH/T sensor platform. Combination of smart textile and electronics (*left*), electronics prototype picture (*top right*), and textile color variations between pH 2 and pH 9 (*bottom right*)

fastness to the cotton fabric and reversibility properties. The fabric color variations are monitored every 5 s, in a reflective way, by using a white LED with a dedicated step-up driver and a high-sensitivity and low-power RGB color sensing. The main measured performances are 205 Hz/lx sensitivity, the quasi-digital output, and a current consumption of only 255 µA (V_{DD} 2.8 V). The temperature sensor is based on reverse-biased Schottky diodes, connected to the same front-end electronics used for the color sensor.

3 Experimental

The pH sensor was first characterized by laboratory experiments performed on different fabric samples, by using buffer solutions and artificial sweat (according to EN ISO 105-E04:2009). The pH estimation model obtained from the data is able to estimate the pH within a range of ±0.4 pH from pH 3 up to pH 9 (Fig. 2), which is compatible with the sweat variations reported in the literature [4]. Additional beneficial features are the immediate color change, an excellent stability, a color reversibility, and washing fastness, which has been proved performing a washing cycle according to EN ISO 6330:2000 and then reverifying the color variations at different pH (Fig. 2).

The temperature sensor exploits the diode reverse current exponential-like dependence with temperature; the sensor characterization has been performed in an oven, comparing the results with those obtained from a calibrated temperature sensor (Sensirion SHT25), having a maximum accuracy of ±0.3 °C. The obtained accuracy on the reverse-biased diode-based sensor, after linearization, lays between −0.3 °C and +0.4 °C in the temperature range 30–40 °C (Fig. 2), and the resolution

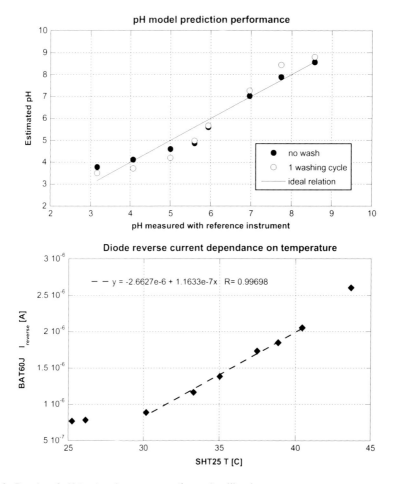

Fig. 2 Results of pH (*top*) and temperature (*bottom*) calibrations

is better than 0.01 °C. The total current consumption of only 85 μA (VDD 2.8 V) and the direct reading with a microcontroller timer are additional attractive features. The ultralow-power microcontroller (STM8L) used in the system manages the pH/T measurements and stores the data in a nonvolatile memory. Every 2 min the onboard Bluetooth module wakes up, in order to send the data to an external device. Recently on-body trials started, by fixing the system on the back with a lumbar elastic band during exercise bike tests; sweat pH and skin temperature were monitored during activity and rest periods (Fig. 3). Data are uploaded with a Bluetooth interface to a PC for analysis and visualization. Preliminary on-body sweat pH sensing showed a slight pH increase during activity, in conformity with what is reported in literature. Additionally the on-body pH resolution is 0.2 pH, a value that can be improved with

Fig. 3 On-body pH and T measurements: at *A* the system has been worn, while at *B* activity session started. pH reference measurements are also reported with *circle markers*

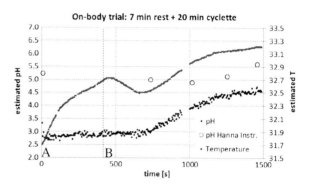

averaging. Finally, smart fabric turned its color about 10 min after sweat production. Concerning on-body temperature monitoring, an excellent resolution is achieved, but a long settling time has been registered, due to the present distance between the diode and the skin.

4 Conclusions

The paper presents the development of a miniaturized sweat pH and skin temperature sensor platform, with processing capabilities and a wireless interface. The low-power consumption (up to 14 h operation with 90 mAh battery), the achieved pH and temperature accuracies, and resolutions and the unobtrusive solution of the smart textile represent attractive features for performing long-monitoring studies. Body hydration assessment will be performed with clinical trials in a sport medicine laboratory.

References

1. M. Chan, D. Estève, J.-Y. Fourniols, C. Escriba, E. Campo, "Smart wearable systems: Current status and future challenges", Artificial Intelligence in Medicine, Volume 56, Issue 3, Pages 137–156, 2012. Available online 1 November 2012, ISSN 0933-3657, doi: 10.1016/j.artmed.2012.09.003
2. C.J. Harvey, R.F. LeBouf, A.B. Stefaniak, "Formulation and stability of a novel artificial human sweat under conditions of storage and use", Toxicology in Vitro, Volume 24, Issue 6, September 2010, Pages 1790–1796, ISSN 0887-2333, doi: 10.1016/j.tiv.2010.06.016.
3. P. Frasco, "Temperature Monitoring", In: J.G. Webster. Encyclopedia of Medical Devices and Instrumentation, 2nd ed., 2006. Wiley, New York. ISBN: 9780471732877, doi: 10.1002/0471732877
4. S. Coyle et al., "BIOTEX—Biosensing Textiles for Personalised Healthcare Management", IEEE Transactions on Information Technology in Biomedicine, Volume 14, Issue 2, March 2010, Pages 364–370, doi: 10.1109/TITB.2009.2038484

Influence Quantity Estimation in Face Recognition Digital Processing Algorithms

G. Betta, D. Capriglione, M. Corvino, C. Liguori, A. Paolillo, and P. Sommella

1 Introduction

The scientific interest in uncertainty estimation in biometric sensor systems is increasing because of their wider and wider diffusion in critical applications such as security and access control. The authors have been involved, in this framework, in the development of general methods for modeling and estimating the uncertainty in biometrical sensor systems based on face recognition algorithms. The aim is providing these systems with the capability of self-estimating the uncertainty of their results, thus giving as output not only the identity of the observed subject but also a confidence level of the identification. The proposed model, presented in [1–4], is based on the estimation of the values assumed by the relevant influence quantities and of their distance from the values assumed under reference conditions. Consequently, the influence quantities have to be identified and estimated starting from the observed image.

2 The Proposed Model

The final goal of the research project is a method for estimating relationships between uncertainty (or confidence level) of final results and influence quantities. The uncertainty model proposed in [1–4] is based on the definition and estimation of a suitable distance function between the N-dimensional vector of the values of

G. Betta (✉) • D. Capriglione • M. Corvino
DIEI, University of Cassino, Via G. Di Biasio 43, Cassino, FR, Italy
e-mail: betta@unicas.it

C. Liguori • A. Paolillo • P. Sommella
DIIn, University of Salerno, via Ponte don Melillo, Fisciano, SA, Italy

C. Di Natale et al. (eds.), *Sensors and Microsystems: Proceedings of the 17th National Conference, Brescia, Italy, 5-7 February 2013*, Lecture Notes in Electrical Engineering 268, DOI 10.1007/978-3-319-00684-0_83, © Springer International Publishing Switzerland 2014

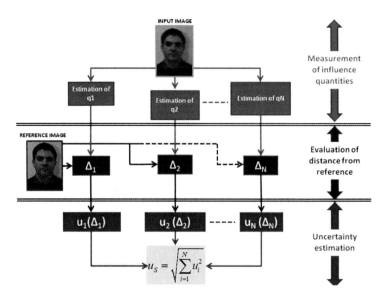

Fig. 1 Model for uncertainty estimation

the influence quantities measured on the image under test and the values of a reference vector containing the values of the same influence quantities in reference conditions. After the measurement phase, the influence quantities are normalized and then the proposed model, schematized in Fig. 1, is applied. In this scheme, u_i is the uncertainty contribution of the ith influence quantity, q_i, and N is the number of relevant influence quantities. Being Δ_i the difference between the values assumed by q_i in the situation under test and those assumed under reference condition, each contribution u_i is estimated with a relation $u_i = f(\Delta_i)$. By this approach, the influence quantities have to be measured only in the situation under test.

3   Measurement of Influence Quantities

The research activity has been then focused on the identification, setup, and characterization of measurement procedures for estimating influence quantities from an acquired image.

Among the influence quantities that could affect an acquired image, the attention was focused on extrinsic uncertainty causes, namely, those related to the sensor systems and not those related to the subject under test. These extrinsic uncertainties are mainly due to the lightening (luminance, wavelength, position, flicker, shadows), the relative position between the camera and subject under test (vibrations, view angle), and camera setup (focal length, focusing, aperture, resolution).

Fig. 2 Performance comparison between two methods for the measurement of the angle between the face and camera

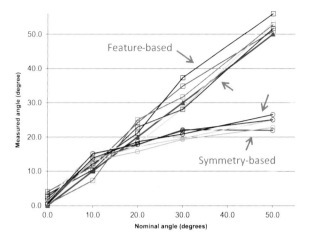

In particular, three relevant influence quantities have been analyzed in detail, setting up suitable measurement procedure for their estimation:

(a) Image luminance. It has been estimated as the average value of gray levels in all the image pixels. It depends on the lightening of the subject and on the camera setup.

(b) Angle between face and camera axes. After having analyzed a method based on an "elastic" model of the face [5], very accurate but requiring a reference image of the subject, or symmetry based [6], a method is adopted that estimates the angle by using a geometrical approach, on the base of the location of specific face features, such as eye center or nose tip, in the acquired image. The main advantage of this approach is that it does not require any reference image. Figure 2 shows a comparison between the proposed feature-based procedure and a symmetry-based one.

(c) Image defocus. It consists in the estimation of the difference between the acquisition optical condition and those of a perfect focus. It cannot be estimated by using known absolute reference (e.g., distance of the subject from the position in which the focus is optimal). The proposed procedure is based on a spectral processing of the acquired image. On the basis of the algorithm described in [7], the proposed solution requires the image decomposition by a three-level wavelet transform in high-pass and low-pass horizontal and vertical components. The higher is the defocusing, and the nearer are second- and third-level coefficients. A $q_{defocus}$ variable was introduced that estimates the defocus value as the sum of the differences between second- and third-level coefficients of the wavelet transform.

4   Conclusions

A proposal for the characterization of face recognition algorithms was formally outlined together with effective solutions for each stage of the uncertainty evaluation. The work is a contribution toward the design of applications based on face recognition provided with automatic capability of evaluating the results reliability. Joined with suitable rules for the subject classification and for the assessment of the uncertainty in conditional statements, the proposed approach can be useful to achieve, as output of face recognition systems, a confidence level for each possible subject (not a simple identity of the classified subject). This important feature will allow the analysis of recognition results to be improved, thus increasing the results reliability in important applications as surveillance and access control to critical sites.

References

1. G. Betta, D. Capriglione, C. Liguori, A. Paolillo. Uncertainty evaluation in face recognition algorithms. In: Proc. of I2MTC 2011, May 2011, Binjiang, Hangzhou, China, 21-26 (2011).
2. G. Betta, D. Capriglione, F. Crenna, G. B. Rossi, M. Gasparetto, E. Zappa, C. Liguori, A. Paolillo. Face-based recognition techniques: proposals for the metrological characterization of global and feature-based approaches. Measurement Science and Technology, **22** (12), 124005-124014 (2011).
3. G. Betta, D. Capriglione, M. Corvino, C. Liguori, A. Paolillo. Estimation of Influence Quantities in Face Recognition. Proc. of Instrumentation and Measurement Technology Conference I2MTC12, May 2012, Graz, Austria, 963-968 (2012).
4. G. Betta, D. Capriglione, M. Corvino, C. Liguori, A. Paolillo. Face based recognition algorithms: a first step toward a metrological characterization. IEEE Transactions on Instrumentation and Measurement, **62** (5), 1008-1016 (2013).
5. S. Zhao, Y. Gao. Automated Face Pose Estimation Using Elastic Energy Models. Proc. Of 18th International Conference on Pattern Recognition (ICPR'06) **4**, 618-621 (2006).
6. V. Pathangay, S. Das, T. Greiner. Symmetry-based Face Pose Estimation from a Single Uncalibrated View. Proc. of Intern. Conf. on Face and Gesture Recognition (FG 2008), art. no. 4813312 (2008).
7. J. Lin, C. Zhang, Q. Shi. Estimating the amount of defocus through a wavelet transform approach. Pattern Recognition Letters, **25**, 407-411 (2004).

Printed Sensors on Textiles for Biomedical Applications

A. Dionisi, E. Sardini, and M. Serpelloni

1 Sensor Implementation Technique

The plane geometry of the plethysmographic sensor had been chosen exploiting the results obtained in [5]. In particular, the sinusoidal geometry is the best configuration for the sensor due to a high impedance value and a low breaking factor. Two plethysmographic sensors, called sensor A and sensor B and shown in Fig. 1a, b, respectively, had been designed and implemented with different geometries: in particular, the sensors differ in the number of meanders and in the height. The sensor had been designed trying to increase the impedance change without losing the elastic features. The trace thickness is measured 5 mm: a lower thickness increased the resistance value, and, on the contrary, an upper thickness increased the quality factor but decreased the sensitivity due to the increase of the stiffness.

The sensor implementation technique is based on the industrial screen printing process, known as serigraphy, which permits the passage of the tight amount of paste through the mesh of the screen. At the beginning, a heat sealable vinyl film was applied on the T-shirt as support, so that it was possible to print on the conductive material. The silver conductive paste DuPont 5000 was printed on the support by using mask with the same sinusoidal form of the heat sealable vinyl film. The conductive paste was subjected to sinterization process positioning the T-shirt into an oven at 125 °C for about 30 min. A protected layer was performed with the same procedure: the dielectric paste DuPont 5018 was printed on the conductive layer, but the sinterization was produced by UV ray. The first paste has an excellent conductivity, less than 15 mΩ/sq/mil, while the dielectric paste has a high insulation resistance, more than 10 GΩ/sq/mil. Both the pastes have allowed to produce a flexible thick film in accordance with the stretching movement of the sensors.

A. Dionisi (✉) • E. Sardini • M. Serpelloni
Department of Information Engineering, University of Brescia,
Via Branze 38, 25123 Brescia, Italy
e-mail: alessandro.dionisi@ing.unibs.it

C. Di Natale et al. (eds.), *Sensors and Microsystems: Proceedings of the 17th National Conference, Brescia, Italy, 5-7 February 2013*, Lecture Notes in Electrical Engineering 268, DOI 10.1007/978-3-319-00684-0_84, © Springer International Publishing Switzerland 2014

Fig. 1 View of the two
plethysmographic sensors
implemented by printed
method, (**a**) sensor A and
(**b**) sensor B. The dimensions
of the two sensors are shown
in the figures, respectively

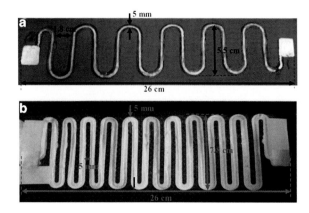

2 Sensor Characterization

The modulus and the phase of the sensor impedance were measured by an impedance analyser (HP4194A) in order to characterize the two sensors. The resulted
trends have represented the typical behaviour of a resistance R_S and an inductance
L_S in series configuration. A capacitor $C_P = 330$ pF was connected to the plethysmographic sensors in parallel in order to create a resonance frequency under 10 MHz
whose quality factor depends on only characteristic parameters of the two sensors
(R_S and L_S), and it is given by the following equation (1):

$$Q = \frac{\omega L_S}{R_S} \tag{1}$$

Figure 2a, b shows, respectively, the modulus and the phase impedance trends of
the two sensors. By means of the impedance analyser, the series resistance and the
series inductance values were measured for both sensors, and they are shown in
Table 1. Furthermore, in the table, there are the values of the resonance frequency
and the peak values of its impedance modulus.

3 Experimental Results

The preliminary tests are related in this preliminary work; in particular, the T-shirt
with the printed sensors was stretched in horizontal direction up to 4 cm with step
of 0.5 cm for sensor A and up to 2.7 cm with step 0.3 cm for the second sensor. The
different stretching movements between the sensors were due to their different
flexibilities. When an elongation was applied to the sensors in horizontal direction,
the resonance peak value decreased with respect to the initial rest value. The results
are shown in Fig. 3, where the impedance modulus close to resonance frequency f_R

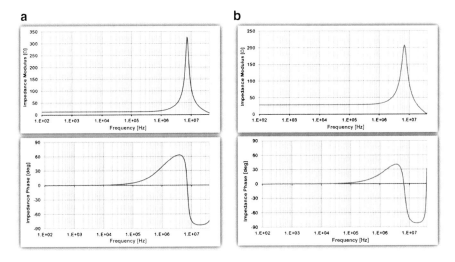

Fig. 2 Modulus and phase of the impedance of the two plethysmographic sensors, (**a**) sensor A and (**b**) sensor B

Table 1 The characteristic parameters of the plethysmographic sensors obtained by the impedance analyser

Parameter	Sensor A	Sensor B
Resistance (R_S)	16.7 Ω	27.5 Ω
Inductance (L_S)	0.62 μH	0.29 μH
Resonance frequency (f_R)	7.42 MHz	6.57 MHz
Impedance amplitude peak ($|Z_{max}|$)	322 Ω	204 Ω

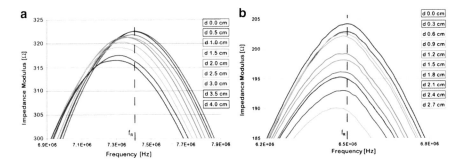

Fig. 3 Amplitude impedance variation as a function of (**a**) sensor A extension and (**b**) sensor B extension

as a function of the extension applied to the sensors is represented. The sensitivity of the implemented plethysmographic sensors measures −1.84 and −5.18 Ω/cm for sensor A and sensor B, respectively.

4   Conclusion

The paper proposes a fabrication technique for biomedical sensors to measure the respiratory rate based on screen printing. Two sensors with different dimensions are compared, and the experimental results show that sensor B, which has more number of meanders, is more sensible than sensor A. In particular, the impedance modulus sensitivities are −1.84 and −5.18 Ω/cm for sensor A and sensor B, respectively. A potential further work may focus on the possibility to involve the print of the sensors around the abdomen or chest wall to improve the sensor functions.

References

1. R. Paradiso, A. Gemignani, E. P. Scilingo, and D. De Rossi (2003), "Knitted bio-clothes for cardiopulmonary monitoring," Proc. 25th Ann. Int. Conf. IEEE EMBS 4: 3720–3723.
2. C. M. Yang, T. Kao, N.N.Y. Chu, C. C. Wu, and T. L. Yang, "A textile-based capacitive breath-sensing system," *2012 IEEE-EMBS International Conference on Biomedical and Health Informatics (BHI)*, 5–7 Jan. 2012, pp. 875-877.
3. E. J. Maarsingh, L. A. van Eykern, A. B. Sprikkelman, *et al.*, "Respiratory muscle activity measured with a noninvasive EMG technique: technical aspects and reproducibility", J Appl Physiol 2000, 88: 1955-1961.
4. M. Di Rienzo, F. Rizzo, G. Parati, G. Brambilla, M. Ferratini, and P. Castiglioni, "MagIC system: A new textile-based wearable device for biological signal monitoring. Applicability in daily life and clinical setting," in Proc. 27th Ann. Int. Conf. IEEE EMBS, Shanghai, Sept. 2005, pp. 7167–7169.
5. T. H Kang, "Textile-embedded sensors for wearable physiological monitoring system", PhD dissertation, Dep Elec Eng, North Carolina State University, USA, 2006.

E-Nose as a Potential Quality Assurance Technology for the Detection of Surface Contamination by Aeronautic Fluids

Paola Di Palma, Saverio De Vito, Mara Miglietta, Ettore Massera, Grazia Fattoruso, Bruno Mastroianni, and Girolamo Di Francia

1 Introduction

The high-efficiency lightweight aircraft realization through the use of composite materials is a primary challenge in the aerospace industry. The estimation of the reduction in aircraft weight is settled at about 15 %, producing considerable savings both during the operative phase, for reduction of fuel consumption and CO_2 emissions, and in aircraft maintenance operations for the structural simplifications.

From the production point of view, adhesive bonding is considered the optimum assembly solution for CFRP components, but the reliability of adhesive bonds strongly depends on the cleanliness of the adherent surfaces [1].

In this field, the goal of our research group, as an alternative to analytical approaches, is to focus on a suitability analysis and on the adaptation of artificial olfaction technologies for quality assessment of adhesive bonds. In order to evaluate detection and discrimination capability also in maintenance scenario, an ad hoc electronic nose prototype is designed and developed.

In this work, current results of this challenge are presented.

P. Di Palma (✉)
Department of Electrical and Information Engineering, University of Cassino and Southern Lazio, via G. Di Biasio, 43, 03043, Cassino, Frosinone, Italy

UTTP-MDB, ENEA C.R. Portici, P.le E. Fermi, 1, 80055 Portici, Naples, Italy
e-mail: dipalmapaola@gmail.com

S. De Vito • M. Miglietta • E. Massera • G. Fattoruso • B. Mastroianni • G. Di Francia
UTTP-MDB, ENEA C.R. Portici, P.le E. Fermi, 1, 80055 Portici, Naples, Italy

C. Di Natale et al. (eds.), *Sensors and Microsystems: Proceedings of the 17th National Conference, Brescia, Italy, 5-7 February 2013*, Lecture Notes in Electrical Engineering 268, DOI 10.1007/978-3-319-00684-0_85, © Springer International Publishing Switzerland 2014

2 Adaptation and Results

Within the EU project described in the beginning of this work, the aim of ENEA UTTP/MDB is to investigate whether the sensor technologies coupled with artificial intelligence subsystems could achieve adequate discrimination and classification capabilities in different CFRP contamination scenarios [2]. The first step towards this aim was made in the identification of main contaminants that could impact with the composite substrates prior to bonding. Different pre-bond contamination scenarios have been considered, namely, the hydraulic fluid (e.g. Skydrol 500-B), the runway de-icing fluid (e.g. Kilfrost) and the release agent (e.g. Frekote) commonly used in composite material production. Also the water can penetrate into the bulk material, acting like an interferent for the behaviour of other contaminants.

In this work, we investigated the use of a commercial hybrid array composed of two semiconducting metal oxide sensors (MOX), one electrochemical sensor (EC), one photoionization detector (PID) and one ion-mobility spectrometry (IMS). With this e-nose platform, with respect to the ENCOMB project, a set of measurements have been carried out by using the following specimens: four CFRP samples contaminated by Frekote upon exposure to four levels of contaminant concentration in solution (1, 5, 10, 20 %), four moisture-contaminated samples exposed to different relative humidity environments (30, 75, 95, 100 % RH) and a Skydrol-contaminated sample and an uncontaminated one. Each measurement cycle has been performed at one sample per second rate by an automatic data acquisition process and has been recorded into a database.

Pattern recognition techniques and multivariate data mapping models such as principal component analysis (PCA, Fig. 1a) and linear discriminant analysis (LDA) have been preliminarily applied to extract relevant insights from the data distribution. The dataset has hence been used, point by point, to train and test linear (fisher discriminant based), quadratic, Parzen-like and feedforward neural network (FFNN) classifiers, by using PRTools Matlab® toolbox [3] with appropriate cross-validation.

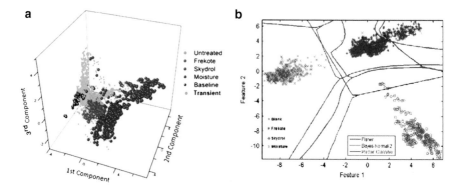

Fig. 1 (**a**) PCA loading map of untreated (uncontaminated), Frekote-, Skydrol- and moisture-contaminated samples; (**b**) 2D LDA mapping with the overlapping of discriminant plot of classifiers

Table 1 Correct classification rate of applied classifiers

Mapping technique	Correct classification rate (%)
Linear classification	87.6
Parzen classifier	98.2
Quadratic Bayes normal classifier	98.1

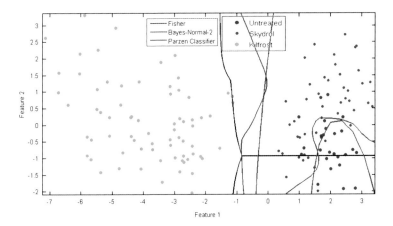

Fig. 2 LDA mapping of the ICARO measure dataset with the overlapping of discriminant plots

In Fig. 1b, discrimination capability of the electronic nose is shown by using a 2D LDA projection that caught the cluster structure of data. Discriminant plots overlapped to the data mapping highlight a good separability, especially between Skydrol data points and the other classes, showing the correct classification rate summarized in Table 1.

Furthermore, a neural network (NN) classifier, characterized by 10-neuron hidden layer, has been trained, for classification purposes, with 30 % of samples randomly chosen from the entire dataset. The remaining samples have been used for testing scopes, leading to a combined index of 99.4 % of correctly classified samples, computed by using a tenfold cross-validation procedure.

Of course, these performed tests made with the commercial hybrid nose have contributed to build an operative experience used as a starting point for the design of a new ENEA electronic nose prototype, cofinanced by the ICARO project. With ICARO prototype, other analyses, which will be presented below, were carried out for surface detection of Kilfrost and Skydrol on contaminated specimens.

The performances of the used classifiers, superimposed on the transformed plane as shown in Fig. 2, show classification capabilities reaching, at best, with Bayes normal classifier, 96.8 % value of correct classification index.

Furthermore, in order to correctly estimate the values of the contaminant concentrations, sensor responses were used as inputs of an FFNN used to fit an input–output relationship in the regression problem, and the mean absolute error (MAE), mean squared error (MSE) and correlation coefficient (CC), related to Kilfrost and Skydrol concentration prediction problems, are summarized in Table 2.

Table 2 Performance indexes of a 5-20-3 FFNN for Skydrol and Kilfrost concentration on predictions

Performance indexes	Skydrol concentration prediction	Kilfrost concentration prediction
MAE	8E–3	0.0102
MSE	5.5E–3	1.2E–3
CC	0.9896	0.9978

3 Conclusions

In conclusion, these preliminary results show that it is reasonable to expect that the electronic nose technology can be a suitable technology to address the problem of surface contaminant detection in adhesive bonding quality assurance.

The ICARO e-nose results indicate that this prototype is capable, thanks to the development calibration methodology, to distinguish and quantify with interesting performance figures among the target analytes.

Acknowledgments The research leading to these results has received funding from the European Union Seventh Framework Programme (FP7/2007–2013) under grant agreement no. 266226 and no. 286786.

References

1. D.N. Markatosa, K.I. Tserpesa, E. Rau, K. Brun, Sp. Pantelakis (2014), Degradation of Mode-I fracture toughness of CFRP bonded joints due to release agent and moisture pre-bond contamination: The Journal of Adhesion, Vol. 90, Issue 2, 156-173.
2. S. De Vito, E. Massera, G. Fattoruso, M.L. Miglietta, G. Di Francia (2012), Developing artificial olfaction techniques for contamination detection on aircraft CFRP surfaces: The Encomb project, Sensors and Microsystems, Lecture Notes in Electrical Engineering, Vol. 109, Chapter 28, 163-166.
3. F. van der Heijden, R. Duin, D. de Ridder, D. M. J. Tax (2005), Classification, Parameter Estimation and State Estimation: An Engineering Approach Using MATLAB, Wiley & Sons Australia.

IMS Development at NRNU MEPhI

N. Samotaev, A. Golovin, V. Vasilyev, E. Malkin, E. Gromov, Y. Shaltaeva, A. Mironov, and D. Lipatov

1 Sources of Ionization

Typical IMS drift tube structure is shown in Fig. 1. Drift tube consists of ionization chamber, ion source, electrostatic gate, drift tube, and collector with signal amplifier. One of the most important IMS parts is ion source. Ionization source affects the sensitivity and selectivity of the spectrometer. The second most important part in the IMS device is the drift tube. Parameter drift tube gives resolution and the warm-up time of the device. The lighter the drift tube mains, the faster the warm-up IMS device, thus reducing power consumption. The longer the drift tube, the higher the resolution. The third most important is the sampling system. Sampling system is most sensitive to the type of application and one construction depending on the type of the detected substance (volatile, liquid, or solid components).

ELNOS research laboratory has developed IMS spectrometers with various type sources of ionization [1]. In the last 10 years in work following technology of sampling ionization, we have used capsule with ^{3}H (Tritium) isotopic as a radioactive ionization source, hydrogen glow discharge lamp (200 V, 1 mA) with maximal quanta energy of 10.5 eV for ultraviolet photoionization, 266 nm source with 1 mJ 20 ns pulse laser ionization, X-ray tube with total size smaller than 20 cm^3, anode voltage 4 kV, and anode current 0.1 mA roentgen ionization and several discharge gaps with optical synchronization for corona discharge. Table 1 presents technical parameters of IMS systems which were created in the last 15 years. From the table, it is possible to see the evolution in the field of replacing the radioactive ionization sources for nonradioactive. Also from the table, a progressive movement towards more compact and cheap methods of ionization is observed. As a last step in our

N. Samotaev (✉) • A. Golovin • V. Vasilyev • E. Malkin • E. Gromov • Y. Shaltaeva
A. Mironov • D. Lipatov
Department of Nano- and microelectronics, National Research Nuclear University MEPhI,
Kashirskoe Highway 31, Moscow 115409, Russian Federation
e-mail: samotaev@mail.ru

C. Di Natale et al. (eds.), *Sensors and Microsystems: Proceedings of the 17th National*   447
Conference, Brescia, Italy, 5-7 February 2013, Lecture Notes in Electrical Engineering 268,
DOI 10.1007/978-3-319-00684-0_86, © Springer International Publishing Switzerland 2014

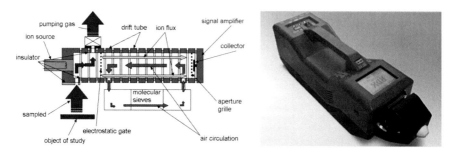

Fig. 1 Structure of IMS drift tube (*left*) and commercially available IMS detector Kerber

Table 1 Technical parameters of different types of IMS developed by ELNOS laboratory

Year of development	1996	1999	2005	2010
Ionization source	³H, 500 MBq	Laser, 266 nm	X-ray, 4 kV	Corona discharge
Length of the drift tube (mm)	79	130	120	120
Inner drift tube diameter (mm)	50	35	30	30
Number of rings	8	13	11	10
Drift voltage (V)	2,000–3,000	2,500–4,000	2,500–4,000	2,000–3,000
Shutter opening times (ms)	0.6	–	0.06–0.6	0.06–0.6
Temperature (°C)	50	23	23–100	23–100
Pressure	Atmospheric pressure	Atmospheric pressure	Atmospheric pressure	Atmospheric pressure
Drift gas	Dry ambient air	Dry ambient air	Dry ambient air	Dry ambient air
Sample gas flow rate (liter/min)	0.5–3	1–10	1–10	1–10
Ion current charge conversion coeff. (mV/pA)	1	3	33	33

modern researches and development, we apply only corona discharge ionization source. This method is simple in fabrication, is inexpensive, and does not contain hazardous substances and radioactive isotopes.

2 Drift Tubes

Because achieving the best measurement properties of IMS device needs to have a system as a combination of ion source and drift tube, scientific staff of the laboratory had made a design of drift tubes for IMS (Fig. 2). Ion source camera is based on PTFE with pushing electrode, electrostatic gate, inlet channel for sampling gas, and

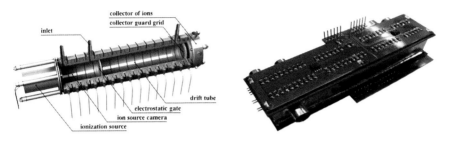

Fig. 2 ELNOS research laboratory drift tube designs. Aluminum (*left*) and PCB (*right*) technologies

outlet channel for used sampling and drift gas. Tundal type [2] electrostatic gate is based on 30 μm thickness nickel (or its alloy) grid with rectangular (0.5×0.5 mm) cells and 50 μm frame. Drift tube consists of ten uniform stainless steels of aluminum guard rings with PTFE gap isolation, nickel aperture grid, and collector of ions. Also at the laboratory, we have designed several extra small and light drift tubes based on PCB technology, which provides warm-up time of IMS device of about 2 min. ELNOS research laboratory alumina and PCB drift tube designs are shown in Fig. 2.

3 Sampling Systems

The main part of spectrometers which was created at ELNOS laboratory is based on classic scheme [3] working at room temperature and using atmosphere air as a drift gas. The important part of IMS detectors is sampling gas system, which depends on the substances to be detected. The research laboratory activity has mainly focused on application which connection with search various types of explosives and drugs. Therefore, for detection of solid particles, different types of sampling systems have been created, the most important of which was system with aluminum sample swab. For investigations over time, vapor sampling for volatile substances, hot air vortex sampling system and swab oven system heated up to 200°C were created. For trace detection of substances with low vapor pressure on human fingers and documents, the method of surface heating by impulse gas-discharge lamp [4] in comparison with the use of a heated air flow or a heat transfer from warm body by thermal conductivity was investigated. The evaporated molecules move from tested surface as narrow bunch to input of ion-mobility spectrometer. Studies were conducted on samples of TNT, PETN, and RDX. Scheme and appearance of the human finger sampling system are shown in Fig. 3.

Today, the current research activity of the laboratory focuses on medical topics, such as the analyses of human breath for noninvasive diagnostics of diseases [5]. The measurement of biomarker concentration in human breath has some problems connected with sampling. In particular, it is the problem of the high-level moisture in expiration air. To eliminate this problem, three methods are investigated. The first method is based on using the molecular sieves for adsorption of the water

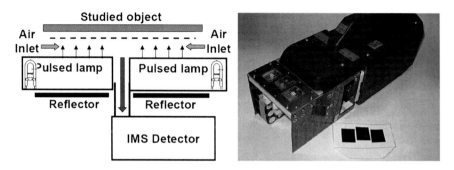

Fig. 3 Human finger sampling system. Sampling setup scheme (*left*). Combined IMS device with human finger sampling system and samples of swab for providing testing (*right*)

molecules. In the second method for creation of compact drying system without an additional sorption loss, the expiration air flow is cooled by using cuprum radiator and Peltier element. The third method is based on the collection of the alveolar part of the breath in special closed volume. In the case of diabetic diseases, acetone smell was registered without the additional air drying. The experiments showed that the acetone concentration in breath of healthy persons is no more than 1 ppm, but persons with diabetic illness have the acetone concentration in their breath more than two times. The sensitivity of ELNOS ion-mobility spectrometers allows to confidently detect the acetone concentration on the level less than 100 ppb.

4 Conclusions

ELNOS research laboratory developed stationary and mobile versions of IMS devices with different sampling systems, drift tubes, and ionization sources for different applications, including built-in versions for air monitoring systems on transport and safety portals for access control on protected objects. Researches and design results of ELNOS laboratory activity are already used in commercially available ion-drift detector Kerber (Fig. 1). Kerber using corona discharge ion source and multifunctional sampling system (air/surface) is able to detect explosives, narcotics, and TICs.

References

1. V. Matsaev, M. Gumerov, L. Krasnobaev, V. Pershenkov, V. Belyakov, A. Chistyakov, V. Boudovitch. IMS Spectrometers with Radioactive, X-ray, UV and Laser Ionization. International Journal for Ion Mobility Spectrometry, 5 (3), 112-114 (2002)
2. H. Borsdorf, G.A. Eiceman, E.G. Nazarov, D. Schultze, H. Schelhom "Influence of Structural Features on Ion Mobility Spectra of Non-polar Hydrocarbons", International Journal for Ion Mobility Spectrometry, 4 (2), 96-99 (2002)

3. A.M. Tyndall. The Mobility of Positive Ions in Gases. (Cambridge University Press. Cambridge, UK 1938)
4. A.V. Golovin, V.V. Belyakov, V.K. Vasilyev, E.K. Malkin, E.A. Gromov, V.S. Pershenkov. The sampling unit of IMS device for detection of trace amounts substances on human fingers and documents. Proc. of the 21th International conference on ion mobility spectrometry, July 22–27, 2012, Orlando, Florida, USA (2012)
5. A.V. Golovin, V.V. Belyakov, V.K. Vasilyev, E.K. Malkin, E.A. Gromov, V.S. Pershenkov. The methods of air sample preparation during human breath detection by ion mobility spectrometry technique. Proc. of the 21th International conference on ion mobility spectrometry, July 22–27, 2012, Orlando, Florida, USA. (2012)

Fiber-Optic Flow Sensor for the Measurement of Inspiratory Efforts in Mechanical Neonatal Ventilation

Luigi Battista, Andrea Scorza, and Salvatore Andrea Sciuto

1 Introduction

Measurements of air flow rate in neonatal mechanical ventilation are carried out mainly for monitoring air flows supplied by a mechanical ventilator to support infants in intensive care units and for the evaluation of lung ventilator performances [1–4]. Flow measurements in infant mechanical ventilation are typically performed by means of electrical and electromagnetic sensors [5, 6] that are exposed to errors due to electromagnetic interferences. The reduction of the above-quoted shortcomings, achieved by means of fiber-optic sensing techniques, is widely reported in scientific literature and is due to optical fiber features, i.e., immunity to electromagnetic interferences and electrical insulation allowing the improvement of electrical safety conditions [7–9].

By the way, even if various optical fiber sensors have been proposed, it is difficult to find an exhaustive investigation on the design and testing of optical fiber air flow meters in mechanical ventilation applications.

L. Battista (✉)
Department of Engineering, University of Rome "ROMA TRE",
Via della Vasca Navale 79/81, 00146 Rome, Italy

CNR – National Institute of Optics, Via Campi Flegrei 34, 80078 Pozzuoli, NA, Italy
e-mail: luigi.battista@uniroma3.it

A. Scorza • S.A. Sciuto
Department of Engineering, University of Rome "ROMA TRE",
Via della Vasca Navale 79/81, 00146 Rome, Italy

C. Di Natale et al. (eds.), *Sensors and Microsystems: Proceedings of the 17th National Conference, Brescia, Italy, 5-7 February 2013*, Lecture Notes in Electrical Engineering 268, DOI 10.1007/978-3-319-00684-0_87, © Springer International Publishing Switzerland 2014

2 Materials and Methods

In former works [10, 11], authors proposed a fiber-optic air flow sensor based on an arrangement that provides immunity to light intensity variations not related to the air flow, developed for monitoring air flows supplied by a neonatal ventilator to support infants in intensive care units. The principle of operation of the proposed sensor is based on the measurement of transversal displacement of an emitting fiber-optic cantilever due to action of air flow acting on it, performed by means of a photodiode linear array.

The aforementioned transducer can detect air flow rates up to 18.0 l/min (i.e., flow rates normally encountered during neonatal ventilation in infants [12]); nevertheless, the minimum detectable air flow rate is 2.0 l/min, in monodirectional configuration, and 3.0 l/min in bidirectional configuration, due to resolution of the photodiode array. The poor lower limit of measurement range and resolution at low flow rates implies that the considered flow sensor cannot discriminate flow variations in the range between 0.5 l/min and 5 l/min that are the typical flow trigger levels set during assist-control ventilation (ACV) [13], i.e., a ventilation mode used when a patient is able to initiate breaths but requires ventilatory assistance.

In order to reduce the minimum detectable flow allowing the measurement of air flow rates lower than 3.0 l/min, an alternative arrangement is proposed here (Fig. 1), in which the displacement of the tip of the optical fiber is detected by means of a photoswitch rather than a photodiode array. Contrary to the typical commercial photoswitches, the through-beam photomicrosensor Omron EE-SX1041 (made of an infrared emitter and an infrared receiver) is characterized by linear transition from the two conditions "dark/light" allowing the detection of displacements up to about 250 μm with a sensitivity of 20 μm/mA; therefore, an amperemeter with a resolution of 1 mA is used with the photomicrosensor Omron EE-SX1041 to detect displacements up to 250 μm with a resolution of 20 μm that overcome the limit of 63.5 μm achieved by means of the photodiode array previously adopted [10, 11].

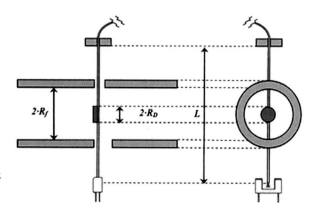

Fig. 1 Schematic of the proposed sensor ($R_f = 10$ mm; $R_D = 5$ mm; $L = 55$ mm) (not in scale)

3 Experimental Setup and Results

In order to perform static calibration of the device, experimental trials have been performed for different air flow rates by adjusting the output pressure from a compressor and comparing the air flow values with a reference sensor (variable area flowmeter Key Instrument Flo-Rite MR3A14, 0.4–5 l/min, accuracy of 4 %). Flowing air produces a drag force applied on the optical fiber and on the flat target disk, so that the emitting optical fiber bends with a fiber tip displacement measured by means of an electronic circuit (Fig. 2) made up of the through-beam photomicrosensor.

Experimental relationship between the output voltage of the electronic circuit V_{OUT} and air flow Q, measured with an accuracy of 5 %, is displayed in Fig. 3, and a parabolic relationship can be deduced, as expected [11], with a coefficient of determination $r^2 = 0.998$.

Results show that the lower limit of the measuring interval is 0.6 l/min, confirming that the novel proposed configuration for the optical fiber air flow sensor is suitable for monitoring flow variations due to infants' inspiratory attempts that typically occur during the ACV in the range between 0.5 l/min and 5 l/min.

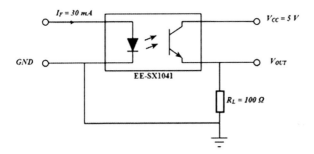

Fig. 2 Electronic circuit for displacement detection and conversion

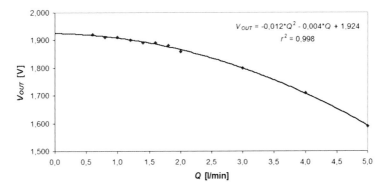

Fig. 3 Measurement results: output voltage of electronic circuit V_{OUT} as a function of flow rate Q

4 Discussion and Conclusion

A novel fiber-optic flow sensor has been developed for monitoring inspiratory efforts during neonatal mechanical ventilation. The device is based on fiber-optic sensing techniques, allowing the reduction of the effects due to electromagnetic interferences and a possible improvement of the electrical safety conditions.

In the arrangement described here, the fiber-optic sensor is able to measure, with an accuracy of 5 %, flow variations in the range between 0.5 l/min and 5 l/min that are the typical flow variations due to infants' inspiratory attempts and typical flow trigger levels set during assist-control ventilation.

Moreover, a good agreement ($r^2 = 0.998$) between experimental data and the parabolic theoretical model can be deduced.

The experimentally obtained metrological characteristics confirm that the novel proposed configuration for the optical fiber air flow sensor is suitable for monitoring flow variations due to infants' inspiratory attempts.

References

1. F. P. Branca, P. Cappa, S. A. Sciuto and S. Silvestri. A novel methodology for the experimental evaluation of pulmonary ventilator performance drift. Journal of Clinical Engineering. **22**, 163–170 (1997)
2. P. Cappa, S. A. Sciuto and S. Silvestri. A novel preterm respiratory mechanics active simulator to test the performances of neonatal pulmonary ventilators. Review of Scientific Instruments. **73**, 2411–2416 (2002)
3. P. Cappa, S. A. Sciuto and S. Silvestri. Experimental evaluation of errors in the measurement of respiratory parameters of the newborn performed by a continuous flow neonatal ventilator. Journal of Medical Engineering & Technology. **30**, 31–40 (2006)
4. P. Saccomandi, E. Schena and S. Silvestri. A novel target-type low pressure drop bidirectional optoelectronic air flow sensor for infant artificial ventilation: measurement principle and static calibration. Review of Scientific Instruments. **82**, 024301 (2011).
5. J. G. Webster and A. M. Cook, Clinical engineering: principles and practices. Englewood Cliffs: Prentice-Hall, NJ (1979)
6. R. C. Baker, Flow measurement handbook: industrial designs, operating principles, performance, and applications (Cambridge University Press, Cambridge, 2000)
7. B. Lee. Review of the present status of optical fiber sensors. Optical Fiber Technology. **9**, 57–79 (2003)
8. L. Battista, A. Scorza, S. A. Sciuto. Preliminary evaluation of a simple optical fiber measurement system for monitoring respiratory pressure in mechanically ventilated infants. Proc. 9th IASTED Int. Conf. of Biomedical Engineering, 443–449 (2012)
9. L. Battista, A. Scorza, S. A. Sciuto. Experimental characterization of a novel fiber-optic accelerometer for the quantitative assessment of rest tremor in parkinsonian patients. Proc. 9th IASTED Int. Conf. of Biomedical Engineering, 437–442 (2012)
10. L. Battista, S. A. Sciuto, A. Scorza. Preliminary evaluation of a fiber-optic sensor for flow measurements in pulmonary ventilators. Proc. 2011 IEEE Int. Symp. on Medical Measurements and Applications, 29–34 (2011)

11. L. Battista, S. A. Sciuto, A. Scorza. An air flow sensor for neonatal mechanical ventilation application based on a novel fiber-optic sensing technique. Review of Scientific Instruments. **84** (3), 035005 (2013). doi: 10.1063/1.4798298

12. U. Frey et al. Specifications for equipment used for infant pulmonary function testing. Eur Respir J. **16**, 731–740 (2000)

13. Ghulam Nabi. Mechanical Ventilation in Infants. JK-Practitioner. **12**, 31–33 (2005)

Determination of Polyphenols in Bakery Food Matrices with New Detection Methods

L. Pigani, R. Seeber, A. Bedini, E. Dalcanale, and M. Suman

1 Introduction

The understanding of the molecular origin and the transduction process of bitter taste on the human tongue represents a challenge for scientists, due to the high number of receptors involved [1]. In bakery commodities, like biscuits, bitter taste is found frequently. The origin may lie, for example, in the common use of cocoa and coffee: the flavonoids present in the cocoa beans are among the main responsible for the typical bitter taste, together with xanthines, recognizable also in the derivative foodstuffs [2, 3]. In particular, the catechins, belonging to the category of the flavan-3-ols, i.e., a subgroup of the flavonoids, are abundantly present in numerous food matrices, cocoa and derivatives included.

The exploitation of dedicated panel tests for bitter taste sensory purpose is useful, but it suffers from limitations related to subjectivity, reproducibility, and number of analysis per day (Profile Attribute Analysis – PAA) [4]. Besides that, the most used methods for the quantification of the phenolic content in the plant products are based upon the reaction of polyphenols with colorimetric reactants, as in the Folin–Ciocalteu method [5]: reduced specificity constitutes the main limit, since other oxidation products interfere, causing an overestimation of the phenolic content. A different analytical perspective investigated here is based on FT-NIR technique,

L. Pigani • R. Seeber
Department of Chemical and Geological Sciences, University of Modena and Reggio Emilia, via G. Campi, 18, 41125 Modena, Italy

A. Bedini (✉) • E. Dalcanale
Department of Chemistry, University of Parma, Parco Area delle Scienze 17/A, 43124 Parma, Italy
e-mail: a.b.85@hotmail.it

M. Suman
Barilla SpA, Food Research Labs, via Mantova 166, 43122 Parma, Italy

C. Di Natale et al. (eds.), *Sensors and Microsystems: Proceedings of the 17th National Conference, Brescia, Italy, 5-7 February 2013*, Lecture Notes in Electrical Engineering 268, DOI 10.1007/978-3-319-00684-0_88, © Springer International Publishing Switzerland 2014

with the aim to effectively monitor the industrial product without any destructive time-consuming pretreatment of the samples. Furthermore, all flavonoids are electroactive, easily subjected to either anodic or cathodic reaction; hence, they can be spontaneously determined by electrochemical methods. Unfortunately, flavonoids present a strong tendency to give adsorption onto the electrode surfaces, often fouling them, even in the absence of any applied potential [6–8]. It is, however, known that in some specific cases the adsorption phenomena can be exploited to a quantitative scope, namely, in the so-called adsorptive stripping voltammetry (AdSV).

2   LC-MS/MS and FT-NIR Methods

Eleven types of different biscuits were chosen to analyze their bitterness. According to a dedicated LC-MS procedure [9, 10], analysis showed that biscuits with high content of cocoa present concentrations of polyphenols of the order of tenths of mg kg^{-1} (Table 1). In particular, in the case of chocolate biscuit 3, the concentration is higher than 80 mg kg^{-1}, and also PAA indicated that this is the most bitter-tasting biscuit of the analyzed series. On the contrary, the biscuits containing the recipe coffee (little percentage) or lemon achieved levels of few mg kg^{-1}, and biscuits containing cream, cereals, and honey did not show any remarkable amount of polyphenols. The comparison between panel test and LC-MS/MS data demonstrated that there is a correlation between the concentration of polyphenols and the biscuit's bitterness, confirming the hypothesis that these two classes of molecules are included within the main compounds responsible for this taste. LC-MS data were exploited for the calibration of the FT-NIR spectrophotometer by using partial least squares (PLS) regression. The values of the standard errors of prediction (lower than 10 %) were comparable to those of the standard errors of cross-validation. Coefficient of determination indicates a good predictive power for the calibration

Table 1 LC-MS results for the 11 categories of biscuits

Sample	Bitter taste index (PAA)	Catechin (mg kg^{-1})	Epicatechin (mg kg^{-1})	Tot. polyph. (mg kg^{-1})
Chocolate biscuit 3	8	28	55	83
Chocolate biscuit 2	7	17	39	56
Nuts biscuit	5	10	7	17
Chocolate and cream biscuit 1	5	16	38	54
Chocolate biscuit	6	10	6	16
Chocolate and cream biscuit 2	5	3	2	5
Coffee biscuit	4	<2	<3	<5
Lemon biscuit	4	2	<3	≤5
Cereal biscuit	<4	<2	<3	<5
Honey biscuit	<4	<2	<3	<5
Cream biscuit	<4	<2	<3	<5

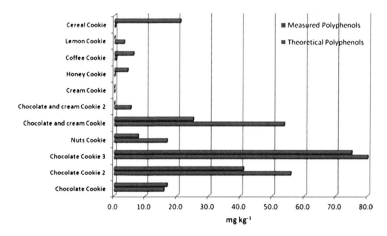

Fig. 1 FT-NIRS validation experiment for polyphenols (measured: FT-NIR; Theo.: LC-MS)

model (R^2 polyphenols = 0.96,) and satisfying discriminating power among different contents in the validation models (R^2 polyphenols = 0.96) [10].

Measurements were made directly on the solid sample with shorter analysis time compared to the LC-MS. Comparison between the techniques shows that the method is able to detect polyphenols when they are present in concentration close to few tenths of mg kg^{-1} (Fig. 1).

3   AdSV Method

We focused the attention on the use of a conducting polymer-modified electrode for the analysis of epicatechin: poly(3,4-ethylenedioxythiophene) (PEDOT) films have attracted special interest, thanks to their high conductivity, electrochemical stability, and possibility to work properly in aqueous media [11]. Three typologies of biscuits were considered with different PAA scores and amounts of polyphenols: chocolate biscuit 3, nuts biscuit, and cream biscuit. The extraction solutions for analysis were prepared as follows: a suitable quantity of biscuit, ranging from 5 to 10 g, was finely crumbled and, after the addition of 25 mL of phosphate buffer (PB) solution, posed under stirring for 1 h. After filtration on Buchner, the solution was directly used for analysis or diluted before electroanalytical tests. AdSV on PEDOT-modified Pt electrodes was therefore investigated together with the previously described FT-NIR approach.

As shown in Fig. 2, the height of the voltammetric peak recorded increases at increasing the percentages of cocoa in the biscuits, and hence with the polyphenolic content of the different food matrices' aqueous extraction solutions. In particular, the effectiveness of the device and procedure proposed has been evaluated for the electrochemical determination of a specific substance [12], namely, epicatechin, largely present in cocoa-containing biscuits (see Table 1).

Fig. 2 AdSV signal for the three categories of biscuits

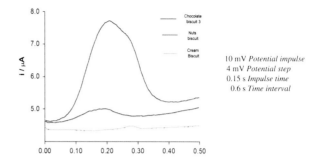

10 mV *Potential impulse*
4 mV *Potential step*
0.15 s *Impulse time*
0.6 s *Time interval*

4   Conclusions

The presence of a relevant category of bitter taste molecular markers in biscuits, i.e., polyphenols, has been explored with different analytical methodologies. LC-MS shows great potentiality as reference method, and AdSV shows its remarkable capabilities in differentiating samples as to their polyphenol content. FT-NIR shows to be the best quantitative solution for fast and reproducible industrial quality control analysis.

References

1. B. P. Trivedi, Nature, Taste Outlook, 486 (7403) S2-S3 (2012)
2. T. Stark, S. Bareuther, T. Hofmann, J. Agric. Food Chem., 53, 5407 (2005)
3. J. Serra Bonvehi, F. Ventura Coil, Food Chem., 60, 365 (1997)
4. J.M. Murray, C.M. Delahunty, I.A. Baxter, Food Res. Int., 34, 461 (2001)
5. ISO TC 34/SC 8 N 444 (1994)
6. N.E. Zoulis, C.E. Efstathiou, Anal. Chim. Acta, 320, 255 (1996)
7. D. Guo, D. Zheng, G. Mo, J. Ye, Electroanal., 21, 762 (2009)
8. K. Fan, X. Luo, J. Ping, W. Tang, Y. Ying, Q. Zhou, J. Agric. Food Chem., 60, 6333 (2012)
9. H. Schneider, L. Ma, H. Glatt, J. Chrom. B 789 (2003) 227.
10. A. Bedini, V. Zanolli, S. Zanardi,U. Bersellini, E. Dalcanale, M. Suman, Food Anal. Meth., 6 (1), 17-27 (2013)
11. L. Groenendaal,F. Jonas, D. Freitag, H. Pielartzik, J. Reynolds, Adv. Mat., 12, 481 (2000)
12. L. Pigani, R. Seeber, A. Bedini, E. Dalcanale, M. Suman, Eletroanal. 2013, submitted

Electronic Nose and Gas Chromatography–Mass Spectrometry for the Philippine Civet Coffee Discrimination

Veronica Sberveglieri, Emelda Ongo, Matteo Falasconi, Isabella Concina, Andrea Pulvirenti, and Fortunato Sevilla III

1 Materials and Methods

Civet coffee ranks as the top most expensive coffee due to its unique aroma and taste [1]. The Philippines is one of the few countries which produce civet coffee. Until now, coffee is considered as one of the Philippines' agricultural top ten crops. The climatic conditions and the soil conditions of the country make it suitable to produce the four varieties of commercial coffee beans among the five varieties (Arabica, Liberica, Excelsa, and Robusta). It comes from berries, which have been eaten and passed through the digestive tract of the Asian palm civet (Paradoxus hermaphrodites). Civet cat eats red coffee cherries containing the fruit and seed, but only the pulp is digested so that the bean ferments in the digestive system of the animal. The unique combination of enzymes in the stomach of the civet breaks down proteins in the bean and imparts to the coffee a unique bitter taste. The beans are defecated, still covered in some inner layers. The droppings are harvested by hand from the forest floor. The beans are then washed, air dried, and given only a light roast so as not to destroy the complex flavors that develop through the process. In order to sustain the strong market value of civet coffee and eliminate adulterated and fraudulent products, quality authentication is essential. In this work, E-nose and GC-MS analysis of Philippine civet coffee were carried out for the first time to determine the potential

V. Sberveglieri (✉) • A. Pulvirenti
Depart. Of life Science, University of Modena and Reggio Emilia, Via Amendola, 2, 42122 Reggio Emilia, Italy

CNR-IDASC SENSOR Lab, via Valotti, 9, 25122 Brescia, Italy
e-mail: veronica.sberveglieri@unimore.it

E. Ongo • F. Sevilla III
Graduate School, University of Santo Tomas, Santo Tomas, Philippines

M. Falasconi • I. Concina
CNR-IDASC SENSOR Lab, via Valotti, 9, 25122 Brescia, Italy

C. Di Natale et al. (eds.), *Sensors and Microsystems: Proceedings of the 17th National Conference, Brescia, Italy, 5-7 February 2013*, Lecture Notes in Electrical Engineering 268, DOI 10.1007/978-3-319-00684-0_89, © Springer International Publishing Switzerland 2014

of the instruments to discriminate civet coffees with their control coffee beans as well as to demonstrate the correlation between headspace aroma attributes and volatile compound composition of civet coffee [2].

1.1 Headspace Analysis of Philippine Civet Coffee Using EN and GC-MS with SPME

EN recently emerged as potential instruments for the detection of microbial contamination and quality selection in food. The EN EOS 835 (SACMI Imola, Italy) consists of the thermally controlled sensor chamber of 20 mL internal volume accommodating 6 thin metal oxide (MOX) semiconductor sensors, an electronic board for controlling the sensor heaters and measuring the sensing layers and software for data acquisition and signal processing [3–5].

MOX sensors are produced by sputtering a thin layer of sensing material over an alumina substrate. When the chemical species interact with the sensing layer, redox reactions occur at the surface causing a measurable change in the thin film conductance. The advantages of MOX sensor technology are good sensitivity toward a large spectrum of volatile compounds, high stability with time, and scalability at the industrial level. The responses of the MOX sensors to the coffee samples varied greatly. However, chemometric analysis through principal component analysis (PCA) and cluster analysis revealed groupings that differentiate civet and non-civet (control) coffee. Four different commercial brands of civet coffee and their corresponding control coffee (not eaten by civet) were used in the analysis.

Measurements were performed by static headspace using an automated sampling unit provided by a 40-loading-position carousel. Headspace analysis of Philippine coffees was performed for the first time using three phases (DVB/CAR/PDMS) 50/30 μm SPME fiber in combination with GC-MS. Chemometric analysis was carried out to visually display the similarities and distinction between civet and control coffee beans. One gram of roasted coffee beans was placed in a 20 mL crimp top-sealed vial which was then heated to a temperature of 70 °C for 10 min to reach sample headspace equilibrium. The volatile compounds were identified by comparing the mass spectral data with the NIST spectral database (MS library) and confirmed using reference standards. A comprehensive view of the PCA score plot of the E-nose data is illustrated in Figs. 1 and 2, and the dendrogram graph is shown in Fig. 3. Data obtained from GC-MS indicates the presence of at least 47 major components in the headspace of civet and non-civet (control) coffees.

The separation in the GC-MS-PCA plot between civet and control coffees is complementary with the E-nose results. The integration of E-nose and GC-MS data of the PCA plot shown in Fig. 4 reveals a good correlation between aroma quality and coffee volatiles of civet coffee, considering that individual civet coffee was successfully discriminated with its control beans also. The distinct data structure of civet coffee reveals that the passage of the beans through the digestive tract of civet

Fig. 1 E-nose PCA score plot of different civet and control coffee beans

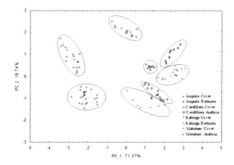

Fig. 2 GC-MS-PCA score plot of different civet and control coffee beans

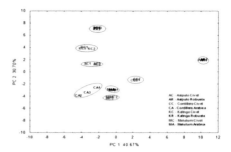

Fig. 3 E-nose cluster analysis (dendrogram) of civet and control coffee beans

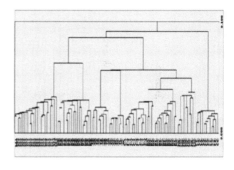

Fig. 4 PCA plot of E-nose and GC-MS data for civet and control coffee beans

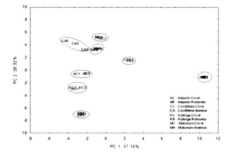

affects the aroma attributes of coffee beans [6, 7]. Cluster separation among civet coffees suggested that the aroma quality of civet coffee is region specific. It was observed that the composition of volatile compounds in civet coffee is almost similar to their controls but of different concentrations.

2 Conclusions

In conclusion EN has proven to be useful in discriminating the aroma quality of the Philippine civet coffee. The finding was supported by the differences between the relative GC-MS traces of the different coffees. Chemometric analysis revealed groupings that differentiate civet and control coffees. Also, cluster separation among civet coffee indicated that the aroma characteristics of civet coffee are varietal and region specific.

Acknowledgments This work was partially supported by CAFIS "Utilizzo di tecniche analitiche per la determinazione di indici di qualità nel caffè verde tostato e macinato," Progetto Operativo FESR 2007–2013–CUP G73F12000120004.

References

1. H.Y. Lee, "Wine and food feature: Most expensive coffee", Forbes Magazine, USA (July 2006)
2. V. Sberveglieri, I. Concina, M. Falasconi, E. Ongo, A. Pulvirenti, P. Fava, "Identification of geographical origin of coffee before and after roasting by electronic nose" AIP Conference proceeding, (2011) 1362, 86-87.
3. G. Sberveglieri, I. Concina, E. Comini, M. Falasconi, M. Ferroni, V. Sberveglieri, "Synthesis and integration of tin oxide nanowires into an electronic nose", Vacuum, 86 (2012) 532-535.
4. V. Sberveglieri, I. Concina, M. Falasconi, E. Gobbi, A. Pulvirenti, P. Fava, "Early detection of fungal contamination on green coffee by MOX sensors based Electronic nose" AIP Conference proceeding, (2011) 119-120.
5. I. Concina, M. Falasconi, V. Sberveglieri, "Electronic nose as a flexible tools to assess food quality and safety: Should we trust them?" IEEE sensors journal (2012) Vol 12, 3232-3237.
6. M.F. Marcone, "Composition and properties of Indonesian palm civet coffee (Kopi Luwak) and Ethiopian civet coffee", Food Research International, 37 (2008) 901–912.
7. E. Ongo, M. Falasconi, G. Sberveglieri, A. Antonelli, G. Montevecchi, V. Sberveglieri, I. Concina, F. Sevilla III, "Chemometric discrimination of Philippine civet coffee using electronic nose and gas chromatography mass spectrometry" Procedia Engineering vol 47 (2012) 977-980.

Simulation of Chlorine Decay in Drinking Water Distribution Systems: Case Study of Santa Sofia Network (Southern Italy)

G. Fattoruso, D. De Chiara, S. De Vito, V. La Ferrara, G. Di Francia, A. Leopardi, E. Cocozza, M. Viscusi, and M. Fontana

1 Introduction

Any drinking water distribution system has to make the water available to the consumer in proper quantity and pressure, with acceptable quality and sanitary security. Preserving the water quality throughout the distribution system is currently one of the most challenging technological issues for suppliers. For providing water that is safe from disease-causing pathogenic microorganisms, it is commonly treated by disinfection. Chlorine is widely used as a disinfectant in drinking water systems for its advantages such as high oxidation potential, long-term disinfection until the water reaches the consumer, excellent disinfection effectiveness, and relatively low cost. However, chlorine disinfection has some drawbacks such as the formation of undesirable by-products and the decay of chlorine concentration along the water distribution network due to three separated mechanisms, namely, the chlorine reactions in bulk fluid, the chlorine reactions with pipe walls and other system elements, and the natural evaporation.

In order to ensure water safety until the point of use (i.e., a desirable level of 0.2 mg/l of residual chlorine at the consumer), the approach commonly used by water suppliers is to carry out periodic chlorine sampling plans in fixed points of the aqueduct. Only recently, in addition, water quality within aqueducts is also investigated by mathematical modeling.

G. Fattoruso (✉) • D. De Chiara • S. De Vito • V. La Ferrara • G. Di Francia
UTTP/Basic Materials and Devices Dept., ENEA RC Portici, P.le E. Fermi,
1-80055 Portici, NA, Italy
e-mail: grazia.fattoruso@enea.it

A. Leopardi
Civil and Mechanical Engineering Dept., Università degli Studi di Cassino e del Lazio
Meridionale, Via G. Di Biasio, 43-03043 Cassino, FR, Italy

E. Cocozza • M. Viscusi • M. Fontana
Feronia srl, Centro Direzionale Isola E/7, Naples, Italy

C. Di Natale et al. (eds.), *Sensors and Microsystems: Proceedings of the 17th National Conference, Brescia, Italy, 5-7 February 2013*, Lecture Notes in Electrical Engineering 268, DOI 10.1007/978-3-319-00684-0_90, © Springer International Publishing Switzerland 2014

Over the last decades, several mathematical models have been investigated and developed in literature. They are mostly based on first-order kinetics, second-order kinetics, power-law (nth order) kinetics, and exponential decay assumption or reacting balance equation. The most used models assume an overall first-order reaction within the water distribution system. This is the case of Clark et al.'s water quality model [1], incorporated in the open source software *EPA-Epanet*, that models water quality behavior of an aqueduct by a simple first-order reaction for a single chemical (i.e., chlorine), with user input first-order reaction rate coefficients k_b and k_w for the bulk and wall reactions, respectively [1]. The quality model constants are generally defined through a calibration procedure based on chlorine point measurements along aqueduct obtained by sampling plans or by lab scale.

The objective of the present research work has been to simulate water quality behavior (i.e., chlorine decay process) along the Santa Sofia aqueduct (Campania, Southern Italy) by a modeling approach based on *live* data (pressures and residual chlorine parameters), gathered continuously by a wireless network of multiparametric probes distributed along the aqueduct.

2 Material and Method

The Santa Sofia aqueduct starts at the San Prisco reservoir, in Caserta City (Campania, Southern Italy), running for 22 km, supplying water to 28 municipalities. The scheme of the Santa Sofia aqueduct (Fig. 1a), built on the basis of GIS layers, is composed by 1 reservoir (San Prisco), 186 junctions node, 186 pipes, and 5 regulation valves.

The Santa Sofia aqueduct is equipped with a wireless network of commercial multi-parametric probes deployed along the system (Fig. 1b). The sensor network allows a distributed (10 points along the aqueduct), continued, and real-time monitoring of physical and chemical parameters such as pressure, residual chlorine,

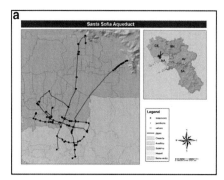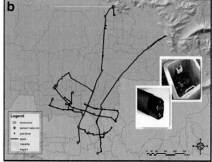

Fig. 1 (**a**) Map of Santa Sofia aqueduct; (**b**) map of the wireless network of commercial multiparametric probes (ENDETEC-KAPTA tm 3000) installed along the Santa Sofia water network

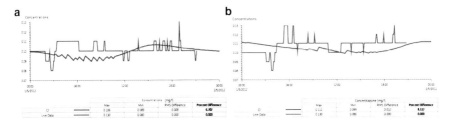

Fig. 2 Comparison between predicted and observed residual chlorine concentrations (mg/l) at (**a**) multi-parametric probe 5 location; (**b**) multi-parametric probe 10 of the WSN

conductivity, temperature, and pH. The *live data* of the pressure and residual chlorine have been used for the hydraulic and quality models calibration.

For simulating the chlorine decay process along the Santa Sofia aqueduct, a step-by-step approach has been followed which includes (1) water system model construction, (2) hydraulic behavior and water quality (chlorine decay) simulation, (3) models calibration, and (4) output scenario mapping.

First of all, the Santa Sofia water facilities have been visited to ensure accurate representation in the hydraulic model. The model has been built using GIS layers and the information gathered from the facility site visits and existing worksheets. The junctions' elevations have been obtained by a digital elevation model (20 m) of the interest area by using spatial analysis tools. Moreover, water demands and controls have manually been inserted into the model. The chlorine decay simulation has been performed on the basis of Clark et al.'s water quality model [1], incorporated in the open source *EPA-Epanet*. This model takes into account the phenomena of chlorine reaction with chemical species at bulk fluid and with pipe walls of the water network by means of specific constants, those required to be calibrated on the real Santa Sofia aqueduct (e.g., [2]). The calibration of this model as well as hydraulic models has been performed through an iterative process by adjusting the principal model parameters on the basis of the *live data* gathered by the probes network along the Santa Sofia aqueduct. In more detail, the chlorine decay model parameters [3] have been adjusted so that the simulated residual chlorine concentrations along the network over a 24 h simulation period at 5 min intervals gave a good match to the chlorine readings, observed by multi-parametric probes network. In Fig. 2, the plots show some results of water quality model calibration based on the *live data* of the selected probes P5 and P10.

3   Results and Conclusion

The water quality model calibration process, based on the *live data* of installed probes, has allowed to develop an effective predictive chlorine decay model along Santa Sofia aqueduct. The scenarios (Fig. 3) performed by the calibrated chlorine decay model show that the predicted residual chlorine concentrations along the

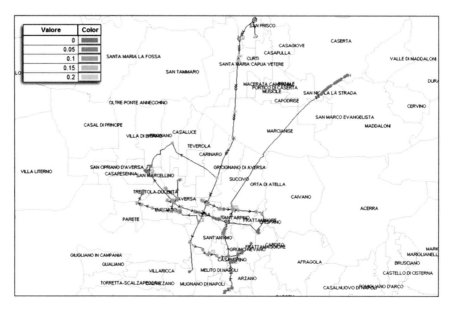

Fig. 3 A predictive spatial chlorine decay scenario along the Santa Sofia aqueduct

Santa Sofia aqueduct meet the acceptable values (i.e., 0.2 mg/l at the consumer) imposed by national water quality laws.

The present research work has allowed to provide the water supplier company with an effective chlorine decay model that enables water quality analysis and management of the Santa Sofia aqueduct, providing support for sustainable decision-making processes. Although there are other chlorine decay models, from which the parallel first-order and the n-order models stand out, the EPANET-incorporated first-order model was found to be satisfactory.

Acknowledgement The research work has been funded by ACQUARETI Project—POR Campania 2000/2006 Misura 3.17.

References

1. Clark, R., Rossman, L., and Wymer, L. Modeling Distribution System Water Quality: Regulatory Implications, Journal Water Resources Planning and Management 138:6, 614-623 (1995).
2. Di Cristo C., Esposito G., Leopardi A. Modelling trihalomethanes formation in water supply systems. ENVIRONMENTAL TECHNOLOGY, vol. 34(1), pp. 61-70, ISSN 0959-3330 (Print), ISSN 1479-487X (Online), DOI:10.1080/09593330.2012.679702, (2013)
3. Shang, F., Uber, J. G., & Rossman, L. A. (2008). Modeling Reaction and Transport of Multiple Species in Water Distribution Systems. *Environmental Science and Technology*, 42 (3), 808-814.

Use of Kinetic Models for Predicting DBP Formation in Water Supply Systems

G. Fattoruso, A. Agresta, E. Cocozza, S. De Vito, G. Di Francia,
M. Fabbricino, C.M. Lapegna, M. Toscanesi, and M. Trifuoggi

1 Introduction

Disinfection processes are used worldwide to reduce the risk related to waterborne diseases. Due to its low cost, stability, and effectiveness, chlorine is the most used agent for drinking water disinfection. Nonetheless chlorine has a main drawback related to its reaction with the natural organic matter (NOM), which is responsible for the formation of the so-called disinfection by-products (DBPs). These by-products include species such as trihalomethanes (THMs) and haloacetic acids (HAAs) that are characterized by recognized toxicity and carcinogenicity. The possibility of predicting their formation as function of water quality and chlorine dose, and even more important the possibility of predicting their evolution in water supply systems as function of water network characteristics, is therefore crucial. Although several predictive models of THM and HAA formation are available in the scientific literature, basically grouped into empirical models and kinetic ones, they are almost exclusively calibrated at lab scale. Very few attempts have been done, instead, to apply these models to existing water supply system.

The objective of our research work is to give a contribution to cover this gap, applying two of the existing kinetic models throughout the Santa Sofia aqueduct, located downstream the reservoir of San Prisco (Campania, Southern Italy), and calibrating them with full-scale data.

G. Fattoruso (✉) • A. Agresta • E. Cocozza • S. De Vito • G. Di Francia
UTTP/Basic Materials and Devices Dept., ENEA RC Portici, P.le E. Fermi,
1-80055 Portici, NA, Italy
e-mail: grazia.fattoruso@enea.it

M. Fabbricino
Civil, Architectural and Engineering Dept., University of Naples, Via Claudio, 21, Naples, Italy

C.M. Lapegna • M. Toscanesi • M. Trifuoggi
Chemical Science Dept, University of Naples, Via Cintia, 21, Naples, Italy

C. Di Natale et al. (eds.), *Sensors and Microsystems: Proceedings of the 17th National Conference, Brescia, Italy, 5-7 February 2013*, Lecture Notes in Electrical Engineering 268, DOI 10.1007/978-3-319-00684-0_91, © Springer International Publishing Switzerland 2014

471

2 Materials and Methods

The Santa Sofia aqueduct stretches about 22 km downstream from the San Prisco reservoir. It is operated by Acqua Campania Spa Company and supplies water to 28 municipalities of the western area of Campania Region. Its hydraulic scheme, built on the basis of GIS layers (Fig. 1), is composed by 1 reservoir (San Prisco), 186 junction nodes, 186 pipes, and 5 regulation valves.

The two kinetic models used to simulate DBP formation in this aqueduct, shortly indicated as DGF and LY, were, respectively, the one proposed by Fabbricino and Della Greca [1] and the one proposed by Lin and Yeh [2]. The DGF one predicts 12 types of DBPs: four THMs and eight HAAs. The master equation of this model, based on the reaction pathway proposed by Nokes et al. [3], is $DBP_i = A_{if}S_0\left(1-e^{-k_{if}t}\right)+A_{is}S_0\left(1-e^{-k_{is}t}\right)$ where S_0 is the dissolved organic carbon (DOC), while k_{if} and k_{is} are DBPi kinetic constants, and A_{if} and A_{is} are equilibrium constants. The model requires the calibration of 52 parameters, 4 for each single DBP species. The LY model, instead, predicts only the total THM concentration. The formation rate of total THMs is assumed to be equal to the chlorine consumption rate. The master equation of the model, therefore, is $\dfrac{\partial C_{THM}}{\partial t} = K_b \cdot C_{HOCL}$. The decay of total residual chlorine is modeled using a first-order decay reaction: $\dfrac{\partial C_{HOCL}}{\partial t} = -K_a \cdot C_{HOCL}$ where t is the time; C_{HOCL} and C_{THM} are, respectively, the HOCL and the total THM concentrations; K_a is a global chorine decay coefficient; and K_b is a kinetic constant.

Model application required of course, as preliminary steps, i) the construction of the water network model and ii) the calibration and the simulation of the hydraulic behavior of the water network. Regarding the hydraulic modeling, an open source

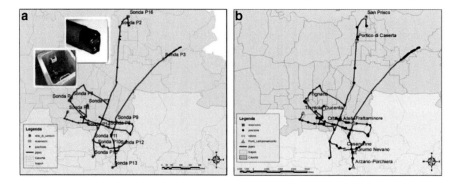

Fig. 1 (**a**) The wireless multi-parametric probe network and the commercial probe ENDETEC-KAPTA tm 3000, installed along the Santa Sofia aqueduct, within POR/Acquareti Project. (**b**) Water quality sampling plan carried out along the Santa Sofia aqueduct

GIS-based application, Net Tools, was used. This application integrates the hydraulic engine code EPA-Epanet in the open source GIS uDig. The hydraulic calibration of the Santa Sofia model was carried out by an iterative process, modifying the principal model parameters on the basis of the *live* data, such as pressure and flow rate, gathered by a continuous and real-time monitoring system based on a wireless network of commercial multi-parametric probes (Fig. 1a), installed on the Santa Sofia aqueduct. The selected kinetic models for DBP formation were hence applied throughout the aqueduct by using the computer program for multispecies modeling Epanet-MSX, integrated as extension in the GIS uDig platform, by ENEA researches. MSX and Epanet software as integrated tools of GIS uDig provides the user with an integrated open source GIS-based platform for water quality analysis.

For the kinetic models calibration, a campaign of water quality measurements (Fig. 1b) was carried out throughout the Santa Sofia aqueduct during the months of November and December 2012. The physical/chemical parameters, sampled in 9 points of the aqueduct, were temperature (T), pH, DOC, residual chlorine concentration (Cl), THM, and HAA concentrations.

Temperature and pH measurements, also gathered by multi-parametric probes, returned constant values in different points of the aqueduct. The model parameter *contact time* of the water with the disinfectant was assumed equal to the *age of the water* by using the related EPANET tools.

The models calibration was achieved by minimizing the sum of squared residuals and the difference between the observed chemical concentration values and the simulated ones (i.e., by least-squares method).

3   Results and Discussion

Mean and maximum relative errors associated with the simulated chlorine and THM concentration values, along the Santa Sofia aqueduct, returned by DGF and LY kinetic models, are shown in the table of Fig. 2a. It is worth noting that the magnitude of these errors is comparable with the one of several kinetic and empirical models, available in literature, with the best performances. Plotting the observed THM concentration values against the simulated ones (Fig. 2b), it can be observed that the DGF model tends to overestimate the lowest THM concentrations and underestimate the highest ones, whereas the LY model tends to underestimate the THM values in most of the measured points. However, the best fitting of the measured THM values is given by DGF model in each sampling point.

The DBP concentrations have been simulated in each node of hydraulic scheme of Santa Sofia aqueduct over a 24 h period. The simulation results are returned as predictive thematic maps.

As a result of this research work, it can be stated that the DBP concentrations throughout the Santa Sofia network, predicted by the DGF and LY kinetic models calibrated on the same aqueduct, may be considered sufficiently reliable.

a

LY's Model	RE$_M$ [%]	RE$_{max}$ [%]
Cl residue	3.5	20.6
Total THMs	15.4	72.7
DGF's Model	RE$_M$[%]	RE$_{max}$[%]
CHCl$_2$Br	16.8	126
CHClBr$_2$	12.5	75.2
CHBr$_3$	10.7	62.1

b

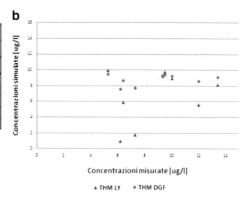

Fig. 2 (**a**) Mean and maximum relative errors of the chlorine and THM concentrations predicted by the calibrated DGF and LY kinetic models. (**b**) Measured THM concentrations [μg/l] against the values predicted by DGF and LY kinetic models, respectively

Acknowledgement The research work has partially been funded by ACQUARETI Project – POR Campania 2000/2006 Misura 3.17. The authors wish to thank Acqua Campania Spa Company for its profitable collaboration.

References

1. Della Greca G. & Fabbricino M. DBP formation in drinking water: kinetics and linear modelling. Water Science & Technology: Water Supply, (2008)
2. Lin Y. & Yeh H. Trihalomethane Species Forecast Using Optimization Methods: Genetic Algorithms and Simulated Annealing. Journal of Computing in Civil Engineering, (2005)
3. Nokes, C.J., Fenton, E., and Randall, C.J. "Modeling the formation of brominated trihalomethanes in chlorinated drinking water." Water Research, (1999)

Development of Electronic-Nose Technologies for Biomedical Applications

S.G. Leonardi, M. Cannistraro, E. Patti, D. Aloisio, N. Donato, G. Neri,
C. Pace, M. Mazzeo, and W. Khalaf

1   Introduction

Today, electronic-nose (e-nose) applications are growing at a rapid rate, thanks to the capability of this device to solve the problem associated with the detection/quantification of gaseous or vapor analytes in complex gas mixtures. An electronic nose is an artificial sensor system that generally consists of an array of chemical sensors and which responds essentially to all gases but produces a distinguishable response pattern for each separate type of analyte or mixture. Pattern recognition algorithms and/or neural network hardware is used on the output signals arising from the electronic nose to classify, identify, and where necessary quantify the gases of concerns.

The focus of this work will be on exploiting the vapor detection technology recently developed by us [1] for a low-power, simple, and easy-to-use "electronic nose" for biomedical applications in order to provide effective solutions to the diagnostic needs of modern medicine. The application here considered is related to the monitoring of volatile biomarkers in the human breath associated with certain disease states. This is a major challenge for modern medicine allowing to achieve effective early disease diagnoses, facilitating rapid cure treatments and, at the same time, reducing the invasiveness of diagnostic treatments.

S.G. Leonardi (✉) • M. Cannistraro • E. Patti • D. Aloisio • N. Donato • G. Neri
Dipartimento di Ingegneria Elettronica, Chimica e Ingegneria Industriale,
Università di Messina, Contrada di Dio, 98166 Messina, Italy
e-mail: leonardis@unime.it

C. Pace • M. Mazzeo
Dipartimento di Ingegneria Informatica, Modellistica, Elettronica e Sistemistica,
Università della Calabria, Via P. Bucci 42C, 87036 Rende, Italy

W. Khalaf
Computer & Software Eng. Dep., Almustansiriya University, Bab Al Muadham,
10047 Baghdad, Iraq

C. Di Natale et al. (eds.), *Sensors and Microsystems: Proceedings of the 17th National*   475
Conference, Brescia, Italy, 5-7 February 2013, Lecture Notes in Electrical Engineering 268,
DOI 10.1007/978-3-319-00684-0_92, © Springer International Publishing Switzerland 2014

Specifically, the performance of e-nose developed was tested in the breath monitoring of renal disease patients with the aim to find differences in the pattern, taken at different time intervals during hemodialysis and correlated to the variation of the clinical parameters due to the treatment [2].

2   Experiments

The e-nose system developed consists of an array of different commercial gas sensors placed in a small volume chamber (Fig. 1a), whose signals are conditioned and sampled by a multifunction board connected to a personal computer. A program, implementing efficient support vector machine and least-square model algorithms (Fig. 1c), executes the gas classification and the concentration estimation and warns about set risk thresholds overcoming. The system training was performed in the laboratory, over a wide range of analyte concentrations in air. Other boundary conditions, such as oxygen and CO_2 concentration, temperature, and RH, are also taken into account.

The overall cost of the system has been made very low, adopting an embedded architecture approach. So, also the size of the analyzer resulted small. For this scope, a simple system for breath sampling has been fabricated and interfaced with the e-nose analyzer (Fig. 1b).

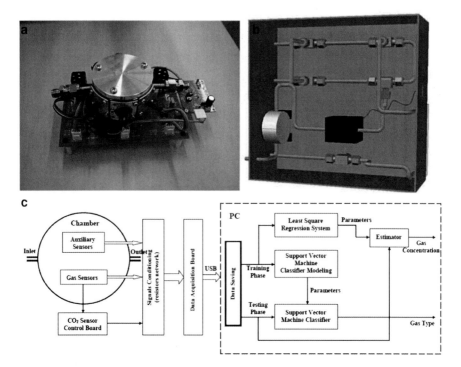

Fig. 1 Photographs of the e-nose (**a**), render of the sampling system (**b**), developed and block diagram of e-nose (**c**)

3 Results and Discussion

First, the developed analyzer was tested in the laboratory. As an example, the responses in the concentration range 0.25–3 % of CO_2 are reported in Fig. 2. The monitoring of carbon dioxide is important for breath analysis because it can be regarded as an internal standard and give information about the correctness of the breath sampling [3].

The analyzer has been then tested, sampling the breath of volunteer subjects. The response patterns of five sensors are reported in Fig. 3. Preliminarily, the stability of the analyzer response during repeated sampling of a same volunteer has been tested (Fig. 3a). The response patterns of three volunteers are reported in Fig. 3b. It can be noted that the analyzer is able to monitor the differences in the breath of these subjects.

After these promising results, the analyzer has been tested in hospital, sampling the breath of patients with renal disease during the hemodialysis treatment. The raw results registered by the analyzer at different time intervals during hemodialysis are reported in Fig. 4.

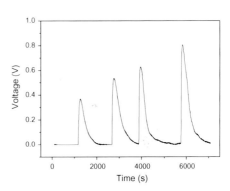

Fig. 2 Responses to different concentrations of CO_2

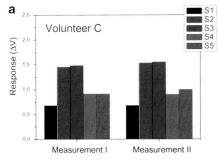

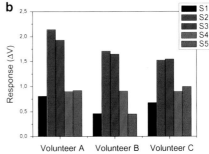

Fig. 3 Stability of the analyzer response during repeated sampling of a same volunteer (**a**), and response patterns of three volunteers (**b**)

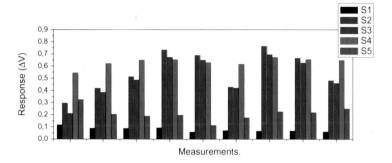

Fig. 4 Analyzer response at different time intervals during hemodialysis

The response patterns change during the treatment (about 4 h). It is well known that during this period, many clinical parameters of the patient vary largely. In particular, the concentration of toxic nitrogen compounds (e.g., urea) decreases in the blood, and consequently a lower concentration of ammonia in the breath can be found [2]. This is well evidenced by our analyzer where sensor S1, specific for ammonia, shows a continuous decrease with dialysis time. Our final aim is to correlate the other differences observed in the response patterns with the variation of the various clinical parameters registered in the blood or urine during the treatment of hemodialysis [2].

4 Conclusions

The development of an electronic nose comprising commercial gas sensors, portable and easy to use for biomedical applications, is reported. Tests carried out in the laboratory have shown the promising features of the analyzer. The real-time breath monitoring of patients under hemodialysis treatment has been performed to validate the developed analyzer.

References

1. C. Pace, W. Khalaf, M. Latino, N. Donato, G. Neri, E-nose development for safety monitoring applications in refinery environment, Procedia Engineering, 47, 1267-1270 (2012).
2. G. Neri, G. Rizzo, N. Donato, M. Latino, A. Laquaniti, M. Buemi, Real-time monitoring of breath ammonia during hemodialysis. Part I—IMS and CRDS techniques. Nephr. Dial. Trans. Nephrology Dialysis Transplantation 27, 2945-2952 (2012).
3. W. Miekisch, A. Hengstenberg, S. Kischkel, U. Beckmann, M. Mieth, J. K. Schubert, Construction and evaluation of a versatile CO$_2$ controlled breath collection device, IEEE Sensors Journal 10, 211-215 (2010).

Printed by Publishers' Graphics LLC
CAMZ131217.15.21.71